Lecture Notes in Computer Science 1448
Edited by G. Goos, J. Hartmanis and J. van Leeuwen

Springer
Berlin
Heidelberg
New York
Barcelona
Budapest
Hong Kong
London
Milan
Paris
Singapore
Tokyo

Martin Farach-Colton (Ed.)

Combinatorial Pattern Matching

9th Annual Symposium, CPM 98
Piscataway, New Jersey, USA, July 20-22, 1998
Proceedings

Series Editors

Gerhard Goos, Karlsruhe University, Germany
Juris Hartmanis, Cornell University, NY, USA
Jan van Leeuwen, Utrecht University, The Netherlands

Volume Editor

Martin Farach-Colton
Bell Labs and
Rutgers University, Department of Computer Science
Piscataway, NJ 08855, USA
E-mail: farach@research.bell-labs.com

Cataloging-in-Publication data applied for

Die Deutsche Bibliothek - CIP-Einheitsaufnahme

Combinatorial pattern matching : 9th annual symposium ; proceedings / CPM 98, Piscataway, New Jersey, USA, July 20 - 22, 1998. Martin Farach-Colton (ed.). - Berlin ; Heidelberg ; New York ; Barcelona ; Budapest ; Hong Kong ; London ; Milan ; Paris ; Singapore ; Tokyo : Springer, 1998
(Lecture notes in computer science ; Vol. 1448)
ISBN 3-540-64739-2

CR Subject Classification (1991): F.2.2, I.5.4, I.5.o, I.7.3, H.3.3, E.4, G.2.1

ISSN 0302-9743
ISBN 3-540-64739-2 Springer-Verlag Berlin Heidelberg New York

Typesetting: Camera-ready by author
SPIN 10638148 06/3142 – 5 4 3 2 1 0 Printed on acid-free paper

Program Committee

Martin Farach-Colton (Committee Chair)
Ricardo Baeza-Yates
Brenda Baker
Gary Benson
Richard Cole
Dannie Durand
Christos Faloutsos
Paolo Ferragina
Paola Merlo
S. Muthukrishnan
Mike Paterson
Prabhakar Raghavan
Jeanette Schmidt
Ron Shamir

Table of Contents

A Fast Bit-Vector Algorithm for Approximate String Matching Based on Dynamic Programming

Gene Myers[1,*]

Dept. of Computer Science, University of Arizona Tucson, AZ 85721

Abstract. The approximate string matching problem is to find all locations at which a query of length m matches a substring of a text of length n with k-or-fewer differences. Simple and practical bit-vector algorithms have been designed for this problem, most notably the one used in *agrep*. These algorithms compute a bit representation of the current state-set of the k-difference automaton for the query, and asymptotically run in $O(nmk/w)$ time where w is the word size of the machine (e.g. 32 or 64 in practice). Here we present an algorithm of comparable simplicity that requires only $O(nm/w)$ time by virtue of computing a bit representation of the *relocatable* dynamic programming matrix for the problem. Thus the algorithm's performance is independent of k, and it is found to be more efficient than the previous results for many choices of k and small m.

Moreover, because the algorithm is not dependent on k, it can be used to rapidly compute blocks of the dynamic programming matrix as in the 4-Russians algorithm of Wu, Manber, and Myers. This gives rise to an $O(kn/w)$ expected-time algorithm for the case where m may be arbitrarily large. In practice this new algorithm, which computes a region of the d.p. matrix in $1 \times w$ blocks using the basic algorithm as a subroutine, is significantly faster than our previous 4-Russians algorithm, which computes the same region in 1×5 blocks using table lookup. This performance improvement yields a code which is superior to *all* existing algorithms except for some filtration algorithms that are superior when k/m is sufficiently small.

1 Introduction

The problem of finding substrings of a text similar to a given query string is a central problem in information retrieval and computational biology, to name but a few applications. It has been intensively studied over the last twenty years. In its most common incarnation, the problem is to find substrings that match the query with k or fewer differences. The first algorithm addressing exactly this problem is attributable to Sellers [Sel80] although one might claim that it was effectively solved by work in the early 70's on string comparison. Sellers

* E-mail: gene@cs.arizona.edu. Partially supported by NLM grant LM-04960

algorithm requires $O(mn)$ time where m is the length of the query and n is the length of the text. Subsequently this was refined to $O(kn)$ expected-time by Ukkonen [Ukk85], then to $O(kn)$ worst-case time, first with $O(n)$ space by Landau and Vishkin [LV88], and later with $O(m^2)$ space by Galil and Park [GP90].

Of these early algorithms, the $O(kn)$ expected-time algorithm was universally the best in practice. The algorithm achieves its efficiency by computing only the region or zone of the underlying dynamic programming matrix that has entries less than or equal to k. Further refining this basic design, Chang and Lampe [CL92] went on to devise a faster algorithm which is conjectured to run in $O(kn/\sqrt{\sigma})$ expected-time where σ is the size of the underlying alphabet. Next, Wu, Manber, and this author [WMM96] developed a practical realization of the 4-Russians approach [MP80] that when applied to Ukkonen's zone, gives an $O(kn/\log s)$ expected-time algorithm, given that $O(s)$ space can be dedicated to a universal lookup table. In practice, these two algorithms are always superior to Ukkonen's zone design, and each faster than the other in different regions of the (k, σ) input-parameter space.

At around the same time, another new thread of practice-oriented results exploited the hardware parallelism of bit-vector operations. Letting w be the number of bits in a machine word, this sequence of results began with an $O(n\lceil m/w \rceil)$ algorithm for the exact matching case by Baeza-Yates and Gonnet [BYG92], and culminated with an $O(kn\lceil m/w \rceil)$ algorithm for the k-differences problem by Wu and Manber [WM92]. These authors were interested specifically in text-retrieval applications where m is quite small, small enough that the expression between the ceiling braces is 1. Under such circumstances the algorithms run in $O(n)$ or $O(kn)$ time, respectively. More recently, Baeza-Yates and Navarro [BYN96] have realized an $O(n\lceil km/w \rceil)$ variation on the Wu/Manber algorithm, implying $O(n)$ performance when $mk = O(w)$.

The final recent thrust has been the development of *filter* algorithms that eliminate regions of the text that cannot match the query. The results here can broadly divided into on-line algorithms (e.g. [WM92,CL94]) and off-line algorithms (e.g. [Mye94]) that are permitted to preprocess a presumably static text before performing a number of queries over it. After filtering out all but a presumably small segment of the text, these methods then invoke one of the algorithms above to verify if a match is actually present in the portion that remains. The *filtration efficiency* (i.e. percentage of the text removed from consideration) of these methods increases as the *mismatch ratio* $e = k/m$ approaches 0, and at some point, dependent on σ and the algorithm, they provide the fastest results in practice. However, improvements in verification-capable algorithms are still very desirable, as such results improve the filter-based algorithms when there are a large number of matches, and also are needed for the many applications where e is such that filtration is ineffective.

In this paper, we present two verification-capable algorithms, inspired by the 4-Russians approach, but using bit-vector computation instead of table lookup. First, we develop an $O(n\lceil m/w \rceil)$ bit-vector algorithm for the approximate string matching problem. This is asymptotically superior to prior bit-vector results, and in practice will be shown to be superior to the other bit-vector algorithms for all but a few choices of m and k. In brief, the previous algorithms use bit-vectors to model and maintain the state set of a non-deterministic finite automaton with $(m+1)(k+1)$ states that (exactly) matches all strings that are k-differences or fewer from the query. Our method uses bit-vectors in a very different way, namely, to encode the list of m arithmetic differences between successive entries in a column of the dynamic programming matrix. Our second algorithm comes from the observation that our first result can be thought of as a subroutine for computing any $1 \times w$ block of a d.p. matrix in $O(1)$ time. We may thus embed it in the zone paradigm of the Ukkonen algorithm, exactly as we did with the 4-Russians technique [WMM96]. The result is an $O(kn/w)$ expected-time algorithm which we will show in practice outperforms both our previous work and that of Chang and Lampe [CL92] for *all* regions of the (k, σ) parameter space.

2 Preliminaries

Assume the query sequence is $P = p_1 p_2 \ldots p_m$, the text is $T = t_1 t_2 \ldots t_n$, and that we are given a positive threshold $k \geq 0$. Further let $\delta(A, B)$ be the unit cost edit distance between strings A and B. The approximate string matching problem is to find all positions j in T such that there is a suffix of $T[1..j]$ matching P with k-or-fewer differences, i.e., j such that $\min_g \delta(P, T[g..j]) \leq k$.

The classic approach to this problem [Sel80] is to compute an $(m+1) \times (n+1)$ *dynamic programming (d.p.) matrix* $C[0..m, 0..n]$ for which $C[i,j] = \min_g \delta(P[1..i], T[g..j])$ after an $O(mn)$ time computation using the well-known recurrence:

$$C[i,j] = \min \left\{ \begin{array}{c} C[i-1,j-1] + (\text{if } p_i = t_j \text{ then } 0 \text{ else } 1) \\ C[i-1,j] + 1 \\ C[i,j-1] + 1 \end{array} \right\} \tag{1}$$

subject to the boundary condition that $C[0,j] = 0$ for all j. It then follows that the solution to the approximate string matching problem is all locations j such that $C[m,j] \leq k$.

A basic observation is that the computation above can be done in only $O(m)$ space because computing column $C_j = < C[i,j] >_{i=0}^{m}$ only requires knowing the values of the previous column C_{j-1}. This leads to the important conceptual realization that one may think of a column C_j as a state of an automaton, and the algorithm as advancing from state C_{j-1} to state C_j as its "scans" symbol t_j of the text. The automaton is started in the state $C_0 =< 0, 1, 2, \ldots, m >$ and any state whose last entry is k-or-fewer is considered to be a final state.

The automaton just introduced has at most 3^m states. This follows because the d.p. matrix C has the property that the difference between adjacent entries in any row or any column is either 1, 0, or -1. Formally, define the *horizontal delta* $\Delta h[i,j]$ at (i,j) as $C[i,j] - C[i,j-1]$ and the *vertical delta* $\Delta v[i,j]$ as $C[i,j] - C[i-1,j]$ for all $(i,j) \in [1,m] \times [1,n]$.

Lemma 1 [MP80,Ukk85]: For all i, j: $\Delta v[i,j], \Delta h[i,j] \in \{-1,0,1\}$.

It follows that to know a particular state C_j it suffices to know the *relocatable* column $\Delta v_j = < \Delta v[i,j] >_{i=1}^{m}$ because $C[0,j] = 0$ for all j.

We can thus replace the problem of computing C with the problem of computing the *relocatable d.p. matrix* Δv. One potential difficulty is that determining if Δv_j is final requires $O(m)$ time if one computes the sum $\Sigma_i \, \Delta v_j[i] = C[m,j]$ explicitly in order to do so. Our algorithm will compute a *block* of vertical deltas in $O(1)$ time, and thus cannot afford to compute this sum. Fortunately, one can simultaneously maintain the value of $Score_j = C[m,j]$ as one computes the $\Delta v_j' s$ using the fact that $Score_0 = m$ and $Score_j = Score_{j-1} + \Delta h[m,j]$. Note that the *horizontal* delta in the last row of the matrix is required, but the horizontal delta at the end of a block of vertical delta's will be seen to be a natural by-product of the block's computation. Figure 1 illustrates the basic dynamic programming matrix and its formulation in relocatable terms.

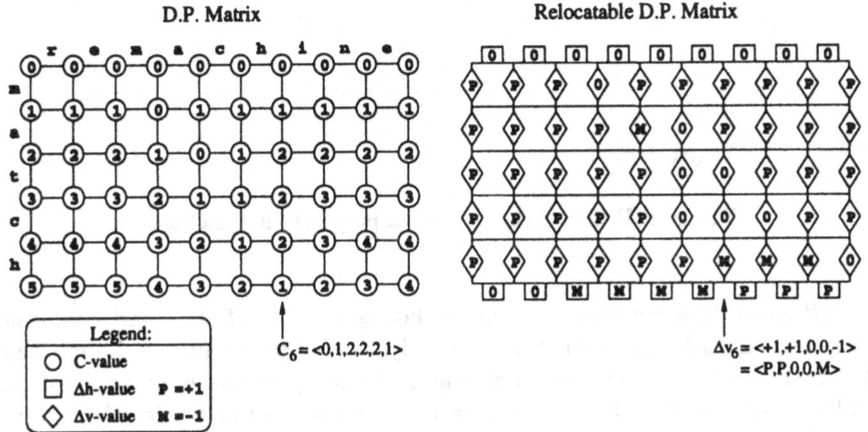

Fig. 1. Dynamic Programming (D.P.) Matrices for P = *match* and T = *remachine*.

3 The Basic Algorithm

Representation. We seek to compute successive $\Delta v_j' s$ in $O(1)$ time using bit-vector operations under the assumption that $m \leq w$. We begin by choosing to represent the column Δv_j with two bit-vectors Pv_j and Mv_j, whose bits are set according to whether the corresponding delta in Δv_j is $+1$ or -1, respectively. Formally, $Pv_j(i) \equiv (\Delta v[i,j] = +1)$ and $Mv_j(i) \equiv (\Delta v[i,j] = -1)$, where the notation $W(i)$ denotes the ith bit of the integer W. Note that $\Delta v[i,j] = 0$ exactly when *not* $(Pv_j(i)$ *or* $Mv_j(i))$ is true.

Cell Structure. Consider an individual *cell* of the d.p. matrix consisting of the square $(i-1, j-1)$, $(i-1, j)$, $(i, j-1)$, and (i,j). There are two vertical deltas, $\Delta v_{out} = \Delta v[i,j]$ and $\Delta h_{in} = \Delta v[i,j-1]$, and two horizontal deltas, $\Delta h_{out} = \Delta h[i,j]$ and $\Delta h_{in} = \Delta h[i-1,j]$, associated with the sides of this cell as shown in Figure 2(a). Further define $Eq = Eq[i,j]$ to be 1 if $p_i = t_j$ and 0 otherwise. Using the definition of the deltas and the basic recurrence for C-values it follows that:

$$\Delta v_{out} = \min\{-Eq, \Delta v_{in}, \Delta h_{in}\} + (1 - \Delta h_{in}) \tag{2a}$$
$$\Delta h_{out} = \min\{-Eq, \Delta v_{in}, \Delta h_{in}\} + (1 - \Delta v_{in}) \tag{2b}$$

It is thus the case that one may view Δv_{in}, Δh_{in}, and Eq as *inputs* to the cell at (i,j), and Δv_{out} and Δh_{out} as its *outputs*.

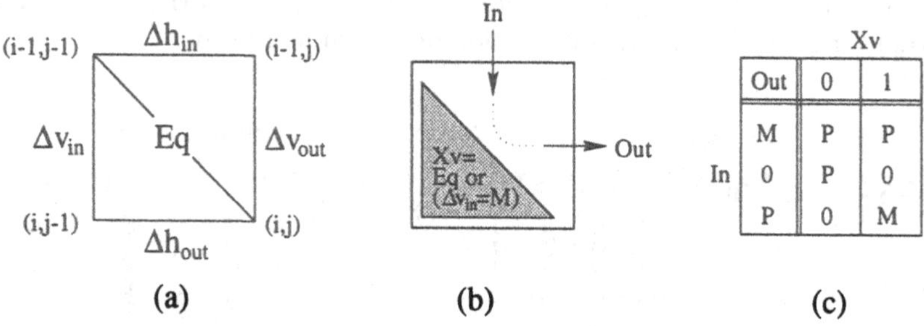

Fig. 2. D.P. Cell Structure and Input/Output Function.

Cell Logic. Observe that there are 3 choices for each of Δv_{in} and Δh_{in} and 2 possible values for Eq. Thus there are only 18 possible inputs for a given cell. This gave rise to the key idea that one could compute the numeric values in a column with boolean logic, in contrast to the previous methods, which use a traditional bit vector implementation of a set over a finite number of elements.

As Figure 2(b) suggests, we find it conceptually easiest to think of Δv_{out} as a function of Δh_{in} modulated by an auxiliary boolean value $Xv \equiv (Eq \textit{ or } (\Delta v_{in} = -1))$ capturing the net effect of both Δv_{in} and Eq on Δv_{out}. With a brute force enumeration of the 18 possible inputs, one may verify the correctness of the table in Figure 2(c) which describes Δv_{out} as a function of Δh_{in} and Xv. In the table, the value -1 is denoted by M and $+1$ by P, in order to emphasize the logical, as opposed to the numerical, relationship between the input and output. Let Px_{io} and Mx_{io} be the bit values encoding Δx_{io}, i.e. $Px_{io} \equiv (\Delta x_{io} = +1)$ and $Mx_{io} \equiv (\Delta x_{io} = -1)$. By definition $Xv = Eq \textit{ or } Mv_{in}$ and from the the table one can verify that:

$$(Pv_{out}, Mv_{out}) = (Mh_{in} \textit{ or not } (Xv \textit{ or } Ph_{in}), Ph_{in} \textit{ and } Xv) \tag{3a}$$

By symmetry, given $Xh = (Eq \textit{ or } Mh_{in})$, it follows that:

$$(Ph_{out}, Mh_{out}) = (Mv_{in} \textit{ or not } (Xh \textit{ or } Pv_{in}), Pv_{in} \textit{ and } Xh) \tag{3b}$$

Alphabet Preprocessing. To evaluate cells according to the treatment above, one needs the boolean value $Eq[i,j]$ for each cell (i,j). In terms of bit-vectors, we will need an integer Eq_j for which $Eq_j(i) \equiv (p_i = t_j)$. Computing these integers during the scan would require $O(m)$ time and defeat our goal. Fortunately, in a preprocessing step, performed before the scan begins, we can compute a table of the vectors that result for each possible text character. Formally, if Σ is the alphabet over which P and T originate, then we build an array $Peq[\Sigma]$ for which: $Peq[s](i) \equiv (p_i = s)$. Constructing the table can easily be done in $O(|\Sigma|m)$ time and it occupies $O(|\Sigma|)$ space (continuing with the assumption that $m \leq w$). We are assuming, or course, that Σ is finite. At a small loss in efficiency our algorithm can be made to operate over infinite alphabets.

The Scanning Step. The central inductive step is to compute $Score_j$ and the bit-vector pair (Pv_j, Mv_j) encodingΔv_j, given the same information at column $j-1$ and the symbol t_j. In keeping with the automata conception, we refer to this step as *scanning* t_j and illustrate it in Figure 3 at the left. The basis of the induction is easy as we know at the start of the scan that $Pv_0(i) = 1$, $Mv_0(i) = 0$, and $Score_0 = m$. A scanning step is accomplished in two stages as illustrated in Figure 3:

1. First, the vertical delta's in column $j-1$ are used to compute the horizontal delta's at the bottom of their respective cells, using formula (3b).
2. Then, these horizontal delta's are used in the cell *below* to compute the vertical deltas in column j, using formula (3a).

In between the two stages, the *Score* in the last row is updated using the last horizontal delta now available from the first stage, and then the horizontal deltas are all *shifted* by one, pushing out the last horizontal delta and introducing a 0-delta for the first row. We like to think of each stage as a pivot, where the pivot of the first stage is at the lower left of each cell, and the pivot of the second stage is at the upper right. The delta's swing in the arc depicted and produce results modulated by the relevant X values.

The logical formulas (3) for a cell and the schematic of Figure 3, lead directly to the formulas below for accomplishing a scanning step. Note that the horizontal deltas of the first stage are recorded in a pair of bit-vectors, (Ph_j, Mh_j), that encodes horizontal deltas exactly as (Pv_j, Mv_j) encodes vertical deltas, i.e., $Ph_j(i) \equiv (\Delta h[i,j] = +1)$ and $Mh_j(i) \equiv (\Delta h[i,j] = -1)$.

$$\begin{aligned} Ph_j(i) &= Mv_{j-1}(i) \textit{ or not } (Xh_j(i) \textit{ or } Pv_{j-1}(i)) \\ Mh_j(i) &= Pv_{j-1}(i) \textit{ and } Xh_j(i) \end{aligned} \qquad \text{(Stage 1)}$$

$$Score_j = Score_{j-1} + (1 \text{ if } Ph_j(m)) - (1 \text{ if } Mh_j(m)) \qquad (4)$$

$$\begin{aligned} Ph_j(0) &= Mh_j(0) = 0^1 \\ Pv_j(i) &= Mh_j(i-1) \textit{ or not } (Xv_j(i) \textit{ or } Ph_j(i-1)) \\ Mv_j(i) &= Ph_j(i-1) \textit{ and } Xv_j(i) \end{aligned} \qquad \text{(Stage 2)}$$

[1] In the more general case where the horizontal delta in the first row can be -1 or $+1$ as well as 0, these two bits must be set accordingly.

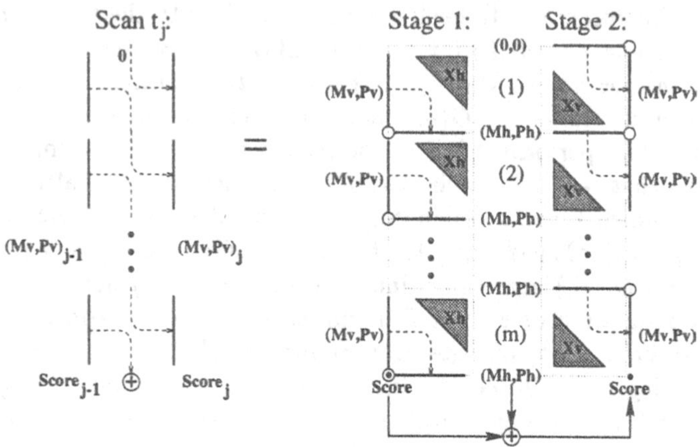

Fig. 3. The Two Stages of a Scanning Step.

At this point, it is important to understand that the formulas above specify the computation of bits in bit-vectors, all of whose bits can be computed in parallel with the appropriate machine operations. For example in *C*, we can express the computation of *all* of Ph_j as '`Ph = Mv | ~ (Xh | Pv)`'.

The X-Factors. The induction above is incomplete as we have yet to show how to compute Xv_j and Xh_j. By definition $Xv_j(i) = Peq[t_j](i)$ *or* $Mv_{j-1}(i)$ and $Xh_j(i) = Peq[t_j](i)$ *or* $Mh_j(i-1)$ where *Peq* is the precomputed table supplying *Eq* bits. The bitvector Xv_j can be directly computed at the start of the scan step as the vector Mv_{j-1} is input to the step. On the other hand, computing Xh_j requires the value of Mh_j which in turn requires the value of Xh_j! We thus have a cyclic dependency which must be unwound. Lemma 2 gives such a formulation of Xh_j which depends only on the values of Pv_{j-1} and $Peq[t_j]$.

Lemma 2: $Xh_j(i) = \exists k \leq i, Peq[t_j](k)$ *and* $\forall x \in [k, i-1], Pv_{j-1}(x)^2$. (5)

Basically, Lemma 2 says that the ith bit of *Xh* is set whenever there is a preceding *Eq* bit, say the kth and a run of set *Pv* bits covering the interval $[k, i-1]$. In other words, one might think of the *Eq* bit as being "propagated" along a run of set *Pv* bits, setting positions in the *Xh* vector as it does so. This brings to mind the addition of integers, where carry propagation has a similar effect on the underlying bit encodings. Figure 4 illustrates the way we use addition to have the desired effects on bits which we summarize as Lemma 3 below, and which we prove precisely in the full paper.

Lemma 3: If $X = (((E\&P) + P)\,\hat{}\,P)|E$ then
$$X(i) \;=\; \exists k \leq i, E(k) \text{ and } \forall x \in [k, i-1], P(x).$$

[2] In the more general case where the horizontal delta in the first row can be -1 or $+1$ as well as 0, $Peq[t_j](1)$ must be replaced with $Peq[t_j](1)$ *or* $Mh_j(0)$.

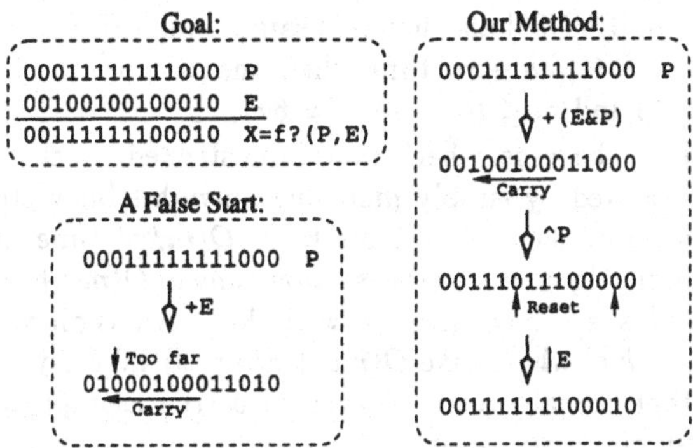

Fig. 4. Illustration of Xv computation.

The Complete Algorithm. It now remains just to put all the pieces together. Figure 5 gives a complete specification in the style of a C program to give one a feel for the simplicity and efficiency of the result.

```
Precompute Peq[Σ]
Pv = 1^m
Mv = 0
Score = m

for j = 1, 2, ...n do

{ Eq = Peq[t_j]
  Xv = Eq | M
  Xh = (((Eq & Pv) + Pv) ^ Pv) | Eq

  Ph = Mv | ~ (Xh | Pv)
  Mh = Pv & Xh

  if Ph & 10^(m-1) then
     Score += 1
  else if Mh & 10^(m-1) then
     Score -= 1

  Ph <<= 1
  Pv = (Mh << 1) | ~ (Xv | Ph)
  Mv = Ph & Xv

  if Score ≤ k then
     print "Match at " · j
}
```

Fig. 5. The Basic Algorithm.

Under theh prevailing assumption that $m \leq w$, the complexity of the algorithm is easily seen to be $O(m\sigma + n)$ where σ is the size of the alphabet Σ. Indeed only 17 bit operations are performed per character scanned. This is to be contrasted with the Wu/Manber bit-vector algorithm [WM92] which takes $O(m\sigma + kn)$ under the same assumption about m. The Baeza-Yates/Navarro bit-vector algorithm [BYN96] also has this same complexity under this assumption, but improves to $O(m\sigma + n)$ time when one assumes $m \leq 2\sqrt{w} - 2$ (e.g., $m \leq 9$ when $w = 32$ and $m \leq 14$ when $w = 64$).

Finally, consider the case where m is unrestricted. Such a situation can easily be accommodated by simply modeling an m-bit bit-vector with $\lceil m/w \rceil$ words. An operation on such bit-vectors takes $O(m/w)$ time. It then directly follows that the basic algorithm of this section runs in $O(m\sigma + nm/w)$ time and $O(\sigma m/w)$ space. This is to be contrasted with the previous bit-vector algorithms [WM92,BYN96], *both* of which take $O(m\sigma + knm/w)$ time asymptotically. This leads us to say that our algorithm represents a true asymptotic improvement over previous *bit-vector* algorithms.

4 The Unrestricted Algorithm

The Blocks Model. Just as we think of the computation of a single cell as realizing an input/output relationship on the four deltas at its borders, we may more generally think of the computation of a $u \times v$ rectangular subarray or *block* of cells as resulting in the output of deltas along its lower and right boundary, given deltas along its upper and left boundary as input. This is the basic observation behind Four Russians approaches to sequence comparison [MP80,WMM96], where the output resulting from every possible input combination is pretabulated and then used to effect the computation of blocks as they are encountered in a particular problem instance. We can similarly modify our basic algorithm to effect the $O(1)$ computation of $1 \times w$ blocks (via bitvector computation as opposed to table lookup), given that we are careful to observe that in this context the horizontal input delta may be -1 or $+1$, as well as 0.

There are several sequence comparison results that involve computing a region or *zone* of the underlying dynamic programming matrix, the first of which was [Ukk85]. Figure 6 depicts such a hypothetical zone and a tiling of it with $1 \times w$ blocks. Provided that one can still effectively delimit the zone while performing a block-based computation, using such a tiling gives a factor w speedup over the underlying zone algorithm. For example, we take Ukkonen's $O(kn)$ expected-time algorithm and improve it to $O(kn/w)$ below. Note that blocks are restricted to one of at most $b_{max} = \lceil m/w \rceil$ levels, so that only $O(\sigma m/w)$ *Eq*-vectors need be precomputed. Further note that any internal boundary of the tiling has a delta of 1.

A Block-Based Algorithm for Approximate String Matching. Ukkonen improved the expected time of the standard $O(mn)$ d.p. algorithm for approximate string matching, by computing only the zone of the d.p. matrix consisting of the prefix of each column ending with the last k in the column. That is, if $x_j = \max\{i : C(i,j) \leq k\}$ then the algorithm takes time proportional to the size of the zone $Z(k) = \cup_{j=0}^{n}\{(i,j) : i \in [0, x_j]\}$. It was shown in [CL92] that the expected size of $Z(k)$ is $O(kn)$. Computing just the zone is easily accomplished with the observation that $x_j = \max\{i : i \leq x_{j-1} + 1 \text{ and } C(i,j) \leq k\}$.

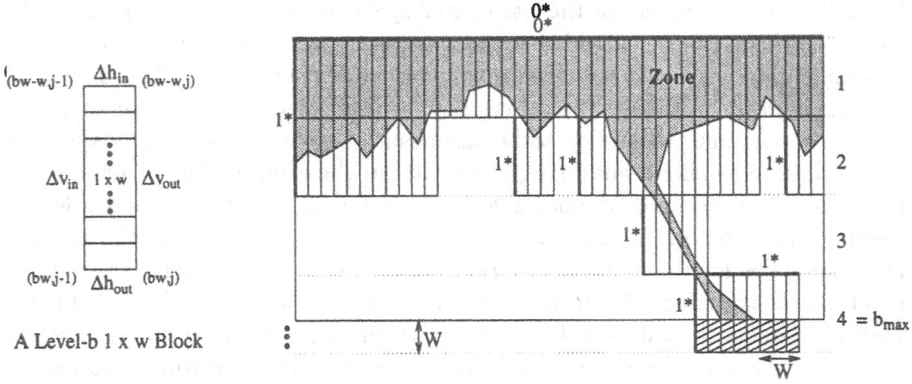

Fig. 6. Block-Based Dynamic Programming.

A block-based algorithm for this $O(kn)$ expected-time algorithm was devised and presented in an earlier paper of ours [WMM96] where the blocks were computed in $O(1)$ time using a 4-Russians lookup table. What we are proposing here, is to do exactly the same thing, except to use our bit-vector approach to compute $1 \times w$ blocks in $O(1)$ time. As we will see in the next section, this results in almost a factor of 4-5 improvement in performance, as the 4-Russians table lookups were limited to 1×5 blocks and the large tables involved result in much poorer cache coherence, compared to the bit-vector approach where all the storage required typically fits in the on-board CPU cache. We describe the small modifications necessary to tile $Z(k)$ in the full paper.

5 Some Empirical Results

We report on two sets of comparisons run on a Dec Alpha 4/233 for which $w = 64$. The first is a study of our basic bit-vector algorithm and the two previous bit-vector results [WM92,BYN96] for approximate string matching when $m \leq w$. The second set of experiments involves all verification-capable algorithms that work when k and m are unrestricted. Experiments to determine the range of k/m for which filter algorithms are superior have not been performed in this preliminary study.

The expected time complexity of each algorithm, A, is of the asymptotic form $\Theta(f_A(m,k,\sigma)n)$ and our experiments are aimed at empirically measuring f_A. In the full paper we will detail the specific trials, noting here only that (a) their design guarantees a measurement error of at most 2.5%, and (b) the search texts were obtained by randomly selecting characters from an alphabet of size σ. with equal probability.

Our first set of experiments compare the three bit-vector algorithms for the case where $m \leq 64$, and the results are shown in Figure 7. At left we show our best estimate for f_A for each algorithm. For our basic algorithm, f_A is a constant as is also true of the Baeza-Yates and Navarro algorithm save that it can only be applied to the region of the parameter space where $(m-k)(k+2) \leq 64$. Their algorithm can be extended to treat a greater range of k and m by linking automata together in a thresholded manner as we did in Section 4. This may lead to some increase in the area of superiority for their algorithm, but we have not had the opportunity to develop and measure such a code in this first study, which we consider to be about what one can do without the complication of multi-word bitvectors. The Wu and Manber algorithm performs linearly in k and a least-squares regression line fits the results of 90 trials very well, save that the fit is off by roughly 9% for the first two values of k (see Figure 7). We hypothesize that this is due to the effect of branch-prediction in the instruction pipeline hardware. Figure 7 depicts the values of k and m for which each method is superior to the others. In the zone where the Baeza-Yates and Navarro algorithm requires no automata linking it is 12% faster than our basic algorithm, and for $k = 0$ the algorithm of Manber and Wu is 29% faster.

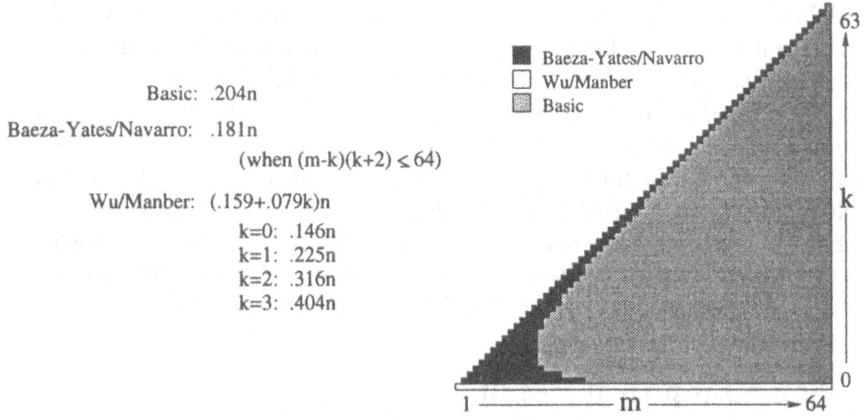

Fig. 7. Performance Summary and Regions of Superiority for Bit-Vector Algorithms.

Our second set of experiments are aimed at comparing verification-capable algorithms that can accommodate unrestricted choices of k and m. In this case, we need only consider our block-based algorithm and the results of [CL92] and [WMM96], as all other competitors are already known to be dominated in practice by these two [WMM96]. These three algorithms are all zone-based and when m is suitably large, the zone never reaches the last row of the d.p. matrix, so that running time does not depend on m. We set $m = 400$ for all trials and ran 107 trials with $(k, \sigma) \in \{0, 1, 2, \ldots 6, 8, 10, \ldots 20, 24, 28 \ldots 60\} \times \{2, 4, 8, 16, 32\} \cup \{64, 68, 72, \ldots 120\} \times \{32\}$. For each of the five choices of σ, Figure 8 has a graph of time as a function of k, one curve for each algorithm. From this figure it is immediately clear that our block-based algorithm is superior to the others for all choices of k and σ we tried. The Change and Lampe algorithm may eventually overtake ours but it did not do so with $\sigma = 95$, the number of printable ASCII characters. Again, one should consider that filtration algorithms (and perhaps a thresholded version of the Baeza-Yates and Navarro algorithm), will definitely outperform our algorithm when k/m is sufficently small.

References

[BYG92] R.A. Baeza-Yates and G.H. Gonnet. A new approach to text searching. *Communications of the ACM*, 35:74–82, 1992.

[BYN96] R.A. Baeza-Yates and G. Navarro. A faster algorithm for approximate string matching. In *Proc. 7th Symp. on Combinatorial Pattern Matching. Springer LNCS 1075*, pages 1–23, 1996.

[CL92] W.I. Chang and J. Lampe. Theoretical and empirical comparisons of approximate string matching algorithms. In *Proc. 3rd Symp. on Combinatorial Pattern Matching. Springer LNCS 644*, pages 172–181, 1992.

[CL94] W.I. Chang and E.L. Lawler. Sublinear expected time approximate matching and biological applications. *Algorithmica*, 12:327–344, 1994.

[GP90] Z. Galil and K. Park. An improved algorithm for approximate string matching. *SIAM J. on Computing*, 19:989–999, 1990.

[LV88] G.M. Landau and U. Vishkin. Fast string matching with k differences. *J. of Computer and System Sciences*, 37:63–78, 1988.

[MP80] W.J. Masek and M. S. Paterson. A faster algorithm for computing string edit distances. *J. of Computer and System Sciences*, 20:18–31, 1980.

[Mye94] E.W. Myers. A sublinear algorithm for approximate keywords searching. *Algorithmica*, 12:345–374, 1994.

[Sel80] P.H. Sellers. The theory and computations of evolutionary distances: Pattern recognition. *J. of Algorithms*, 1:359–373, 1980.

[Ukk85] E. Ukkonen. Finding approximate patterns in strings. *J. of Algorithms*, 6:132–137, 1985.

[WM92] S. Wu and U. Manber. Fast text searching allowing errors. *Communications of the ACM*, 35:83–91, 1992.

[WMM96] S. Wu, U. Manber, and G. Myers. A subquadratic algorithm for approximate limited expression matching. *Algorithmica*, 15:50–67, 1996.

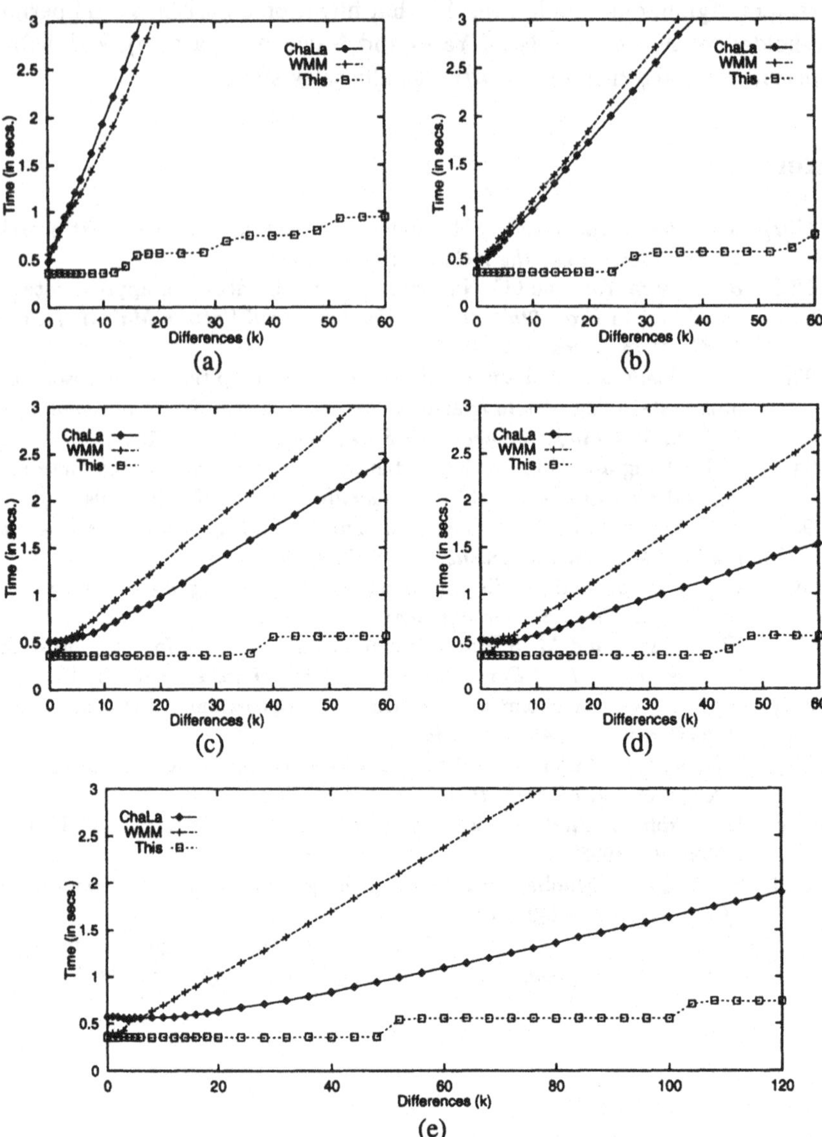

Fig. 8. Timing curves for the $O(kn/w)$ block-based algorithm ("This") versus Chang/Lampe ("ChaLa") and Wu/Manber/Myers ("WMM"), with alphabet sizes (a) $\sigma = 2$, (b) $\sigma = 4$, (c) $\sigma = 8$, (d) $\sigma = 16$, and (e) $\sigma = 32$.

A Bit-Parallel Approach to Suffix Automata: Fast Extended String Matching

Gonzalo Navarro[13] Mathieu Raffinot[2]

[1] Dept. of Computer Science, University of Chile. Blanco Encalada 2120, Santiago, Chile. gnavarro@dcc.uchile.cl.

[2] Institut Gaspard Monge, Cité Descartes, Champs-sur-Marne, 77454 Marne-la-Vallée Cedex 2, France. raffinot@monge.univ-mlv.fr.

[3] Partially supported by Chilean Fondecyt grant 1-950622.

Abstract. We present a new algorithm for string matching. The algorithm, called BNDM, is the bit-parallel simulation of a known (but recent) algorithm called BDM. BDM skips characters using a "suffix automaton" which is made deterministic in the preprocessing. BNDM, instead, simulates the nondeterministic version using bit-parallelism. This algorithm is 20%-25% faster than BDM, 2-3 times faster than other bit-parallel algorithms, and 10%-40% faster than all the Boyer-Moore family. This makes it *the fastest algorithm in all cases* except for very short or very long patterns (e.g. on English text it is the fastest between 5 and 110 characters). Moreover, the algorithm is very simple, allowing to easily implement other variants of BDM which are extremely complex in their original formulation. We show that, as other bit-parallel algorithms, BNDM can be extended to handle classes of characters in the pattern and in the text, multiple patterns and to allow errors in the pattern or in the text, combining simplicity, efficiency and flexibility. We also generalize the suffix automaton definition to handle classes of characters. To the best of our knowledge, this extension has not been studied before.

1 Introduction

The string-matching problem is to find all the occurrences of a given pattern $p = p_1p_2 \ldots p_m$ in a large text $T = t_1t_2 \ldots t_n$, both sequences of characters from a finite character set Σ.

Several algorithms exist to solve this problem. One of the most famous, and the first having linear worst-case behavior, is Knuth-Morris-Pratt (KMP) [14]. A second algorithm, as famous as KMP, which allows to skip characters, is Boyer-Moore (BM) [6]. This algorithm leads to several variations, like Hoorspool [12] and Sunday [20], forming the fastest known string-matching algorithms.

A large part of the research in efficient algorithms for string matching can be regarded as looking for automata which are efficient in some sense. For instance, KMP is simply a deterministic automaton that searches the pattern, being its main merit that it is $O(m)$ in space and construction time. Many variations of the BM family are supported by an automaton as well.

Another automaton, called "suffix automaton" is used in [9, 10, 11, 15, 19], where the idea is to search a substring of the pattern instead of a prefix (as KMP), or a suffix (as BM). Optimal sublinear algorithms on average, like "Backward DAWG Match" (BDM) or Turbo_BDM [10, 11], have been obtained with this approach, which has also been extended to multipattern matching [9, 11, 19] (i.e. looking for the occurrences of a set of patterns).

Another related line of research is to take those automata in their nondeterministic form instead of making them deterministic. Usually the nondeterministic versions are very simple and regular and can be simulated using "bit-parallelism" [1]. This technique uses the intrinsic parallelism of the bit manipulations inside computer words to perform many operations in parallel. Competitive algorithms have been obtained for exact string matching [2, 22], as well as approximate string matching [22, 23, 3]. Although these algorithms work well only on relatively short patterns, they are simpler, more flexible, and have very low memory requirements.

In this paper we merge some aspects of the two approaches in order to obtain a fast string matching algorithm, called Backward Nondeterministic Dawg Matching (BNDM), which we extend to handle classes of characters, to search multiple patterns, and to allow errors in the pattern and/or in the text, like Shift-Or [2]. BNDM uses a nondeterministic suffix automaton that is simulated using bit-parallelism. This new algorithm has the advantage of being faster than previous ones which could be extended in such a way (typically 2-3 times faster than Shift-Or), faster than its deterministic-automaton counterpart BDM (20%-25% faster), using little space in comparison with the BDM or Turbo_BDM algorithms, and being very simple to implement. It becomes the **fastest** string matching algorithm, beating all the Boyer-Moore family (Sunday included) by 10% to 40%. Only for very short (up to 2-6 letters) or very long patterns (past 90-150 letters), depending on $|\Sigma|$ and the architecture, other algorithms become faster than BNDM (Sunday and BDM, respectively). Moreover, we define a new suffix automaton which handles classes of characters and we simulate its nondeterministic version using bit-parallelism. This extension has not been considered for the BDM or Turbo_BDM algorithms before.

We introduce some notation now. A word $x \in \Sigma^*$ is a *factor* (i.e. substring) of p if p can be written $p = uxv$, $u, v \in \Sigma^*$. We denote Fact(p) the set of factors of p. A factor x of p is called a *suffix* of p is $p = ux$. The set of suffixes of p is called Suff(p).

We denote as $b_\ell ... b_1$ the bits of a mask of length ℓ. We use exponentiation to denote bit repetition (e.g. $0^3 1 = 0001$). We use C-like syntax for operations on the bits of computer words: "|" is the bitwise-or, "&" is the bitwise-and, "^" is the bitwise-xor and "~" complements all the bits. The shift-left operation, "<<", moves the bits to the left and enters zeros from the right, i.e. $b_m b_{m-1} ... b_2 b_1 << r = b_{m-r} ... b_2 b_1 0^r$. We can interpret bit masks as integers also to perform arithmetic operations on them.

An expanded version of this work can be found in [17].

2 Searching with Suffix Automata

We describe in this section the BDM pattern matching algorithm [10, 11]. This algorithm is based on a suffix automaton. We first describe such automaton and then explain how is it used in the search algorithm

2.1 Suffix Automata

A *suffix automaton* on a pattern $p = p_1p_2 \ldots p_m$ (frequently called DAWG(p) - for Deterministic Acyclic Word Graph) is the minimal (incomplete) deterministic finite automaton that recognizes all the suffixes of this pattern. By "incomplete" we mean that some transitions are not present.

The nondeterministic version of this automaton has a very regular structure and is shown in Figure 1. We show now how the corresponding deterministic automaton is built.

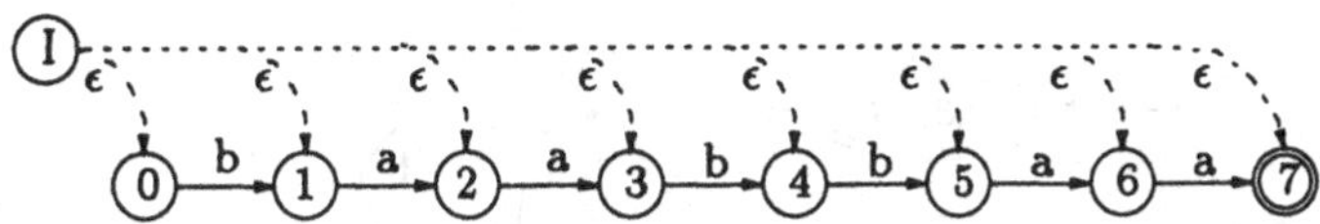

Fig. 1. A nondeterministic suffix automaton for the pattern $p = baabbaa$. Dashed lines represent epsilon transitions (i.e. they occur without consuming any input). I is the initial state of the automaton.

Given a factor x of the pattern p, *endpos(x)* is the set of all the pattern positions where an occurrence of x ends (there is at least one, since x is a factor of the pattern, and there are as many as repetitions of x inside p). Formally, given $x \in \text{Fact}(p)$, we define *endpos(x)* $= \{i \;/\; \exists u,\ p_1p_2...p_i = ux\}$. We call each such integer a *position*. For example, *endpos*(baa) $= \{3, 7\}$ in the word *baabbaa*. Notice that *endpos*(ϵ) is the complete set of possible positions (recall that ϵ is the empty string). Notice that for any u, v, *endpos*(u) and *endpos*(v) are either disjoint or one contained in the other.

We define an equivalence relation $\equiv$ between factors of the pattern. For $u, v \in$ Fact(p), we define

$$u \equiv v \text{ if and only if } endpos(u) = endpos(v)$$

(notice that one of the factors must be a suffix of the other for this equivalence to hold, although the converse is not true). For instance, in our example pattern $p = baabbaa$, we have that $baa \equiv aa$ because in all the places where aa ends in the pattern, baa ends also (and vice-versa).

The nodes of the DAWG correspond to the equivalence classes of $\equiv$, i.e. to sets of positions. A state, therefore, can be thought of a factor of the pattern already recognized, except because we do not distinguish between some factors. Another way to see it is that the set of positions is in fact the set of active states in the nondeterministic automaton.

There is an edge labeled σ from the set of positions $\{i_1, i_2, \ldots i_k\}$ to $\gamma_p(i_1 + 1, \sigma) \cup \gamma_p(i_2 + 1, \sigma) \cup \ldots \cup \gamma_p(i_k, \sigma)$, where

$$\gamma_p(i,\sigma) = \begin{cases} \{i\} \text{ if } i \leq m \text{ and } p_i = \sigma \\ \emptyset \text{ otherwise} \end{cases}$$

which is the same to say that we try to extend the factor that we recognized with the next text character σ, and keep the positions that still match. If we are left with no matching positions, we do not build the transition. The initial state corresponds to the set $\{0..m\}$. Finally, a state is *terminal* if its corresponding subset of positions contains the last position m (i.e. we matched a suffix of the pattern). As an example, the deterministic suffix automaton of the word *baabbaa* is given in Figure 2.

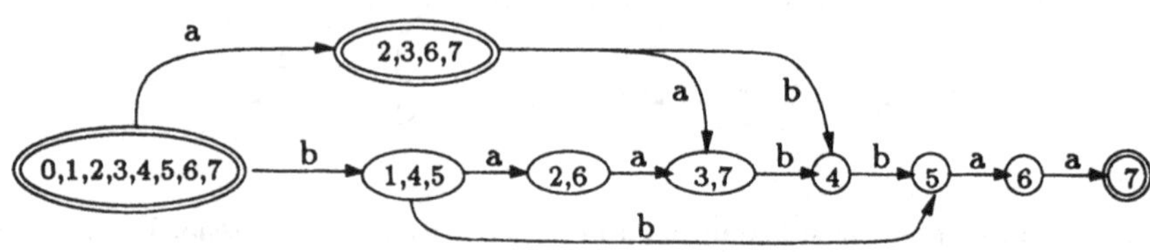

Fig. 2. Deterministic suffix automaton of the word *baabbaa*. The largest node is the initial state.

The (deterministic) suffix automaton is a well known structure [8, 5, 11, 18], and we do not prove any of its properties here (neither the correctness of the previous construction). The size of DAWG(p) is linear in m (counting both nodes and edges), and can be built in linear time [8]. A very important fact for our algorithm is that this automaton can not only be used to recognize the suffixes of p, but also factors of p. By the suffix automaton definition, there is a path labeled by x form the initial node of DAWG(p) if and only if x is a factor of p.

2.2 Search Algorithm

The suffix automaton structure is used in [10, 11] to design a simple pattern matching algorithm called BDM. This algorithm is $O(mn)$ time in the worst case, but optimal in average ($O(n \log m/m)$ time)[4]. Other more complex variations such as Turbo_BDM[10] and MultiBDM[11, 19] achieve linear time in the worst

[4] The lower bound of $O(n \log m/m)$ in average for any pattern matching algorithm under a Berbouilli model is from A. C. Yao in [24].

case. To search a pattern $p = p_1p_2\ldots p_m$ in a text $T = t_1t_2\ldots t_n$, the suffix automaton of $p^r = p_mp_{m-1}\ldots p_1$ (i.e the pattern read backwards) is built. A window of length m is slid along the text, from left to right. The algorithm searches backwards inside the window for a factor of the pattern p using the suffix automaton. During this search, if a terminal state is reached which does not correspond to the entire pattern p, the window position is remembered (in a variable *last*). This corresponds to finding a *prefix* of the pattern starting at position *last* inside the window and ending at the end of the window (since the suffixes of p^r are the reverse prefixes of p). The last recognized prefix is the *longest* one. The backward search ends because of two possible reasons:

1. We fail to recognize a factor, i.e we reach a letter σ that does not correspond to a transition in DAWG(p^r). Figure 3 illustrates this case. We then shift the window to the right in *last* characters (we cannot miss an occurrence because in that case the suffix automaton would have found its prefix in the window).

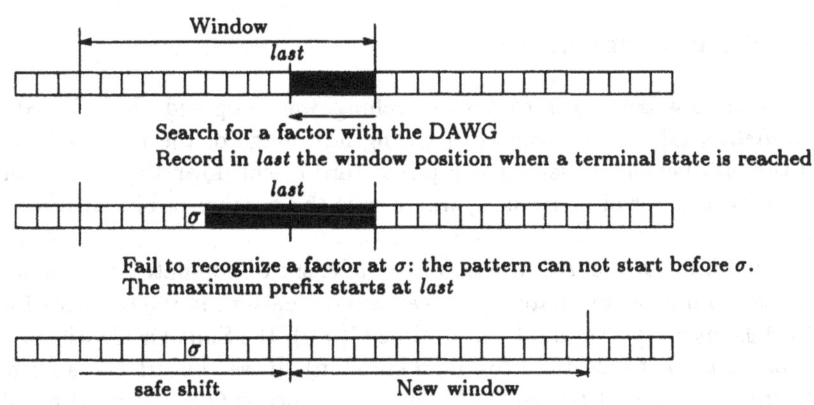

Fig. 3. Basic search with the suffix automaton

2. We reach the beginning of the window, therefore recognizing the pattern p. We report the occurrence, and we shift the window exactly as in the previous case (notice that we have the previous *last* value).

Search example: we search the pattern *aabbaab* in the text

$$T = a\ b\ b\ a\ b\ a\ a\ b\ b\ a\ a\ b.$$

We first build DAWG(p^r =*baabbaa*), which is given in Figure 2. We note the current window between square brackets and the recognized prefix in a rectangle. We begin with
$T =\ [a\ b\ b\ a\ b\ a\ a\ ]b\ b\ a\ a\ b$, $m = 7$, $last = 7$.

1. $T = [\,a\,b\,b\,a\,b\,a\,\boxed{a}\,]\,b\,b\,a\,a\,b$.
   a is a factor of p^r and a reverse prefix of p. $last = 6$.
2. $T = [\,a\,b\,b\,a\,b\,\boxed{a\,a}\,]\,b\,b\,a\,a\,b$.
   aa is a factor of p^r and a reverse prefix of p. $last = 5$.
3. $T = [\,a\,b\,b\,a\,\boxed{b\,a\,a}\,]\,b\,b\,a\,a\,b$.
   aab is a factor of p^r.
   We fail to recognize the next a. So we shift the window to *last*. We search again in the position: $T = a\,b\,b\,a\,b\,[\,a\,a\,b\,b\,a\,a\,b\,]$, $last = 7$.
4. $T = a\,b\,b\,a\,b\,[\,a\,a\,b\,b\,a\,a\,\boxed{b}\,]$.
   b is a factor of p^r.
5. $T = a\,b\,b\,a\,b\,[\,a\,a\,b\,b\,a\,\boxed{a\,b}\,]$.
   ba is a factor of p^r.
6. $T = a\,b\,b\,a\,b\,[\,a\,a\,b\,b\,\boxed{a\,a\,b}\,]$.
   baa is a factor of p^r.
7. $T = a\,b\,b\,a\,b\,[\,a\,a\,b\,b\,\boxed{a\,a\,b}\,]$.
   baa is a factor of p^r, and a reverse prefix of p. $last = 4$.
8. $T = a\,b\,b\,a\,b\,[\,a\,a\,b\,\boxed{b\,a\,a\,b}\,]$.
   $baab$ is a factor of p^r.
9. $T = a\,b\,b\,a\,b\,[\,a\,a\,\boxed{b\,b\,a\,a\,b}\,]$.
   $baabb$ is a factor of p^r.
10. $T = a\,b\,b\,a\,b\,[\,a\,\boxed{a\,b\,b\,a\,a\,b}\,]$.
    $baabba$ is a factor of p^r.
11. $T = a\,b\,b\,a\,b\,[\,\boxed{a\,a\,b\,b\,a\,a\,b}\,]$.
    We recognize the word *aabbaab* and report an occurrence.

3 Bit-Parallelism

In [2], a new approach to text searching was proposed. It is based on *bit-parallelism* [1], which consists in taking advantage of the intrinsic parallelism of the bit operations inside a computer word to cut down the number of operations by a factor of at most w, where w is the number of bits in the computer word.

The Shift-Or algorithm uses bit-parallelism to simulate the operation of a nondeterministic automaton that searches the pattern in the text (see Figure 4). As this automaton is simulated in time $O(mn)$, the Shift-Or algorithm achieves $O(mn/w)$ worst-case time (optimal speedup). If we convert the automaton to deterministic we get a version of KMP [14], which is $O(n)$ search time, although twice as slow in practice for $m \leq w$.

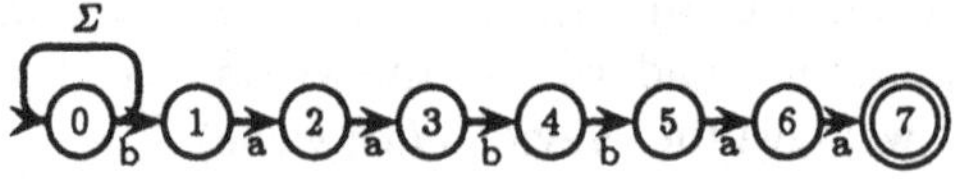

Fig. 4. A nondeterministic automaton to search the pattern $p = baabbaa$ in a text. The initial state is 0.

We explain now a variant of the Shift-Or algorithm (called Shift-And). The algorithm builds first a table B which for each character stores a bit mask $b_m ... b_1$. The mask in $B[c]$ has the i-th bit set if and only if $p_i = c$. The state of the search is kept in a machine word $D = d_m ... d_1$, where d_i is set whenever the state numbered i in Figure 4 is active. Therefore, we report a match whenever d_m is set.

We set $D = 0$ originally, and for each new text character T_j, we update D using the formula

$$D' \leftarrow ((D << 1) \mid 0^{m-1}1) \ \& \ B[T_j]$$

which mimics what occurs inside the nondeterministic automaton for each new text character: each state gets the value of the previous one, provided the text character matches the corresponding arrow. The "$\mid 0^{m-1}1$" corresponds to the initial self-loop. For patterns longer than the computer word (i.e. $m > w$), the algorithm uses $\lceil m/w \rceil$ computer words for the simulation (not all them are active all the time).

This algorithm is very simple and can be extended to handle classes of characters (i.e. each pattern position matches a set of characters), and to allow mismatches. This paradigm was later enhanced to support wild cards, regular expressions, approximate search, etc. yielding the fastest algorithms for those problems [22, 3]. Bit-parallelism became a general way to simulate simple nondeterministic automata instead of converting them to deterministic. This is how we use it in our algorithm.

4 Bit-Parallelism on Suffix Automata

We simulate the BDM algorithm using bit-parallelism. The result is an algorithm which is simpler, uses less memory, has more locality of reference, and is easily extended to handle more complex patterns. We first assume that $m \leq w$ and show later how to extend the algorithm for longer patterns.

4.1 The Basic Algorithm

We simulate the reverse version of the automaton of Figure 1. Just as for Shift-And, we keep the state of the search using m bits of a computer word $D = d_m...d_1$.

The BDM algorithm moves a window over the text. Each time the window is positioned at a new text position just after pos, it searches backwards the window $T_{pos+1}..T_{pos+m}$ using the DAWG automaton, until either m iterations are performed (which implies a match in the current window) or the automaton cannot perform any transition.

In our case, the bit d_i at iteration k is set if and only if $p_{m-i+1}..{}_{m-i+k} = T_{pos+1+m-k}..T_{pos+m}$. Since we begin at iteration 0, the initial value for D is 1^m. There is a match if and only if after iteration m it holds $d_m = 1$. Whenever $d_m = 1$, we have matched a prefix of the pattern in the current window. The longest prefix matched corresponds to the next window position.

The algorithm is as follows. Each time we position the window in the text we initialize D and scan the window backwards. For each new text character we update D. Each time we find a prefix of the pattern ($d_m = 1$) we remember the position in the window. If we run out of 1's in D then there cannot be a match and we suspend the scanning (this corresponds to not having any transition to follow in the automaton). If we can perform m iterations then we report a match.

We use a mask B which for each character c stores a bit mask. This mask sets the bits corresponding to the positions where the pattern has the character c (just as in Shift-And). The formula to update D follows

$$D' \leftarrow (D \ \& \ B[T_j]) \ << \ 1$$

The algorithm is summarized in Figure 5. Some optimizations done on the real code, related to improved flow of control and bit manipulation tricks, are not shown for clarity.

```
BNDM (p = p1p2...pm, T = t1t2...tn)
    Preprocessing
        For c ∈ Σ do B[c] ← 0^m
        For i ∈ 1..m do B[p_{m-i+1}] ← B[p_{m-i+1}] | 0^{m-i}10^{i-1}
    Search
        pos ← 0
        While pos <= n - m do
            j ← m, last ← m
            D = 1^m
            While D != 0^m do
                D ← D & B[T_{pos+j}]
                j ← j - 1
                if D & 10^{m-1} != 0^m then
                    if j > 0 then last ← j
                             else report an occurrence at pos + 1
                D ← D << 1
            End of while
            pos ← pos + last
        End of while
```

Fig. 5. Bit-parallel code for **BDM**. Some optimizations are not shown for clarity.

Search example: we search the pattern *aabbaab* in the text $T =$ *a b b a b a a b b a a b*. Immediately after each step number (1 to 11) we show the text and note the current window between square brackets, as well as the recognized prefix in a rectangle. We begin with

$T =$ [*a b b a b a a*] *b b a a b*, $D = 1\,1\,1\,1\,1\,1\,1$, $B =$

a	1 1 0 0 1 1 0
b	0 0 1 1 0 0 1

, $m = 7$, $last = 7$, $j = 7$.

1. [*abbaba* *a*] *bbaab*.

	1 1 1 1 1 1 1
&	1 1 0 0 1 1 0
$D =$	1 1 0 0 1 1 0

$j = 6$, $last = 6$

2. [*abbab* *aa*] *bbaab*.

	1 0 0 1 1 0 0
&	1 1 0 0 1 1 0
$D =$	1 0 0 0 1 0 0

$j = 5$, $last = 5$

3. [*abba* *baa*] *bbaab*.

	0 0 0 1 0 0 0
&	0 0 1 1 0 0 1
$D =$	0 0 0 1 0 0 0

$j = 4$, $last = 5$

4. [*abb* *abaa*] *bbaab*.

	0 0 1 0 0 0 0
&	1 1 0 0 1 1 0
$D =$	0 0 0 0 0 0 0

$j = 3$, $last = 5$

We fail to recognize the next *a*. So we shift the window to *last*. We search again in the position: *abbab* [*aabbaab*], $last = 7$, $j = 7$.

5. *abbab* [*aabbaa* *b*].

	1 1 1 1 1 1 1
&	0 0 1 1 0 0 1
$D =$	0 0 1 1 0 0 1

$j = 6$, $last = 7$

6. *abbab* [*aabba* *ab*].

	0 1 1 0 0 1 0
&	1 1 0 0 1 1 0
$D =$	0 1 0 0 0 1 0

$j = 5$, $last = 7$

7. *abbab* [*aabb* *aab*].

	1 0 0 0 1 0 0
&	1 1 0 0 1 1 0
$D =$	1 0 0 0 1 0 0

$j = 4$, $last = 4$

8. *abbab* [*aab* *baab*].

	0 0 0 1 0 0 0
&	0 0 1 1 0 0 1
$D =$	0 0 0 1 0 0 0

$j = 3$, $last = 4$

9. *abbab* [*aa* *bbaa b*].

	0 0 1 0 0 0 0
&	0 0 1 1 0 0 1
$D =$	0 0 1 0 0 0 0

$j = 2$, $last = 4$

10. *abbab* [*a* *abbaab*].

	0 1 0 0 0 0 0
&	1 1 0 0 1 1 0
$D =$	0 1 0 0 0 0 0

$j = 2$, $last = 4$

11. *abbab* [*aabbaab*].

	1 0 0 0 0 0 0
&	1 1 0 0 1 1 0
$D =$	1 0 0 0 0 0 0

$j = 0$, $last = 4$

Report an occurrence at 6.

4.2 Handling Longer Patterns

We can cope with longer patterns by setting up an array of words D_t and simulating the work on a long computer word. We propose a different alternative which was experimentally found to be faster.

If $m > w$, we partition the pattern in $M = \lceil m/w \rceil$ subpatterns s_i, such that $p = s_1\ s_2\ ...\ s_M$ and s_i is of length $m_i = w$ if $i < M$ and $m_M = m - w(M-1)$. Those subpatterns are searched with the basic algorithm.

We now search s_1 in the text with the basic algorithm. If s_1 is found at a text position j, we verify whether s_2 follows it. That is, we position a window at $T_{j+m_1}..T_{j+m_1+m_2-1}$ and use the basic algorithm for s_2 in that window. If s_2 is in the window, we continue similarly with s_3 and so on. This process ends either because we find the complete pattern and report it, or because we fail to find a subpattern s_i.

We have to move the window now. An easy alternative is to use the shift $last_1$ that corresponds to the search of s_1. However, if we tested the subpatterns s_1 to s_i, each one has a possible shift $last_i$, and we use the maximum of all shifts.

4.3 Analysis

The preprocessing time for our algorithm is $O(m+|\Sigma|)$ if $m \leq w$, and $O(m(1+|\Sigma|/w))$ otherwise.

In the simple case $m \leq w$, the analysis is the same as for the BDM algorithm. That is, $O(mn)$ in the worst case (e.g. $T = a^n$, $p = a^m$), $O(n/m)$ in the best case (e.g. $T = a^n$, $p = a^{m-1}b$), and $O(n \log_{|\Sigma|} m/m)$ on average (which is optimal). Our algorithm, however, benefits from more locality of reference, since we do not access an automaton but only a few variables which can be put in registers (with the exception of the B table). As we show in the experiments, this difference makes our algorithm the fastest one.

When $m > w$, our algorithm is $O(nm^2/w)$ in the worst case (since each of the $O(mn)$ steps of the BDM algorithm forces to work on $\lceil m/w \rceil$ computer words). The best case occurs when the text traversal using s_1 always performs its maximum shift after looking one character, which is $O(n/w)$. We show, finally, that the average case is $O(n \log_{|\Sigma|} w/w)$. Clearly these complexities are worse than those of the simple BDM algorithm for long enough patterns. We show in the experiments up to which length our version is faster in practice.

The search cost for s_1 is $O(n \log_{|\Sigma|} w/w)$. With probability $1/|\Sigma|^w$, we find s_1 and check for the rest of the pattern. The check for s_2 in the window costs $O(w)$ at most. With probability $1/|\Sigma|^w$ we find s_2 and check s_3, and so on. The total cost incurred by the existence of $s_2 ... s_M$ is at most

$$\sum_{i=1}^{M-1} \frac{w}{|\Sigma|^{wi}} \;\leq\; \varepsilon \;=\; \frac{w}{|\Sigma|^w}\,(1 + O(w/|\Sigma|^w)) \;=\; O(1)$$

which therefore does not affect the main cost to search s_1 (neither in theory since the extra cost is $O(1)$ nor in practice since ε is very small). We consider the shifts now. The search of each subpattern s_i provides a shift $last_i$, and we take the maximum shift. Now, the shift $last_i$ participates in this maximum with probability $1/|\Sigma|^{wi}$. The longest possible shift is w. Hence, if we *sum* (instead of taking the maximum) the *longest possible* shifts w with their probabilities, we get into the same sum above, which is $\varepsilon = O(1)$. Therefore, the average shift is longer than $last_1$ and shorter than $last_1 + \varepsilon = last_1 + O(1)$, and hence the cost is that of searching s_1 plus lower order terms.

5 Further Improvements

5.1 A Linear Algorithm

Although our algorithm has optimal average case, it is not linear in the worst case even for $m \leq w$, since we can traverse the complete window backwards and advance it in one character. Our aim now is to reduce its worst case from $O(nm^2/w)$ to $O(nm/w)$, i.e. $O(n)$ when $m = O(w)$.

Improved variations on BDM already exist, such as Turbo_BDM and Turbo_RF [10, 15], the last one being linear in the worst case and still sublinear on average. The main idea is to avoid retraversing the same characters in the backward window verification using the fact that when we advance the window in *last* positions, we already know that $T_{i+last}..T_{i+m-1}$ is a prefix of the pattern (recall Figure 3). The ending position of the prefix in the window is usually called the *critical position*. The main problem if this area is not retraversed is how to determine the next shift, since among all possible shifts in $T_{i+last}..T_{i+m-1}$ we remember only the first one.

One strategy adds a kind of BM machine to the BDM algorithm. It works as follows: let u be the pattern prefix before the critical position. If we reach the critical position after reading (backwards) a factor z with the DAWG, it is possible to know whether z^r is a suffix of the pattern p: if z^r is a suffix, (i.e. $p = uz^r$) we recognize the whole pattern p, and the next shift corresponds to the longest *border* of p (i.e. the longest proper prefix that is also a suffix), which can be computed in advance. If z^r is not a suffix, it appears in the pattern in a set of positions which is given by the state we reached in the suffix automaton. We shift to the rightmost occurrence of z^r in the pattern.

It is not difficult to simulate this idea in our BNDM algorithm. To know if the factor z we read with the DAWG is a suffix, we test whether $d_{|z|} = 1$. To get the rightmost occurrence, we seek the rightmost 1 in D, which we can get (if it exists) in constant time with $\log_2(D \,\&\, \sim (D-1))$ [5]. We implemented this algorithm under the name BM_BNDM in the experimental part of this paper.

This algorithm remains quadratic, because we do not keep a prefix of the pattern after the BM shift. We do that now. Recall that u is the prefix before the critical position. The Turbo_RF (second variation) [10] uses a complicated preprocessing phase to associate in linear time an occurrence of z^r in the pattern to a border b_u of u, in order to obtain the maximal prefix of the pattern that is a suffix of uz^r. Moreover, the Turbo_RF uses a suffix tree, and it is quite difficult to use this preprocessing phase on DAWGs. With our simulation, this preprocessing phase becomes simple. To each prefix u_i of the pattern p, we associate a mask $Bord[i]$ that registers the starting positions of the borders of u_i (ϵ included). This table is precomputed in linear time. Now, to join one occurrence of z^r to a border of u, we want the positions which start a border of u *and* continue with an occurrence of z^r. The first set of positions is $Bord[i]$, and the second one is precisely the current D value (i.e. positions in the pattern where the recognized factor z ends). Hence, the bits of $X = Bord[i] \,\&\, D$ are the positions satisfying both criteria. As we want the rightmost such occurrence (i.e. the maximal prefix), we take again $\log_2(X \& \sim (X-1))$. We implemented this algorithm under the name Turbo_BNDM in the experimental part of this paper.

5.2 A Constant-Space Algorithm

It is also interesting to notice that, although the algorithm needs $O(|\Sigma| m/w)$ extra space, we can make it constant space on a binary alphabet $\Sigma_2 = \{0,1\}$.

[5] It is faster and cleaner to implement this $\log_2$ by shifting the mask to the right until it becomes zero. Using this technique we can use the simpler expression $D \,\hat{}\, (D-1)$ and get the same result.

The trick is that in this case, $B[1] = p$ and $B[0] = \sim B[1]$. Therefore, we need no extra storage apart from the pattern itself to perform all the operations. In theory, any text over a finite alphabet Σ could be searched in constant space by representing the symbols of Σ with bits and working on the bits (the misaligned matches have to be later discarded). This involves an average search time of

$$\frac{n \log_2 |\Sigma|}{m \log_2 |\Sigma|} \log_2(m \log_2 |\Sigma|) \quad = \quad \text{Normal time} \times \log_2 |\Sigma| \left(1 + \frac{\log_2 \log_2 |\Sigma|}{\log_2 m}\right)$$

which if the alphabet is considered of constant size is of the same order of the normal search time.

6 Extensions

We analyze now some extensions applicable to our basic scheme, which form a successful combination of efficiency and flexibility.

6.1 Classes of Characters

As in the Shift-Or algorithm, we allow that each position in the pattern matches not only a single character but an arbitrary set of characters. We call "extended patterns" those that are more complex than a simple string to be searched. In this work the only extended patterns we deal with are those allowing a class of characters at each position.

We denote $p = C_1C_2 \ldots C_m$ such extended patterns. A word $x = x_1x_2 \ldots x_r$ in Σ^* is a factor of an extended pattern $p = C_1C_2 \ldots C_m$ if there exists an i such that $x_1 \in C_{i-r+1}, x_2 \in C_{i-r+2}, \ldots, x_r \in C_i$. Such an i is called a *position* of x in p. A factor $x = x_1x_2 \ldots x_r$ of $p = C_1C_2 \ldots C_m$ is a *suffix* if $x_1 \in C_{m-r+1}, x_2 \in C_{i-r+2}, \ldots, x_r \in C_m$.

Similarly to the first part of this work, we design an automaton which recognizes all suffixes of an extended pattern $p = C_1C_2 \ldots C_m$. This automaton is not anymore a DAWG. We call it Extended_DAWG. To our knowledge, this kind of automaton has never been studied. We first give a formal construction of the Extended_DAWG (proving its correctness) and later present a bit-parallel implementation.

Construction The construction we use is quite similar to the one we give for the DAWG, except for the new definition of suffixes. For any x factor of p, we denote *L-endpos(x)* the set of positions of x in p. For example, *L-endpos*$(baa) = \{3, 7\}$ in the extended pattern *b[a,b]abbaa*, and *L-endpos*$(bba) = \{3, 6\}$ (notice that, unlike before, the sets of positions can be not disjoint and no one a subset of the other). We define the equivalence relation $\equiv_E$ for u, v factors of p by

$$u \equiv_E v \text{ if and only if } L\text{-}endpos(u) = L\text{-}endpos(v).$$

We define $\gamma_p(i,\sigma)$ with $i \in \{0,1,\ldots,m,m+1\}, \sigma \in \Sigma$ by

$$\gamma_p(i,\sigma) = \begin{cases} \{i\} \text{ if } i \leq m \text{ and } \sigma \in C_i \\ \emptyset \text{ otherwise} \end{cases}$$

Lemma 1 *Let p be an extended pattern and $\equiv_E$ the equivalence relation on its factors (as previously defined). The equivalence relation $\equiv_E$ is compatible with the concatenation on words.*

This lemma allows us to define an automaton from this equivalence class. States of the automaton are the equivalence classes of $\equiv_E$. There is an edge labeled by σ from the set of positions $\{i_1, i_2, \ldots i_k\}$ to $\gamma_p(i_1+1,\sigma) \cup \gamma_p(i_2+1,\sigma) \cup \ldots \cup \gamma_p(i_k+1,\sigma)$, if it is not empty. The initial node of the automaton is the set that contains all the positions. Terminals nodes of the automaton are the set of positions that contain m. As an example, the suffix automaton of the word *[a,b]aa[a,b]baa* is given in Figure 6.

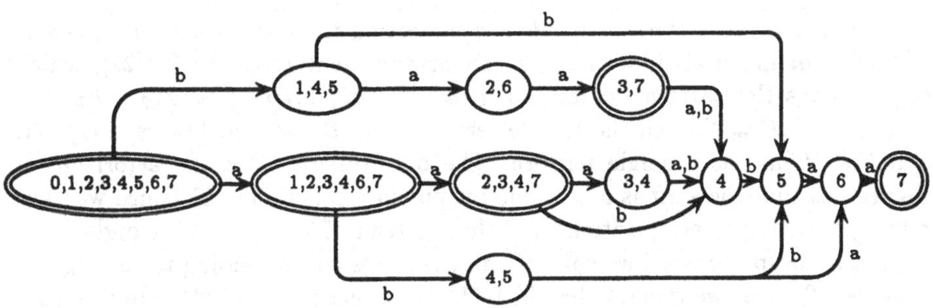

Fig. 6. Extended_DAWG of the extended pattern $^0[a,b]^1a^2a^3[a,b]^4b^5a^6a^7$

Lemma 2 *The Extended_DAWG of an extended pattern $p = C_1C_2\ldots C_m$ recognizes the set of suffixes of p.*

We can use this new automaton to recognize the set of suffixes of an extended pattern p. We do not give an algorithm to build this Extended_DAWG in its deterministic form, but we simulate the deterministic automaton using bit-parallelism.

A bit-parallel implementation: from the above construction, the only modification that our algorithm needs is that the B table has the i-th bit set for all characters belonging to the set of the i-th position of the pattern. Therefore we simply change line 3 (part of the preprocessing) in the algorithm of Figure 5 to

$$\textbf{For } i \in 1..m, c \in \Sigma \textbf{ do if } c \in p_i \textbf{ then } B[c] \leftarrow B[c] \mid 0^{m-i}10^{i-1}$$

such that now the preprocessing takes $O(|\Sigma|m)$ time but the search algorithm does not change.

We combine the flexibility of extended patterns with the efficiency of a Boyer-Moore-like algorithm. It should be clear, however, that the efficiency of the shifts can be degraded if the classes of characters are significantly large and prevent long shifts. However, this is much more resistant than some simple variations of Boyer-Moore since it uses more knowledge about the matched characters.

We point out now another extension related to classes of characters: the text itself may have basic characters as well as other symbols denoting sets of basic characters. This is common, for instance, in DNA databases. We can easily handle such texts. Assume that the symbol C represents the set $\{c_1, ..., c_r\}$. Then we set $B[C] = B[c_1] \mid ... \mid B[c_r]$. This is much more difficult to achieve with algorithms not based on bit-parallelism.

6.2 Multiple Patterns

To search a set of patterns $P1...Pr$ (i.e. reporting the occurrences of all them) of length m in parallel, we can use an arrangement proposed in [22], which concatenates the patterns as follows: $P = P1_1\ P2_1\ ...Pr_1\ P1_2\ P2_2\ ...Pr_2\ .....\ P1_m\ P2_m\ ...Pr_m$ (i.e. all the first letters, then all the second letters, etc.) and searches P just as a single pattern. The only difference in the algorithm of Figure 5 is that the shift is not in one bit but in r bits in line 15 (since we have r bits per multipattern position) and that instead of looking for the highest bit d_m of the computer word we consider all the r bits corresponding to the highest position. That is, we replace the old 10^{m-1} test mask by $1^r0^{r(m-1)}$ in line 12.

This will automatically search for r words of length m and keep all the bits needed for each word. Moreover, it will report the matches of any of the patterns and will not allow shifting more than what all patterns allow to shift.

An alternative arrangement is: $P = P1\ P2\ ...\ Pr$ (i.e. just concatenate the patterns). In this case the shift in line 15 is for one bit, and the mask for line 12 is $(10^{m-1})^r$. On some processors a shift in one position is faster than a shift in $r > 1$ positions, which could be an advantage for this arrangement. On the other hand, in this case we must clear the bits that are carried from the highest position of a pattern to the next one, replacing line 15 for $D = (D << 1)\ \&\ (1^{m-1}0)^r$. This involves an extra operation. Finally, this arrangement allows to have patterns of different lengths for the algorithm of Wu and Manber [22] which is not possible in their current proposal.

Clearly these techniques cannot be applied to the case $m > \lfloor w/2 \rfloor$. However, if $m \leq \lfloor w/2 \rfloor$ and $r \times m > w$ we divide the set of patterns into $\lceil r / \lfloor w/m \rfloor \rceil$ groups, so that the patterns in each group fit in w bits. Since this skips characters, it is better on average than [22]. As we show in the experiments, this is also better than sequentially searching each pattern in turn, even given that the shifts are the most conservative among all the r patterns.

6.3 Approximate String Matching

Approximate string matching is the problem of finding all the occurrences of a pattern in a text allowing at most k "errors". The errors are insertions, deletions and replacements to perform in the pattern so that it matches the text. In [22], an efficient filter is proposed to determine that large text areas cannot contain an occurrence. It is based on dividing the pattern in $k+1$ pieces and searching all the pieces in parallel. Since k errors cannot destroy the $k+_1$ pieces, some of the pieces must appear with no errors close to each occurrence. They use the multipattern search algorithm mentioned in the previous paragraph. In [4, 3], a multipattern Boyer-Moore strategy is preferred, which is faster but does not handle classes of characters and other extensions. This algorithm is the fastest one for low error levels.

Our multipattern search technique presented in the previous section combines the best of both worlds: our performance is comparable to Boyer-Moore algorithms and we keep the flexibility of bit-parallelism handle classes of characters. We show in the experiments how our algorithm performs in this setup.

7 Experimental Results

We ran extensive experiments on random and natural language text to show how efficient are our algorithms in practice. The experiments were run on a Sun UltraSparc-1 of 167 MHz, with 64 Mb of RAM, running SunOS 5.5.1. We measure CPU times, which are within ±2% with 95% confidence. We used random texts and patterns with $\sigma = 2$ to 64, as well as natural language text and DNA sequences.

We show in Figure 7 some of the results for short ($m \leq w$) and long ($m > w$) patterns. The comparison includes the best known algorithms: BM, BM-Sunday, KMP (very slow to appear in the plots, close to 0.14 sec/Mb), Shift-Or (not always shown, close to 0.07 sec/Mb), classical BDM, and our three bit-parallel variants: BNDM, BM_BNDM and Turbo_BNDM.

Our bit-parallel algorithms are always the fastest for short patterns, except for $m \leq$ 2-6. The fastest algorithm is BM_BNDM, though it is very close to simple BNDM. Classical BDM, on the other hand, is sometimes slower than the BM family. Turbo_BNDM is competitive with simple BNDM and has linear worst case. Our algorithms are especially good for small alphabets since they use more information on the matched pattern than others. The only good competitor for small alphabets is Boyer-Moore, which however is slower because the code is more complex (notice that Boyer-Moore is faster than BDM, but slower than BNDM). For larger alphabets, on the other hand, another very simple algorithm gets very close: BM-Sunday. However, we are always at least 10% faster.

On longer patterns[6] our algorithm ceases to improve because it basically

[6] We did not include the more complex variations of our algorithm because they have already been shown very similar to the simple one. We did not include also the algorithms which are known not to improve, such as Shift-Or and KMP.

searches for the first w letters of the pattern, while classical BDM keeps improving. Hence, our algorithm ceases to be the best one (beaten by BDM) for $m \geq$ 90-150. This value would at least duplicate in a 64-bit architecture.

We show also some illustrative results using classes of characters, which were generated manually as follows: we select from an English text an infrequent word, namely "`responsible`" (close to 10 matches per megabyte). Then we replace its first or last characters by the class $\{a..z\}$. This will adversely affect the shifts of the BNDM algorithm. We compare the efficiency against Shift-Or. The result is presented in Table 1, which shows that even in the case of three initial or final letters allowing a large class of characters the shifts are significant and we double the performance of Shift-Or. Hence, our goals of handling classes of characters with improved search times are achieved.

Pattern	Shift-Or	BNDM
`responsible`	6.58	2.71
`responsibl?`	6.51	2.96
`responsib??`	6.52	3.23
`responsi???`	6.49	3.40
`?esponsible`	6.46	2.93
`??sponsible`	6.55	3.42
`???ponsible`	6.51	3.78

Table 1. Search times with classes of characters, in 1/100-th of seconds per megabyte on English text. The question mark '`?`' represents the class $\{a..z\}$.

We present in Figure 8 some results on our multipattern algorithm, to show that although we take the minimum shift among all the patterns, we can still do better than searching each pattern in turn. We take random groups of five patterns of length 6 and compare our multipattern algorithm (in its two versions, called Multi-BNDM (1) and (2) attending to their presentation order), against five sequential searches with BNDM (called BNDM in the legend), and against the parallel version proposed in [22] (called Multi-WM). As it can be seen, our first arrangement is slightly more efficient than the second one, they are always more efficient than a sequential search (although the improvement is not five-fold but two- or three-fold because of shorter shifts), and are more efficient than the proposal of [22] provided $\sigma \geq 8$.

Finally, we show the performance of our multipattern algorithm when used for approximate string matching. We include the fastest known algorithms in the comparison [4, 3, 7, 13, 16, 23, 22]. We compare those algorithms against our version of [4] (where the Sunday algorithm is replaced by our BNDM), while we consider [22] not as the bit-parallel algorithm presented there but their other proposal, namely reduction to exact searching using their algorithm Multi-WM for multipattern search (shown in the previous experiment). Figure 9 shows the results for different alphabet sizes and $m = 20$.

Since BNDM is not very good for very short patterns, the approximate search algorithm ceases to be competitive short before the original version [4]. This is because the length of the patterns to search for is $O(m/k)$. Despite this drawback, our algorithm is quite close to [4] (sometimes even faster) which makes it a reasonably competitive yet more flexible alternative, while being faster than the other flexible candidate [22].

8 Conclusions

We present a new algorithm (called BNDM) based on the bit-parallel simulation of a nondeterministic suffix automaton. This automaton has been previously used in deterministic form in an algorithm called BDM. Our new algorithm is experimentally shown to be very fast on average. It is the fastest algorithm in all cases for patterns from length 5 to 110 (on English; the bounds vary depending on the alphabet size and the architecture). We present also some variations called Turbo_BNDM and BM_BNDM which are derived from the corresponding variants of BDM. These variants are much more simply implemented using bit-parallelism and become practical algorithms. Turbo_BNDM has average performance very close to BNDM, though $O(n)$ worst case behavior, while BM_BNDM is slightly faster than BNDM. The BNDM algorithm can be extended simply and efficiently to handle classes of characters, multiple pattern matching and approximate pattern matching, among others.

The new suffix automaton we introduce and simulate for classes of characters has never been studied. Its study should permit to extend the BDM and Turbo_RF to handle classes of characters.

The Agrep software [21] is in many cases faster than BNDM. However, Agrep is just a BM algorithm which uses pairs of characters instead of single ones. This is an orthogonal technique that can be incorporated in all algorithms, and a general study of this technique would permit to improve the speed of pattern matching softwares. We plan to work on this idea too.

References

1. R. Baeza-Yates. Text retrieval: Theory and practice. In *12th IFIP World Computer Congress*, volume I, pages 465–476. Elsevier Science, September 1992.
2. R. Baeza-Yates and G. Gonnet. A new approach to text searching. *CACM*, 35(10):74–82, October 1992.
3. R. Baeza-Yates and G. Navarro. A faster algorithm for approximate string matching. In *Proc. of CPM'96*, pages 1–23, 1996.
4. R. Baeza-Yates and C. Perleberg. Fast and practical approximate pattern matching. In *Proc. CPM'92*, pages 185–192. Springer-Verlag, 1992. LNCS 644.
5. A. Blumer, A. Ehrenfeucht, and D. Haussler. Average sizes of suffix trees and dawgs. *Discrete Applied Mathematics*, 24(1):37–45, 1989.
6. R. S. Boyer and J. S. Moore. A fast string searching algorithm. *Communications of the ACM*, 20(10):762–772, 1977.

7. W. Chang and J. Lampe. Theoretical and empirical comparisons of approximate string matching algorithms. In *Proc. of CPM'92*, pages 172–181, 1992. LNCS 644.
8. M. Crochemore. Transducers and repetitions. *Theor. Comput. Sci.*, 45(1):63–86, 1986.
9. M. Crochemore, A. Czumaj, L. Gasieniec, S. Jarominek, T. Lecroq, W. Plandowski, and W. Rytter. Fast practical multi-pattern matching. Rapport 93-3, Institut Gaspard Monge, Université de Marne la Vallée, 1993.
10. M. Crochemore, A. Czumaj, L. Gasieniec, S. Jarominek, T. Lecroq, W. Plandowski, and W. Rytter. Speeding up two string-matching algorithms. *Algorithmica*, (12):247–267, 1994.
11. M. Crochemore and W. Rytter. *Text algorithms*. Oxford University Press, 1994.
12. R. N. Horspool. Practical fast searching in strings. *Softw. Pract. Exp.*, 10:501–506, 1980.
13. P. Jokinen, J. Tarhio, and E. Ukkonen. A comparison of approximate string matching algorithms. *Software Practice and Experience*, 26(12):1439–1458, 1996.
14. D. E. Knuth, J. H. Morris, Jr, and V. R. Pratt. Fast pattern matching in strings. *SIAM Journal on Computing*, 6(1):323–350, 1977.
15. T. Lecroq. *Recherches de mot*. Thèse de doctorat, Université d'Orléans, France, 1992.
16. G. Navarro. A partial deterministic automaton for approximate string matching. In *Proc. of WSP'97*, pages 112–124. Carleton University Press, 1997.
17. G. Navarro and M. Raffinot. A bit-parallel approach to suffix automata: Fast extended string matching. Technical Report TR/DCC-98-1, Dept. of Computer Science, Univ. of Chile, Jan 1998. `ftp://ftp.dcc.uchile.cl/pub/users/-gnavarro/bndm.ps.gz`.
18. M. Raffinot. Asymptotic estimation of the average number of terminal states in dawgs. In R. Baeza-Yates, editor, *Proc. of WSP'97*, pages 140–148, Valparaiso, Chile, November 12-13, 1997. Carleton University Press.
19. M. Raffinot. On the multi backward dawg matching algorithm (MultiBDM). In R. Baeza-Yates, editor, *Proceedings of the 4rd South American Workshop on String Processing*, pages 149–165, Valparaiso, Chile, November 12-13, 1997. Carleton University Press.
20. D. Sunday. A very fast substring search algorithm. *CACM*, 33(8):132–142, August 1990.
21. S. Wu and U. Manber. Agrep – a fast approximate pattern-matching tool. In *Proc. of USENIX Technical Conference*, pages 153–162, 1992.
22. S. Wu and U. Manber. Fast text searching allowing errors. *CACM*, 35(10):83–91, October 1992.
23. S. Wu, U. Manber, and E. Myers. A sub-quadratic algorithm for approximate limited expression matching. *Algorithmica*, 15(1):50–67, 1996.
24. A. C. Yao. The complexity of pattern matching for a random string. *SIAM Journal on Computing*, 8(3):368–387, 1979.

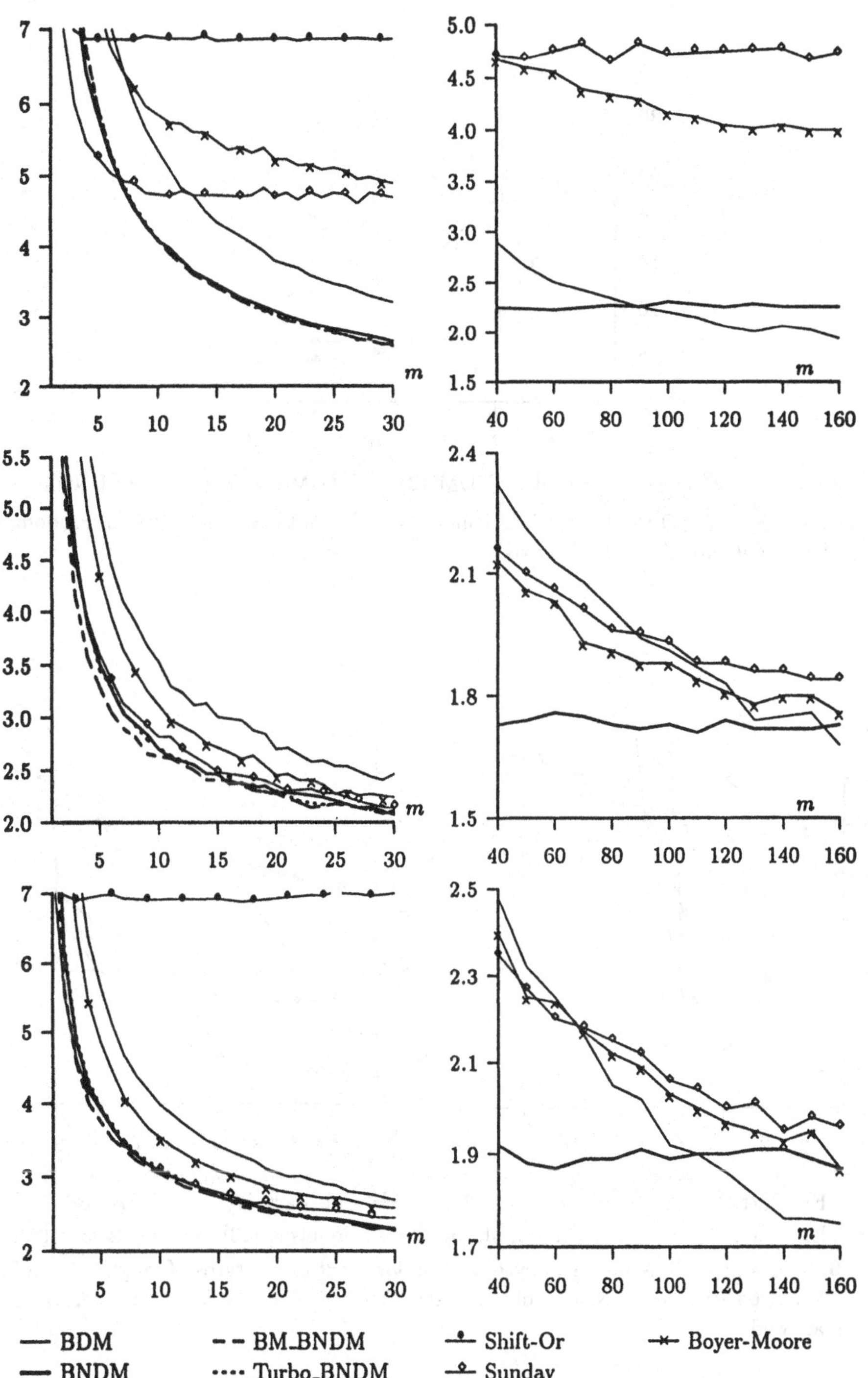

Fig. 7. Times in 1/100-th of seconds per megabyte. For first to third row, random text with $\sigma = 4$, random text with $\sigma = 64$ and English text. Left column shows short patterns, right column shows long patterns.

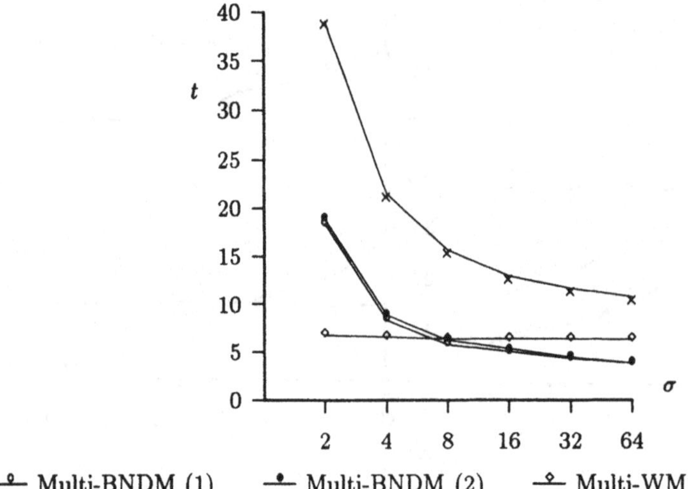

Fig. 8. Times in 1/100-th of seconds per megabyte, for multipattern search on random text of different alphabet sizes (x axis).

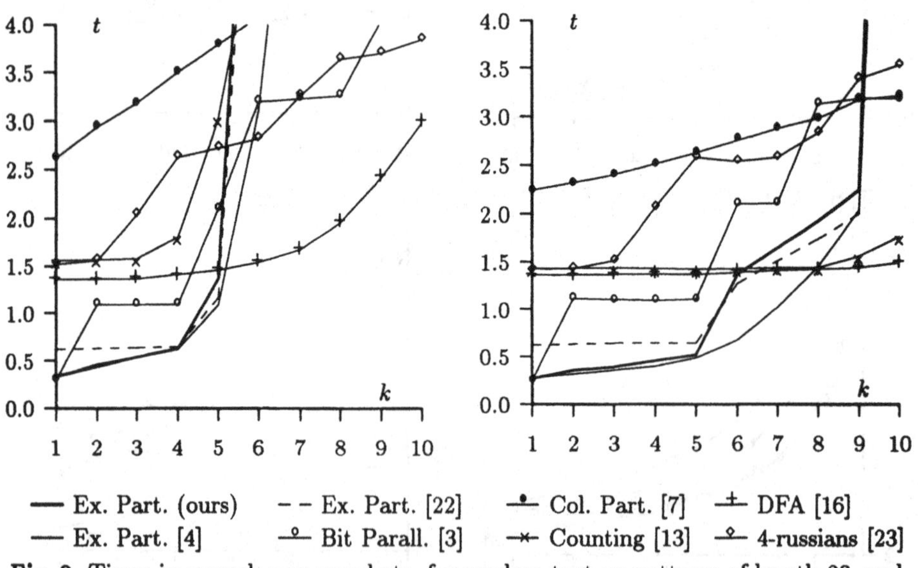

Fig. 9. Times in seconds per megabyte, for random text on patterns of length 20, and $\sigma = 16$ and 64 (first and second column, respectively). The x axis is the number of errors allowed.

A Dictionary Matching Algorithm Fast on the Average for Terms of Varying Length

Michal Ziv-Ukelson[1] and Aaron Kershenbaum[2]

[1] IBM T.J.W Research Center, Hawthorne, N.Y ukelson@watson.ibm.com

[2] Dept of C.S., Polytechnic University, Brooklyn, N.Y akershen@poly.edu

Abstract. We examine the exact dictionary matching problem with dynamic text and static terms and propose a simple but efficient algorithm with sublinear (in size of text) average performance for a wide range of practical problems. The algorithm is based on the Commentz-Walter-Horspool algorithm (CWH), presented by Baeza-Yates and Re'gnier [10]. Typically, our refinement will prune out more than 30% of characters scanned by CWH, when searching for all occurrences of tags, which are of varying lengths and members of a set of moderate size, in natural language text. This problem arises frequently in practice in scanning text downloaded from the internet, and accounts for a major portion of the preprocessing time associated with indexing such text for later retrieval. Our approach, which we refer to as layering, keeps track of an upper bound on the maximal length of potential term prefixes ending at each given position in the text. This information is then used to mask out some of the terms and filter out unnecessary character comparisons during the search. A practical implementation is described, which increases the size of the existing data structures as well as the preprocessing cost only by a factor of the size of the longest term in the set.

1 Introduction

String matching means finding all occurrences of a term p of length m in a text T. The *multi string matching* problem consists of finding all occurrences of any string from a static set $P = \{p1, p2,...pi\}$ in a given dynamic text T of size n. The set of target strings P is called a *dictionary*. Let $mmax$ denote the size of the longest term in the dictionary, $mmin$ the size of the shortest term in the dictionary, and M the sum of the dictionary term lengths. Let σ denote the size of the alphabet.

Several algorithms have been suggested for the string matching problem[1]. The linear algorithm by KMP [28] solves the problem in o(n) time and o(m) preprocessing time and memory space. Another string matching algorithm, which has an average behavior sublinear in n, is the Boyer-Moore algorithm [8], which searches for the pattern from right to left, and thus only a portion of the text needs to be scanned.

Aho and Corasick [2] extended the KMP algorithm to the multi-string search problem and show how to build an automaton of size O(M) and use it to scan the text in O(n) time on bounded alphabets. Various BM type algorithms were proposed for the multi-term case. In the Commentz-Walter [16, 17] algorithm, a reversed trie is constructed from the set of terms. Each node in the trie is then updated with its shift value. Also, an occurrence heuristic shift-table is constructed from all the terms in the set. "The algorithm aligns the root of the trie with the first possible matching position in the text, and starts comparing the text with the set of strings, from right to left. This is continued until we find all possible strings matching that position, or realize that none of them are there. The knowledge of these read characters yields the next possible matching position to the right, etc.."[10] Another version of the CW algorithm is described in [18], where the tables hold a shift value for each possible combination of "good suffix" and "bad character". Another BM type multi pattern matching algorithm which inspects at most 2*n characters is described in [13]. In [10], an algorithm is proposed which is based on the Commentz-Walter approach, combined with the Horspool [24] Boyer-Moore variant. We use CWH whenever referring to this algorithm. Other approaches to dictionary matching are presented in [6, 11, 26, 27], and for dynamic dictionaries in [3, 4, 5, 25].

Both the Boyer-Moore algorithm and its multi-term versions keep no history of the characters already processed in the scan, and may examine a character more than once, leading to a worst case behavior which is O(n*m) (or O(n*mmax) in the multi-term case). Tighter bounds on the complexity of the BM algorithm can be found in [14, 15, 20, 28].Various papers were written on the subject of improving Boyer-Moore type algorithms by retaining some of the information the algorithm forgets. These use more memory, more preprocessing, and do fewer real time calculations. Many of them are guided by the motivation to improve the worst case and optimize the amortized cost [7, 19]. These algorithms try to cut down on the repetitive processing of the same character and increase the size of the shift.[19] shows how to use prefix memorization to make the BM worst case linear. [17] describes a version of the algorithm with a linear worst case, extending Galil's improvement to the multi term case. In their famous paper, KMP[28] introduced the Boyer-Moore-Automaton (BMA for short), which keeps track of the characters already matched for the most recent m characters in the text. The states memorize information about the characters already matched in order to avoid repetitive examination. However, the upper bound for the number of states in such an automaton may be 2^m, except for certain families of words [9,12]. Another variation on the BM Automaton whose number of states is bounded by m^3 is described in [29].

We extend the concept of history memorization to the multi term case in a space-economic way. We also address the problem of handling variable length terms, rather than assuming all terms are of the same length. Our approach is practical, and is designed to improve the average performance of typical natural language text searches. Our algorithm is related to the BMA family in that it memorizes some of the history from previously examined characters. Yet, rather then keeping track of all matching characters examined so far, we utilize the mismatching characters as well as the matching characters in the skip loop as delimiters to the size of the running prefix length upper bound, cutting down on the search space. This is very effective for non periodic languages with large finite alphabets, since most of the skip loop comparisons result in rejections, and the majority of the match cycles are spent comparing characters at the root of the trie.

For simplicity as well as efficiency, we will demonstrate our filter as an extension to the CWH algorithm, even though the idea of layering can be adapted to enhance other Boyer-Moore type multi-string matching algorithms as well.

The paper is organized as follows. Section 2 summarises the CWH algorithm. In Section 3 we describe the new data structures and present our algorithm. Section 4 describes an efficient way to construct the data structures during the preprocessing phase. In Section 5 we show a faster version of the search using Boyer-Moore's fast skip loop structure. Section 6 includes a discussion of the "good suffix" heuristic, and empirical analysis which demonstrates why the good suffix heuristic would be redundant when using our extention of the algorithm for natural language queries. In Section 7 we give empirical results and analyze them.

2 Baeza-Yates's and Re'gnier's CWH Algorithm

We start with Algorithm 0, a simple version of the basic CWH[10] prior to our modifications. The basic data structures include an occurrence shift table, and a trie built from the reversed strings. (See *Figure 0*)

Fig. 0. CWH shift table and trie.

Algorithm 0 can be reverse-derived from the code given for Algorithm 1 in Section 3.3, by deleting the highlighted text, that is - removing line 9 and keeping the values of trie_layer and occurrence_layer set to 1.

As can be seen from line 8, the Horspool version of occurrence shift lookup is used[24], which means the rightmost scanned character is always used to look up the occurrence shift, whether it matches or not, rather then using the rightmost mismatching character, as in the standard BM and CW algorithms. Also, in line 8 only the occurrence heuristic is used to look up the next shift, instead of computing the maximum between the occurrence shift and the "good suffix" heuristic shift. Simplified Boyer Moore Horspool without the "good suffix" heuristic for single term search (SBM) is described in [24] and is shown to be at least as effective as the full BM for natural language applications, due to the non periodic nature of the text.
[10] shows how to use the "good suffix" heuristic as well, stating that it is important when σ is small. We will later include empirical analysis of the multi term case to

demonstrate why the "good suffix" heuristic would be redundant when using our extension to the algorithm for natural language queries.

In the best case, each comparison with the root of the trie results in a mismatch, and the number of characters inspected will be n/mmin. In the worst case (very rare), the number of characters inspected will be n*mmax. For average case complexity analysis see [10].

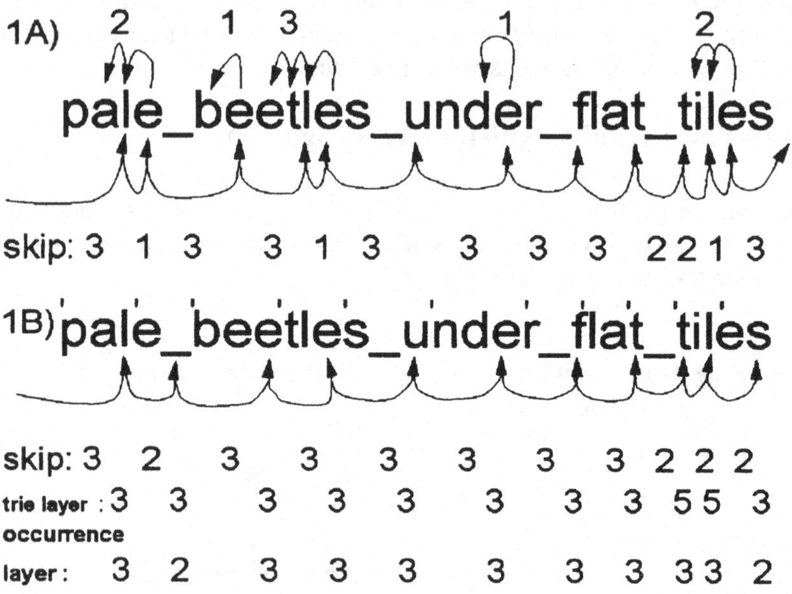

Fig. 1. (A) A trace of CWH algorithm (B) A trace of CWH + Layering algorithm

Figure 1A traces the simple CWH(Algorithm 0) search for the term-set {"log", "title"} over the text "pale beetles under flat tiles.. ". For CWH data structures representing the terms, see *Figure 0*. The route of the forward shift loop is traced underneath the text in *Figure 1A*, while the route of the backward match loop is traced above the text. The shift value is stated per forward skip underneath the forward route, and the maximal depth of trie is stated once per shift/match cycle above the backtrack route.

The CWH algorithm begins by aligning the root of the trie made of both terms with position 3(mmin) in the text, and starts matching the text from right to left following the corresponding path in the trie. The letter "l" in the 3rd position will be rejected, and the trie will be moved 1 character to the right, since shift[l] = 1.

We are now ready to begin the second match/shift cycle. The letter "e" in the 4th position will successfully match the last character of "title", resulting in a backtrack

step. The letter "l" in the 3rd position is examined again, successfully matched against the "l" from "title", resulting in another backtrack step. Comparison of the letter "a" in the 2nd position results in failure, breaking the backtrack loop.
To calculate the next shift, we go back to letter "e" in the 4th position. The shift value for "e" is 3, so the text position is moved forward by 3 to the first "e" in "beetles", and so on. Altogether, 21 characters comparisons are made, 3 of them repetitive examinations of the same character. (Text size is 29).

3 The Layering Mechanism

3.1 Informal outline of our approach

The Commentz-Walter algorithm's sublinearity is accomplished by jumping over some of the text characters. The length of the jumps depends on the density of the occurrence table, and is bounded by the length of the shortest term in the set. The complexity of backtrack comparisons depends on the frequency in the text of the suffixes of the terms.

To explain the intuition behind our algorithm, try to picture a divider wall being pulled along as we traverse the text. We know that all potential term occurrences in the text must start to the right of this wall, and therefore any term longer than the distance between the current text position and the wall gets masked out of the trie, and any character appearing in a term at an index which is greater than this distance gets masked out of the occurrence table. We start the search with the divider wall located prior to the first position in the text, and with the text pointer at position mmin. Therefore, the first upper bound to prefix length (the distance between the divider wall and the search position in the text) is of value mmin. As the search progresses through the text, we use the scanned characters and a data structure compiled at preprocessing time to relocate the divider wall with minimal lag behind the advancing position in the text.

To demonstrate the new heuristic, we go back to the CWH trace in *Figure 1A*: Upon rejection of the letter "l" in the 3rd position we were able to shift the trie 1 character to the right, since shift["l"] is 1. However, if we know that the upper bound on any length of a prefix ending at third position is 3(mmin), and since the character "l" only appears in the term "title" at position 4, we could safely move the trie 2 characters to the right (so we don't miss the term "log"), thus increasing the shift by 1 character.
Upon encountering the letter "e" at the 4th position, the letter was successfully matched against the root of the trie, resulting in backtracking to the left and 2 additional character comparisons for the characters "l" and "a" before the trie is moved to the right again. However, if we know that the upper bound on prefix length at this point is 4, then there is no point in traversing the branch of the trie that leads to a term of length 5 ("title"). The only relevant branch in the trie to be traversed is that for "log", which would result in an immediate mismatch, thus saving 2 extra character comparisons.

In this paper we show how to pull along this "divider wall" as we traverse the text, using matching as well as mismatching characters to delimit the prefix length upper bound, and how to use it as a filter via a mechanism called "Layering by Max Prefix Length". For each layer representing a potential prefix length from 1 to mmax, we keep its relevant occurrence table and sub-trie. (See *Figure* 2)

By layering the occurrence table we reduce its density, resulting in longer jumps. By layering the term trie we reduce the branching factor, thus filtering out unnecessary backtrack match chains.

3.2 The new data structures

The structures used to guide the new search include a forest of tries, and a state machine with an occurrence table per state. (See *Figure* 2)

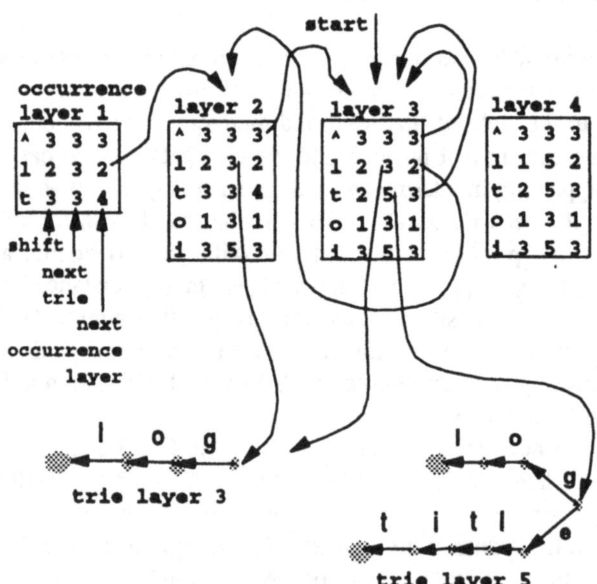

Fig. 2. The state machine for the terms "log" and "title" consists of 4 occurrence tables, and 2 tries. The number of occurrence layers in the PLA is (mmax - 1). (The last occurrence layer state corresponding to mmax is redundant). It is important to note that the symbol "^" in the diagram was used to represent any character alien to the layer, that is, any character not appearing in any of the prefixes up to length L. For such characters, all fields are set to mmin.

Definition 1: we use the term *layer* to define the running upper bound to the maximum length over all potential term prefixes ending at a given position in the text.

In simpler words, we use *layering* to represent information about the value of the current right-offset from the running "divider wall". A different offset value is used for the match and shift phases of the cycle, and therefore both trie and occurrence table are layered.
The *trie layering* is represented by a forest of tries, where a trie of index k (0 < k <= mmax) includes all reversed terms of length smaller than or equal to k.
The *occurrence table layering* is represented by the Prefix Length Automaton(PLA for short). Each state j in the PLA (1 <= j <= mmax) represents occurrence layer j, and contains an *occurrence table* with the following four columns:
(See Section 4 for formal definitions of the field values).
1. *char* - just like in any BM occurrence shift table, there is an entry for each character in the alphabet.
2. *occurrence shift* - is equivalent to BM's delta1 - specifies how many characters to jump forward to the next text position upon encountering the *char* key from the first column.
3. *next_trie_layer* - indicates the trie to be used for the match phase of the next match/shift cycle.
4. *next_occurrence_layer* - indicates the next PLA state, to which a transition should be made. The occurrence table of the new state will be used for the shift phase of the next match/shift cycle.

Lemma 1. *The total size of the new data structures is linearly bounded by O(mmax*(σ + M)), which increases the size of the existing CWH data structures only by a factor of mmax.*

proof: The data structures consist of the PLA and the trie forest. The PLA consists of an occurrence table per layer. By definition, the number of layers is bounded by the size of the longest term in the set. Each occurrence table has a line per character in the alphabet, and therefore the size of the PLA is O(mmax * σ). The trie forest, in the worst case, consists of a trie per layer. The size of a trie is linearly bounded by the sum of term lengths. So the trie forest is O(mmax * M).

3.3 An algorithm for Layered CWH multi-pattern search

The following procedures are used by Algorithm 1:
MakeTrieTransition(q, char, trie_layer) makes a transition on the sub-trie indicated by *trie_layer*, from state *q*, using the edge labeled *char*, and returns the new trie state.
MakePLATransition(char, occurrence_layer) uses entry *char* in the table defined for the current *occurrence_layer* to retrieve the *next trie_layer*, and then to make a transition to the *next occurrence layer*. The *next trie_layer* and the *next occurrence layer* are returned.
Remark: A state q in a trie state machine is labeled *accepting*, if the path from the root of the trie to q spells a word in the dictionary. The sink state and all leaf states are labeled *final*.

```
/* initialization */   i <- mmin
                       occurrence_layer <- mmin
                       trie_layer <- mmin

while ( i <= text_length )                         /* the search   */
  {
    q   <- q0                                      /* match phase */
    j   <- 0
    while( (j < i) and (q is not final) )
    {
      q <- MakeTrieTransition(q, text[i - j], trie_layer)
      if q is accepting print "term found at location i - j"
      j <- j + 1
    }
    i += shift[ text[i] ] [ occurrence_layer ]    /* shift phase   */
    (trie_layer,occurrence_layer)<-MakePLATransition(text[i],occurrence_layer)
  }
```

Algorithm 1. Layered CWH. The highlighted text represents the extensions to CWH. The text indices for this implementation run from 1 to text_length.

Since the complexity of MakePlaTransition is linear, both the best and the worst case complexities of this algorithm are similar to those of Algorithm 0(CWH). However, the average performance improves, as will be demonstrated in Section 7.

Figure 1B traces the same text as *Figure 1A*, this time using the layered CWH(Algorithm 1).

In addition to the occurrence shift values underneath the forward loop trace, we added the trie and occurrence layer values underneath the arrows indicating those positions in the text where the match phase of each cycle begins. Also, a dot above the text indicates the running position of the divider wall.

Just like the CWH algorithm, we start with the text pointer at position 3 (mmin) of the text, and the trie and occurrence layer states are set to 3 accordingly. Comparing the letter "l" with the root of the layer 3 trie results in a mismatch. As can be seen from the row for entry "l" in the table for occurrence_layer 3 in *Figure 2*, we have a new *occurrence shift* of 2, instead of the value 1 in the occurrence table for plain CWH(*Figure 0*). This is because "l" from "title" only appears in layer 4(position 4 of "title") and the only "l" which is candidate for alignment when in layer 3 is the first letter in "log", which is 2 positions away from term-end. The *next trie_layer* entry for the "l" row is 3, pointing to the trie made of terms up to length 3. This is because after skipping 2 characters from an "l" at layer 3 (which could only be the first "l" in "log", since the "l" in "title" was ruled out for layer 3), the only term possibly ending in the new position would be "log". The *next occurrence_layer* state for the row is 2, since after the match phase of the cycle ends with either a match or mismatch, the "l" from the text will no longer be a candidate as a first character of "log"(it has already been ruled out as a candidate for position 4 of "title"), and can now serve as a delimiter to the running prefix length upper bound. The "divider wall" is moved up to the position of the "l" in the text, and after a shift of size 2, we will be at most 2 characters into the prefix of a new term.

The shift of size 2 brings us to the white space at position 5 of the text. We now start a new match/shift cycle, using the layer 3 trie and the layer 2 occurrence_table. Comparison of " " against the root of the trie results in a mismatch. We now look at the row for the "^" key in the layer 2 table, since " " is alien to layer 2. The new values are *occurrence shift* = 3, *next trie_layer* = 3, *next occurrence_layer* = 3.

After skipping 3 characters, we are now pointing to the 2nd "e" of "beetles" in the text. We start the 3rd match/shift cycle at text position 8, using the layer 3 trie and occurrence table. If we were using the full, unlayered trie, we would now get a successful match resulting in a backtrack scan of the previous text character. However, the trie for layer 3 has only the letter "g" as a branch from the root, so comparison with the "e" results in a mismatch, avoiding an extra character comparison.

As can be seen from the example in *Figure 1*, by using the layering mechanism the same text was traversed via 11 forward steps instead of 13, and all 9 backtrack character comparisons were avoided. Of the text of size 29, only 11 characters were examined, instead of 21. For most cycles, the offset from the running "divider wall" was 3, which is the length of the shortest term in the set, and the suffix of the longer term in the set was masked out. This is typical behavior for layer visitation distribution, as is demonstrated below:

mmin	1	2	3	4	5	6	7	8	9	10	11	12
3	0.71	3.22	**42.56**	15.5	14.01	16.65	5.33	1.59	0.33	0.08	0.02	0
4	0.6	2.45	4.17	**40.83**	17.94	15.81	9.73	5.58	1.82	0.92	0.1	0.05
5	0.62	1.3	2.4	6.76	**41.81**	18.99	12.52	7.04	5.35	2.08	0.81	0.21
6	0.14	0.84	2.23	3.86	4.72	**43.09**	20.4	11.64	6.78	4	1.78	0.43

Table 1a. Distribution of cycles spent at each occurrence layer.

mmin	1	2	3	4	5	6	7	8	9	10	11	12
3	0	0	**39.5**	13.65	15.01	17.72	9.95	3.34	0.67	0.15	0.02	0
4	0	0	0	**37.43**	20.71	14.56	16.33	7.73	1.69	1.15	0.31	0.05
5	0	0	0	0	**35.16**	21.95	22.23	10.67	6.37	2.11	0.77	0.6
6	0	0	0	0	0	**33.94**	30.14	20.83	8.87	3.27	2.39	0.49

Table 1b. Distribution of cycles spent at each trie layer.

For both tables, columns 2 to 13 represent the percent of cycles spent at layers 1 to 12. All layers greater than 12 had values smaller then one tenth of a percent for *Table 1a*, and two tenths of a percent for *Table 1b*. From *Table 1a* as well as *Table 1b*, we see that most cycles are spent at a layer state close to the length of the shortest term in the set. This means that if you have very long terms mixed with the short ones, the larger layers will rarely be visited, and the suffixes of the long terms, even if they appear very frequently in the data, will rarely break the skip loop via a successful match. The term-set size was kept constant at 14 terms per set for this experiment. Additional tests show that as the number of terms in the dictionary grows, we spend more and more cycles at larger layers.

3.4 Discussion of the special case when all terms are of similar length

In the special case when all terms are of similar length, or when the set includes a single term, we have only one trie layer, since all terms are of size mmin. However, layering the occurrrence table will result in larger skips, and therefore fewer character comparisons than simple CWH (see *Table 4* in Section 7). This partial gain is achieved by longer shifts from tables of occurrence layers smaller than term_length. Here we utilize the fact that the state machine keeps track of the first character scanned in the previous cycle, and uses it to delimit the next layer upon completion of the current cycle's match phase.

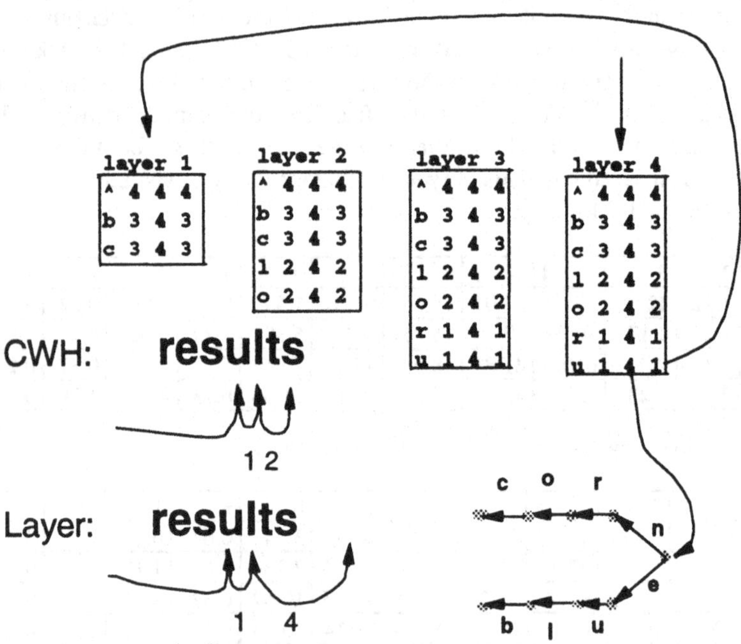

Fig. 3. A trace comparing both algorithms for the case when all terms are of similar length.

To illustrate this, we use the terms "blue" and "corn" and the text "results" from *Figure 3*. The tables in the diagram represent the PLA for Algorithm 1. For simple CWH occurrence_table we use the first two columns from the table for occurrence layer 4. The trie is similar for both algorithms.

We start with the character "u" in text position 4. Since the "u" can be the 3rd character of "blue", the next shift is of size 1. An attempt to compare the next scanned "l" with the root of the trie results in a mismatch. The previously scanned "u" is no longer a candidate as the 3rd character of "blue", and becomes a delimiter to the running max_prefix_length, cutting the value of the next occurrence layer to 1. The entry for character "l" in the table for occurrence layer 1 has a shift value of 4. Had we used standard CWH without layering, we would now be making a much smaller shift of size 2.

4 Construction of Data Structures During the Preprocessing Phase

We are now ready for the formal definitions for setting the occurrence table fields, extending [10].

Let *Pnum* be the number of terms p_i with sizes m_i in set P, and let $p_i[j]$ denote the character at index j of p_i.

Definition 2: A character c is considered a *member of layer L* if it appears in any of the terms in set P at an index smaller than or equal to L.

Definition 3: The *Maximal Index* for character c at layer L and tail offset s

$$\text{MaxIndex[c][L][s]} = \max_{k=1}^{Pnum}\{\, j \mid j = 0 \text{ or } ((1 <= j <= L) \text{ and } p_k[j] = c \text{ and } m_k - j = s)\,\}$$

For each layer state L (1 <= L <= mmax), for each character c which is a member of layer L, we calculate the following three fields:

1) The *Occurrence Shift* for character c at layer L

$$\text{shift[c][L]} = \min_{k=1}^{Pnum}\{\, s \mid s = \text{mmin}, \text{ or } (1 <= s <= \text{mmin} \text{ and } (\exists j \mid 1 <= j <= L \text{ and } p_k[j] = c \text{ and } m_k - j = s))\}$$

2) The *Next Trie Layer* for character c at layer L

next_trie_layer[c][L] = MaxIndex[c][L][shift[c][L]] + shift[c][L]

3)The *Next Occurrence Layer* for character c at layer L

$$\text{next\_occurrence\_layer[c][L]} = \max_{t=d[c][L]+1}^{mmax}\{\, \text{MaxIndex[c][L][t]} \,\} + \text{shift[c][L]}$$

Algorithm 2: creation of prefix layer automaton tables
input: The set of dictionary terms
output: The occurrence tables PLA

```
1  for each character c in the alphabet do      /* initializing the first layer */
   {
         shift[c][1] <– mmin
         max_index[c] <– 0
         next_max_index[c] <– 0
         next_trie[c][1] <– mmin
         next_occurrence_layer[c][1] <– mmin
   }
2  for (L <– 1, L <= mmax, L ++)
   {
3     If ( L > 1)
      {
4        initialize all fields in the table for layer L by copying the values from
         layer  L - 1
      }
5     for each term pi in the set term such that mi > L do
      {
6           c <– pi[L]
7           left_offset <– mi - L    /* the distance of char-index from end of term */
8           if (left_offset < shift[c][L])
            {
9             shift[c][L] <– left_offset
10            next_max_index[c] <– max(next_max_index[c], max_index[c])
11            max_index[c] <– L
            }
12          else /* left_offset >= shift[c][L] */
            {
13               if (left_offset > shift[c][L])
                 {
14                  next_max_index[c] <– L
                 }
15               else /* left_offset = = shift[c][L] */
                 {
16                  max_index[c] <– L
                 }
            }
17          next_trie_layer[c] <– max_index[c] + shift[c][L]
18          next_occurrence_layer[c] <– next_max_index[c] + shift[c][L]
      }
   }
```

Algorithm 2. A preprocessing algorithm for the construction of the PLA tables

Remark: When analyzing Algorithm 2 you will notice that for the field value calculations, each character in each term in the dictionary gets processed only once (except for last characters, which are ignored). This may seem surprising at first glance, since the formal definition for the *next occurrence layer* field includes a two level max calculation. The "trick" to the single pass is in the incremental order by which we process the characters in the dictionary, and the fact that the field settings, once completed for one layer, can serve as the initialization basis for the next layers' field calculations.

Lemma 2. *The time required to construct the occurrence tables is O(mmax*σ + M).*

proof: The time to construct the occurrence tables can be viewed as the sum of two terms: initialization time, plus field-values calculation time. During initialization time (line 1 and then line 4 of Algorithm 2) the first occurrence table is created and initialized(O(σ)) in line 1, and then we incrementally initialize each layer (in line 4) by copying over the values from the previous layer's table (O(σ)) at the beginning of each iteration. (We iterate through all layers from 1 to mmax). The complexity of this is O(mmax*σ).
For the field-value computation process we have one external loop (line 2) iterating over all layers, each time examining 1 character per term (The character at index L). So practically, each character appearing in a term gets examined once. This is O(M).

Remark: The first term can theoretically be further reduced from mmax*σ to mmax*M, using the stamping technique described in [23] for avoiding initialization of large arrays.

Creating the sub tries. We grow a forest of tries, one for each trie_layer. A trie for trie_layer L is made of all reversed strings of length <= L. This is O(mmax*M). See [2] for a trie construction algorithm.

5 The Fast Search Description and Implementation Issues

In the previous sections we have demonstrated how to layer the occurrence shift table and the term trie. In this section we show how to layer the work associated with breaks from the skip loop.

We now present Algorithm 3, a more efficient version of Algorithm 1, based on Boyer-Moore's fast skip loop structure[8, 22]. Algorithm 3 is similar to Algorithm 1 in terms of the number of characters scanned from the text. However, for shift/match cycles resulting in mismatch of the first character examined immediately after a shift, which is the most common case (See the third column of *Table 2* in Section 6 for frequency of the case *good_suffix_backtrack_depth = 0*), we spare the effort of a match against the root of a trie, as well as the control cost of breaking from the skip loop. Here, again, we benefit from the layering mechanism which allows us to reduce the frequency of skip loop breaks. Algorithm 3 also simplifies the data structure navigation control to one pointer transition per skip.

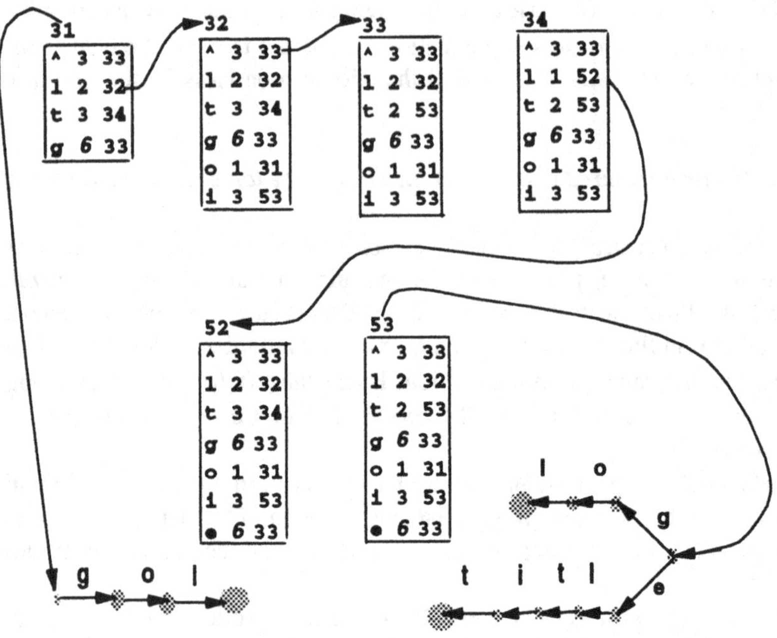

Fig. 4. The relevant members of a 2 by 4 matrix of PLA states, representing the terms "log", "title". Each column in the state matrix corresponds to an occurrence layer (1..4), and each row to a trie layer (3,5). Each table contains an entry per character in the alphabet. Each entry has 2 fields: an *occurrence_shift* value and a *next_state* value. The *next_state* field of each character entry in a state table is represented in the diagram via 2 digits, the first stands for the *next_trie_layer* and the second for the *next_occurrence_layer*.(In the actual implementation, this value is replaced by a pointer to the corresponding states' table.)

Figure 4 demonstrates the PLA structure for the fast version. We chose to represent the layer tables in a more space costly way, even though the fast search can be implemented in the previous model, in a somewhat less efficient way. In the new design each state represents a (next trie_layer, next occurrence_layer) pair. The tables for each state become smaller: instead of the two next layer pointers per character entry (next trie_layer, next occurrence_layer), only one field is needed to point to the next layer state. For each state, one pointer to the relevant trie is kept. This gives us a PLA state machine with number of states bounded by the number of occurrence layers times the number of trie layers, that is (mmax)∗(mmax - mmin). For natural language applications, mmax is usually no bigger than 11.

In this version of the search, we rearrange the order of the shift and match phases in a cycle, so that the shift phase, which is now dominant as far as execution frequency is considered, becomes the external loop, embedding the match phase loop. (We now have shift/match cycles instead of match/shift cycles.) To each occurrence_shift field in an entry corresponding to a last character in any of the terms included in the trie associated with the table we are modifying, we add mmin (which is also the maximal shift allowed). We also do so for end_of_file character entries. (In case our text ends before EOF, we place an EOF symbol in the text buffer after the last character to be scanned). This way, we can use the comparison statement in line 5 to do both EOF detection and skip-loop exit detection.

When observing the shift values for the state tables in the first row of *Figure 5*, (trie layer = 3), notice that the entries for "g" have a shift value of 6, since "g" is the last character of "log", and the only character label on an edge from the root of the trie for layer 3. The PLA state tables in row 2 (trie layer = 5) have an occurrence shift of 6 for both "g" and "e". This value of 6 will cause a break from the skip loop, since it is greater than mmin, which is 3. Upon returning to the next skip loop, the skip value will be reduced back to 3 by subtracting mmin from it.

Algorithm 3: Fast Layered CWH multi-pattern search

```
/* initialization */      i           <- 0
                          skip        <- mmin
                          pla_state   <- (mmin, mmin)

1  do forever
   {
2     i += skip                                        /* skip loop */
3     skip = shift [ text[i] ] [ pla_state ]
4     pla_state <- MakePLATransition( text[i], pla_state )
5     if ( skip > mmin )
      {
6        if ( text[i] is EOF ) return "search terminated"
7        do the match loop as in Algorithm 1          /* match loop */
8        skip <- skip - mmin
      }
   }
```

Algorithm 3. Fast Layered CWH. Procedures *MakePlaTransition* and the *match loop* are similar to those of Algorithm 1.

We start with a skip loop which scans through the text, skipping past immediate mismatches, traversing the PLA but not accessing the tries. The loop is broken upon scanning a character which is either a last character in a term from the set, or an EOF. That is, upon an occurrence shift of value greater than mmin.

We then check if the last character scanned is an EOF symbol. If it is, the search terminates. Otherwise, a possible match position has been found, and we begin the match phase. Now the text is scanned consecutively from right to left and compared to the term-set via trie traversal. After the match phase ends with either any number of matches or a mismatch, the skip phase is resumed, starting with text position

being the character aligned with the root of the trie and a skip forward whose value is obtained by subtracting mmin from the occurrence shift which caused the exit from the previous skip loop.
The benefits of this model during search-time are that we do not need to keep a pointer to the next trie during the search, and can access it from the PLA state upon a break from the skip loop. Also, it allows us to keep full information of the rare exits from the skip loop, and we can stay in this loop until we encounter a character corresponding to a relevant root of a trie. Going back to *Figure 1*, we can see that using Algorithm 0 (*Figure 1A*) the skip loop is broken 5 times, while Layering (*Figure 1B*) allows us to stay in the skip loop throughout the text example. As can be seen from the third column of *Table 2* in Section 6, when layering is applied to content descriptor queries over natural language text, we can avoid breaking the skip loop and accessing the tries for more than 90% of the cycles for dictionaries of 6 terms or less, and for more than 80% of the cycles for sets smaller than 22 terms.(Even at 46 terms, more than 66% of the cycles did not break the skip loop).

6 Discussion of the Good Suffix Heuristic

The unsimplified Commentz-Walter algorithm uses the "good suffix" heuristic as well as the occurrence heuristic to calculate the value of the next shift [10, 16, 17]. The "good suffix" heuristic uses the matching portion of the text scanned in the last match phase to look up a shift value stored per trie node. The maximum value between this shift and the occurrence shift value is then used for the next shift.
The "good suffix" heuristic requires some complicated preprocessing [21], plus the overhead of the arithmetic operation for computing the maximal shift value during search time.

We only demonstrated the layering of the occurrence heuristic, even though the "good suffix" heuristic can be layered as well by applying the [21] preprocessing algorithm to each trie in the layered forest. To understand why the "good suffix" heuristic would not be beneficial in our implementation, we use the results in *Table 2* for distribution of "good suffix" depth frequency, with the following definitions:
"Good Suffix" depth 0 occurs when a comparison of the first character examined in a match phase (immediately after a shift) with the root of a trie results in mismatch.
"Good Suffix" depth 1 occurs when we have a successful comparison of a first character examined in a match phase with the root of a trie, followed by a mismatch upon the next comparison.
Clearly, "good suffixes" of depth 0 are represented by the shift column of the occurrence tables in our model.
Also , "good suffixes" of depth 1 are special cases in the Horspool variant of CW, since their periodicity is already represented by the occurrence shift. (The first character examined in the match phase is used for shift value lookup in the occurrence table, regardless of whether it matched or not).

As can be seen from the 7th column of *Table 2*, when layering is used, most cycles (over 98% for sets smaller than 7 terms, over 96% for sets smaller than 15 terms) result in backtrack to suffix length of depth 0 or 1, which are already represented in our shift calculation model. When layering is applied, "good suffixes" of depth greater than 1 are so rare in this type of data, that it may not be worthwhile to pay the overhead of representing the "good suffix" heuristic in our model.

	depth 0:		**depth 1:**		**depth > 1:**	
set size	**CWH**	**Layer**	**CWH**	**Layer**	**CWH**	**Layer**
2	**92.82**	**96.17**	**5.34**	**2.89**	**1.84**	**0.94**
6	**78.77**	**91.86**	**16.59**	**6.5**	**4.64**	**1.64**
10	**69.95**	**89.34**	**23.38**	**8.28**	**6.67**	**2.39**
14	**63.48**	**84.26**	**27.6**	**11.97**	**8.92**	**3.76**
18	**58.54**	**82.06**	**30.57**	**13.14**	**10.9**	**4.81**
22	**55.75**	**79.73**	**32.47**	**14.91**	**11.78**	**5.36**

Table 2. Analysis of "good suffix" behavior: percent of cycles resulting in backtrack to "good suffix" depth of 0, 1 and > 1, for CWH(Alg0) as well as Layered CWH(Alg1). We continued to measure up to a term set size of 46, which was the sublinearity ceiling for CWH. Even at 46 terms, the distribution of the case *good suffix > 1* was 18.67% for CWH(Alg0), and only 9.65% with Layering(Alg1).

7 Empirical Results

The data for testing was taken from a file of Reuters online news text. The search patterns were taken from the topics content tags. Then the text body was separated from the headers, and lines containing the topics removed (in order to isolate the "Tocc" term - which is the number of occurrences of terms in the text). We ended up with a text search file of over 600,000 characters, and 178 search patterns of length 3 to 19.(Examples: "tin", "jobs", "housing", "interest","cornglutenfeed", etc). The text to be searched was from Reuters body, the alphabet being of size 128. The patterns alphabet was Uppercase and Lowercase, and some of the longer patterns included spaces, since they contained more than one word. The algorithms compared are Algorithm 0(CWH) and Algorithm 1(CWH + Layering).

For each run, the test program is given as input a specification for the number of terms in the dictionary, as well as the range of term lengths allowed. The test program then randomly selects a set of terms complying with the given criteria from the search topics pool, verifying that none of the terms appears more than once in the set, and runs the simulations of both search algorithms over the text. For each given dictionary construction specification, corresponding to a line in a table, the test program was run 25 times and the mean value over all runs calculated.
Various experiments were conducted, differing by the term set construction criteria specification. We only present some of the results in this paper, due to lack of space.

For the first experiment, the length of the terms in the set was kept within the range of 3 to 19 characters. We incrementally increased the size of the pattern set by 4 terms at a time, from 2 terms to 46 terms, at which point the linearity coefficient of 1 was reached for the CWH algorithm. Some of the results for the first experiment are presented in *Table 3* and in *Figure 5*.

set size (Pnum)	CWH	Layer	relative % improvement by Layering
2	25.59	23.23	9.27
6	43.47	32.53	25.19
10	55.69	38.46	31.02
14	65.88	45.02	31.7
18	71.3	48.54	31.96
22	80.38	53.55	33.41

Table 3. (char comparisons*100/size of text) versus number of terms in dictionary. The relative improvement by layering remained over 30% for the larger set sizes. The full results are plotted in *Figure 5*.

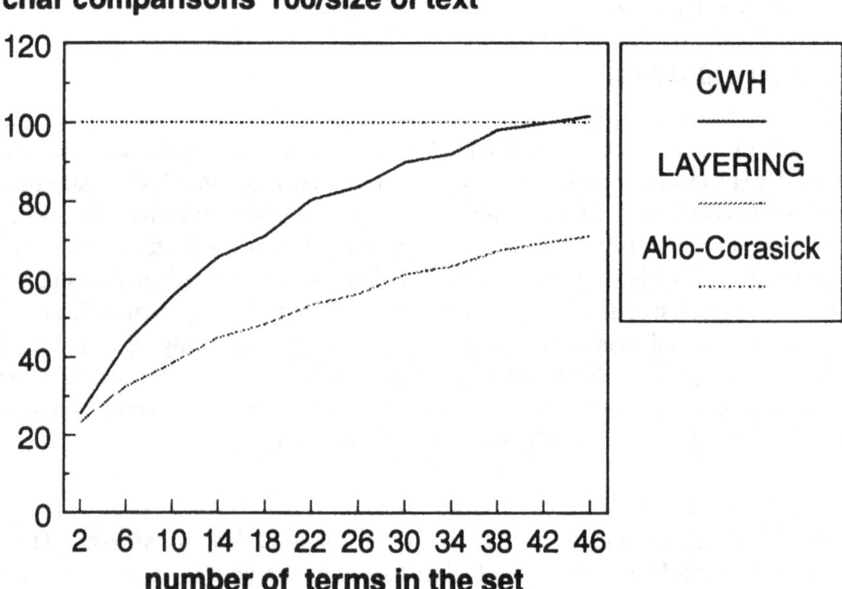

Fig. 5. (char comparisons*100/size of text) versus number of terms in dictionary

In the second experiment, we measured gain by layering for dictionaries of *uniform length* terms. The term set size was kept constant at 14 terms per set. Some of the results for the second experiment are presented in *Table 4*.

	% char comparisons			Mean shift length		
term length	CWH	Layer	% imp	CWH	Layer	% imp
6	51.79	46.05	11.06	3	3.34	11.53
7	44.18	39.78	9.97	3.43	3.77	10.01
8	40.7	36.79	9.61	3.81	4.16	9.1

Table 4. (char comparisons*100/size of text) and mean shift length, versus number of terms in set for *uniform length* dictionaries. The 4th and 7th columns represent relative improvement by layering for the number of characters scanned, and for the mean term length.

From *Table 4* we see that even when the terms in the set are of uniform length, layering still gives some reduction in the number of characters scanned, which is directly proportional to the gain in mean shift length. The improvement is clearly not enough to consider layering for dictionaries of uniform length. However, the table is intended to demonstrate the factor of occurrence table layering contribution, which is independent of variance in term length.(For a set of 14 terms of *variable lengths*, we found that layering will improve the mean shift length by up to 20%)

8 Conclusions

For many Web search engines, the average time passed between consecutive updates of indexed data bases is about a week. During this time, queries by users may get incorrect or misleading results. Therefore, speeding up the Index-Generator performance is of utmost importance. For scanning the text for content descriptors, the preprocessing engine should choose the algorithm to use for the search based on properties of the term set, as well as the data. A Commentz-Walter type algorithm should be chosen over an Aho-Corasick type algorithm when the number of terms in the set is small enough, and the shortest term in the set long enough, the decision threshold depending on system and data specific constants.
Implementing the Layering algorithm can extend the selection cutoff threshold in favor of the Commentz-Walter sublinear type, allowing more terms in the set and a shorter mmin.

Our results show that when searching for all occurrences of tags, which are of varying lengths, in natural language text, our refinement will typically prune out more than 25% of characters scanned by CWH for a set of size greater than 5 tags, and more than 30% for a set of size greater than 9 tags. Layering is recommended when the terms in the dictionary end with a variety of final characters, especially when the suffixes of the longer terms in the set are more frequent in the text than the suffixes of the shortest terms.

9 Acknowledgments

We would like to thank Gad M. Landau for his helpful comments and assistance with the references.

References

[1] A.V.Aho. Algorithms for Finding Patterns in Strings. In *J. van Leeuwen, editor, Handbook of Theoretical Computer Science*. Pages 257-300. Elsevier Science Publishers B.V., Amsterdam, The Netherlands.1990.

[2] A.V.Aho and M.Corasick. Efficient string matching: An aid to bibliographic search. *Comm. of the ACM*, 18(6):333-340, June 1975.

[3] A. Amir and M. Farach. Adaptive dictionary matching. *Proc. 32nd IEEE FOCS*, pages 760-766, 1991.

[4] A. Amir, M. Farach, R. Giancarlo, Z. Galil, and K. Park. Dynamic dictionary matching. *Journal of Computer and System Sciences*, 49(2):208-222, 1994

[5] A. Amir, M.Farach, R.M. Idury, J.A. La Poutre', and A.A Schaffer. Improved dynamic dictionary matching. *Information and Computation*, 119(2):258-282, 1995

[6] A.Amir, M.Farach, and Y.Matias. Efficient ramdomized dictionary matching algorithms, *Proc. of 3rd Combinatorial Pattern Matching Conference*, pages 259-272, 1992. Tucson, Arizona.

[7] A.Apostolico and R.Giancarlo. The Boyer-Moore-Galil string searching strategies revisited, *SIAM J.Comput.* 15,(1), 98-105(1986)

[8] R.S. Boyer and J.S. Moore. A fast string matching algorithm. *Comm. of the ACM*, 20:762-772, 1977.

[9] R. Baeza-Yates and G.H. Gonnet. On Boyer-Moore automata, *Research report, university of Waterloo*,1989.

[10] R. Baeza-Yates and M. Re'gnier. Fast Algorithms for Two Dimensional and Multiple Pattern Matching(Preliminary version). *SWAT 90, In Proc. 2nd Scandinavian Workshop on Algorithm Theory*. Number 447 in Lecture Notes in Computer Science, pages 332-347, Springer-Verlag, Bergen, Sweden, July 1990.

[11] D. Breslauer. Dictionary-Matching on Unbounded Alphabets: Uniform-Length Dictionaries. *Proc. of 5th Combinatorial Pattern Matching Conference*, pages 184-197, 1994, Asilomar, CA, USA

[12] V. Bruye're, R. Baeza-Yates, O. Delgrange and R. Scheihing. On the size of Boyer Moore Automata. *Proceedings of Third South American Workshop on String Processing*. Recife, Brazil, August 1996, 31-46

[13] M.Crochemore,A.Czumaj,L.Gasieniec,S.Jarominek,T.Lecroq,W.Plandowski, and W.Rytter. Fast Practical Multi-Pattern Matching. *Technical Report 93-3, Institut Gaspard Monge, Universit'e de Marne la Vall'ee, Marne la Vall'ee, France,* 1993.

[14] L.Colussi, Z.Galil, and R.Giancarlo.The exact complexity of string matching. *'31st Symposium on foundations of Computer Science I*, 135-143, IEEE(October 22-24 1990)

[15] R.Cole. Tight bounds on the complexity of the Boyer-Moore pattern Matching algorithm, *Technical Report 512, Computer Science Dept, New York University* (June 1990).

[16] B.Commentz-Walter. A string matching algorithm fast on the average. *Technical Report 79.09.007, IBM Wissenchaftliches Zentrum. Heidelberg, Germany,* 1979.

[17] B.Commentz-Walter. A string matching algorithm fast on the average. *Proc,6th International Colloquium on Automata, Languages, and Programming,* Lecture notes in Computer Science. Pages 118-132. Springer-Verlag. Berlin, Germany 1979.

[18] J.J. Fan and K.Y. Su. An efficient algorithm for matching multiple patterns. *IEEE Transactions on Knowledge and Data Engineering.* 5(2):339-351, April, 1993

[19] Z.Galil. On improving the worst case running time of the Boyer-Moore string matching algorithm, *Comm. of the ACM,*22(9) 505-508,(1979)

[20] L.J. Guibas and A.M. Odlyzko. A new proof of the linearity of the Boyer-Moore string searching algorithm, *Siam J. Comput.* 9(1980) 672-682

[21] D.Gusfield.Algorithms on strings, trees and sequences, *published by the press syndicate of the University Of Cambridge,* (1997) 157-164

[22] A. Hume and D. Sunday. Fast String Searching. *Software-Practice and experience,* Vol.21(11). 1221-1248 (November 1991).

[23] T.Hagerup. On saving space in parallel computation, *Information Processing Letters,* Vol.29, 1988, pages 327 -329

[24] R.N. Horspool. Practical fast searching in strings. *Software- Practice and Experience,* 10:501-506, 1980.

[25] R.M. Idury and A.A Schaffer. Dynamic dictionary matching with failure functions. *Proc. 3rd Annual Symposium on Combinatorial Pattern Matching,* pages 273-284,1992.

[26] J.Y. Kim and J. Shawe-Taylor. Fast Multiple Keyword Searching. *Proc. of 3rd Combinatorial Pattern Matching Conference,* pages 41-51,1992. Tucson, Arizona.

[27] G.Kowalski and A. Meltzer. New Multi-Term high speed text search algorithms. *1st conference on computers and applications, IEEE(1984)*

[28] D.E.Knuth, J.H. Morris, and V.R. Pratt. Fast pattern matching in strings. *SIAM J. Comput.,* 6:322-350, 1977.

[29] T.Lecroq. A variation on the Boyer-Moore algorithm. *Theoretical Computer Science 92* (119-144), Elsevier.

A Very Fast String Matching Algorithm for Small Alphabets and Long Patterns
(Extended Abstract)

Christian Charras[1], Thierry Lecroq[1], and Joseph Daniel Pehoushek[2]

[1] LIR (Laboratoire d'Informatique de Rouen) and ABISS (Atelier Biologie Informatique Statistique et Socio-Linguistique), Faculté des Sciences et des Techniques, Université de Rouen, 76128 Mont Saint-Aignan Cedex, France. {Christian.Charras,Thierry.Lecroq}@dir.univ-rouen.fr ***

[2] JDPeh@aol.com

Abstract. We are interested in the exact string matching problem which consists of searching for all the occurrences of a pattern of length m in a text of length n. Both the pattern and the text are built over an alphabet Σ of size σ. We present three versions of an exact string matching algorithm. They use a new shifting technique. The first version is straightforward and easy to implement. The second version is linear in the worst case, an improvement over the first. The main result is the third algorithm. It is very fast in practice for small alphabet and long patterns. Asymptotically, it performs $O(\log_\sigma m(m + n/(m - \log_\sigma m)))$ inspections of text symbols in the average case. This compares favorably with many other string searching algorithms.

1 Introduction

Pattern matching is a basic problem in computer science. The performance of many programs is determined by the work required to match patterns, most notably in the areas of text processing, speech recognition, information retrieval, and computational biology. The kind of pattern matching discussed in this paper is exact string matching.

String matching is a special case of pattern matching where the pattern is described by a finite sequence of symbols. It consists of finding one or more generally all the occurrences of a pattern $x = x_0x_1 \cdots x_{m-1}$ of length m in a text $y = y_0y_1 \cdots y_{n-1}$ of length n. Both x and y are built over the same alphabet Σ.

String matching algorithms use a "window" that shifts through the text, comparing the contents of the window with the pattern; after each comparison, the window is shifted some distance over the text. Specifically, the window has the same length as the pattern x. It is first aligned with the left end of the text y, then the string matching algorithm tries to match the symbols of the pattern

*** The work of these authors was partially supported by the project "Informatique et Génomes" of the french CNRS.

with the symbols in the window (this specific work is called an *attempt*). After each attempt, the window *shifts* to the right over the text, until passing the end of the text. A string matching algorithm is then a succession of attempts and shifts.

The aim of a good algorithm is to minimize the work done during each attempt and to maximize the length of the shifts. After positioning the window, the algorithm tries to quickly determine if the pattern occurs in the window. To decide this, most string matching algorithms have a preprocessing phase, during which a data structure, z, is constructed. z is usually proportional to the length of the pattern, and its details vary in different algorithms. The structure of z defines many characteristics of the search phase.

Numerous solutions to string matching problem have been designed (see [3] and [10]). The two most famous are the Knuth-Morris-Pratt algorithm [5] and the Boyer-Moore algorithm [1].

We use a new shifting technique to construct a basic and straightforward algorithm. This algorithm has a quadratic worst case time complexity though it performs well in practice. Using the shift tables of Knuth-Morris-Pratt algorithm [5] and Morris-Pratt algorithm [8], we make this first algorithm linear in the worst case. Then using a trie, we present a simple and very fast algorithm for small alphabets and long patterns.

We present the basic algorithm in Sect. 1. Section 2 is devoted to the linear variation of the basic algorithm. In the Sect. 3 we introduce the third algorithm which is very fast in practice. Results of experiments are given in Sect. 4.

2 The basic algorithm

2.1 Description

We want to solve the string matching problem which consists in finding all the occurrences of a pattern x of length m in a text y of length n. Both x and y are build over the same finite alphabet Σ of size σ. We are interested in the problem where the pattern x is given before the text y. In this case, x can be preprocessed to construct the data structure z.

The idea of our first algorithm is straightforward. For each symbol of the alphabet, a bucket collects all of that symbol's positions in x. When a symbol occurs k times in the pattern, there are k corresponding positions in the symbol's bucket. When the pattern is much shorter than the alphabet, many buckets are empty.

The main loop of the search phase consists of examining every mth text symbol, y_j (so there will be n/m main iterations). For y_j, use each position in the bucket $z[y_j]$ to obtain a possible starting point p of x in y. Perform a comparison of x to y beginning at position p, symbol by symbol, until there is a mismatch, or until all match.

The entire algorithm is given Fig. 1.

```
SKIPSEARCH(x, m, y, n)
  ▷ Initialization
  for all symbols s in Σ
       do z[s] ← ∅
  ▷ Preprocessing phase
  for i ← 0 to m − 1
       do z[x_i] ← z[x_i] ∪ {i}
  ▷ Searching phase
  j ← m − 1
  while j < n
       do for all i in z[y_j]
                 do if y_{j−i} ··· y_{j−i+m−1} = x
                         then REPORT(j − i)
            j ← j + m
```

Fig. 1. The basic algorithm.

2.2 Complexity

The space and time complexity of this preprocessing phase is in $O(m+\sigma)$. The worst case time complexity of this algorithm is $O(mn)$. This bound is tight and is reached when searching for $a^{m-1}b$ in a^n.

The expected time complexity when the pattern is small, $m < \sigma$, is $O(n/m)$. As m increases with respect to σ, the expected time rises to $O(n)$.

3 A linear algorithm

3.1 Description

It is possible to make the basic algorithm linear using the two shift tables of Morris-Pratt [8] and Knuth-Morris-Pratt [5].

For $1 \le i \le m$, $mpNext[i]$ = length of the longest border of $x_0x_1\cdots x_{i-1}$ and $mpNext[0] = -1$.

For $1 \le i < m$, $kmpNext[i]$ = length of the longest border of $x_0x_1\cdots x_{i-1}$ followed by a character different from x_i, $kmpNext[0] = -1$ and $kmpNext[m] = m - period(x)$.

The lists in the buckets are explicitly stored in a table (see algorithm PRE-KMPSKIPSEARCH in Fig. 3).

A general situation for an attempt during the searching phase is the following (see Fig. 2):

- j is the current text position;
- $i = z[y_j]$ thus $x_i = y_j$;
- $start = j - i$ is the possible starting position of an occurrence of x in y;
- $wall$ is the rightmost scanned text position;
- $y_{start}y_{start+1}\cdots y_{wall-1} = x_0x_1\cdots x_{wall-start-1}$;
- $y_{wall} \neq x_{wall-start}$.

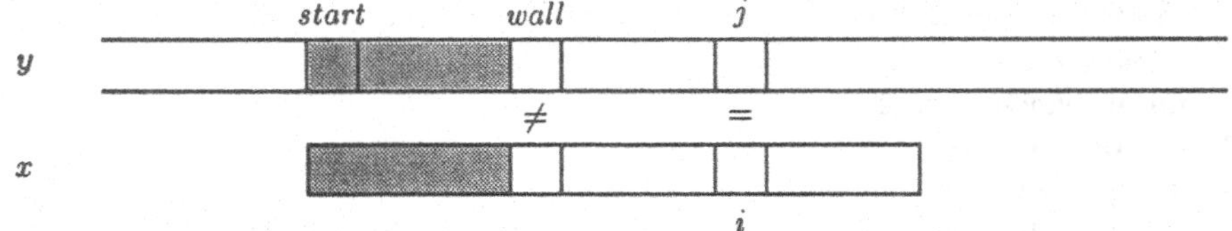

Fig. 2. General situation during the searching phase of the linear algorithm.

The comparisons will be performed from left to right between

$$y_{wall}y_{wall+1}\cdots y_{start+m-1}$$

and

$$x_{wall-start}x_{wall-start+1}\cdots x_{m-1}$$

Let $k \geq wall - start$ be the smallest integer such that $x_k \neq y_{start+k}$ or $k = m$ if an occurrence of x starts at position *start* in y.

Then *wall* takes the value of $start + k$.

Then compute two shifts (two new starting positions): the first one according to the skip algorithm (see algorithm ADVANCESKIP in Fig. 5 for details), this gives us a starting position *skipStart*, the second one according to the shift table of Knuth-Morris-Pratt, which gives us another starting position *kmpStart*.

Several cases can arise:

case 1 $skipStart < kmpStart$ then a shift according to the skip algorithm is applied which gives a new value for *skipStart*, and we have to again compare *skipStart* and *kmpStart*;

case 2 $kmpStart < skipStart < wall$ then a shift according to the shift table of Morris-Pratt is applied which gives a new value for *kmpStart*, and we have to again compare *skipStart* and *kmpStart*;

case 3 $skipStart = kmpStart$ then another attempt can be performed with $start = skipStart$;

case 4 $kmpStart < wall < skipStart$ then another attempt can be performed with $start = skipStart$.

The complete algorithm is given in Fig. 6. When an occurrence of x is found in y, it is, of course, possible to shift by the length of the period of x.

3.2 Complexity

The time complexity of the second algorithm can be easily computed with the following arguments:

```
PRE-KMPSKIPSEARCH(x, m)
▷ Initialization
for all symbols s in Σ
      do z[s] ← −1
list[0] ← −1
z[x[0]] ← 0
for i ← 1 to m − 1
      do list[i] ← z[x[i]]
         z[x[i]] ← i
```

Fig. 3. Preprocessing of KMPSKIPSEARCH algorithm.

```
ATTEMPT(start, wall)
k ← wall − start
while k < m and x_k = y_{wall+k}
      do k ← k + 1
return k
```

Fig. 4. Return $k \geq wall - start$ the smallest integer such that $x_k \neq y_{start+k}$ or $k = m$ if an occurrence of x starts at position *start* in y when comparing $y_{wall} y_{wall+1} \cdots y_{start+m-1}$ and $x_{wall-start} x_{wall-start+1} \cdots x_{m-1}$.

```
ADVANCESKIP()
repeat j ← j + m
  until j ≥ n or z[y_j] ≥ 0
if j < n
  then i ← z[y_j]
```

Fig. 5. Compute the shift using the buckets.

```
KmpSkipSearch(x, m, y, n)
wall ← 0
i ← −1
j ← −1
AdvanceSkip()
start ← j − i
while start ≤ n − m
     do wall ← max(wall, start)
        k ← Attempt(start, wall)
        wall ← start + k
        if k = m
          then Report(start)
               i ← i − period(x)
          else  i ← list[i]
        if i < 0
          then AdvanceSkip()
        skipStart ← j − i
        kmpStart ← start + k − kmpNext[k]
        k ← kmpNext[k]
        while not (case 3 or case 4)
▷ case 1    do if skipStart < kmpStart
                 then i ← list[i]
                      if i < 0
                        then AdvanceSkip()
                      skipStart ← j − i
▷ case 2         else  kmpStart ← kmpStart + k − mpNext[k]
                       k ← mpNext[k]
        start ← skipStart
```

Fig. 6. Searching phase of the KmpSkipSearch algorithm.

- during each attempt no comparison can be done to the left of the position *wall*, this implies that a text symbol can be compared positively only once. This gives n positive symbol comparisons;
- each negative symbol comparison implies a shift of length at least one, there can be at most $n - m + 1$ shifts;
- the total number of text symbols accessed to look into their bucket is $\lfloor n/m \rfloor$.

This gives us a total of $2n + \lfloor n/m \rfloor - m + 1$ symbol comparisons.

The expected time of the search phase of this algorithm is similar to the first algorithm.

4 A very fast practical algorithm

4.1 Description

Instead of having a bucket for each symbol of the alphabet. We can build a trie $T(x)$ of all the factors of the length $\ell = \log_\sigma m$ occurring in the pattern x. The leaves of $T(x)$ represent all the factors of length ℓ of x. There is then one bucket for each leaf of $T(x)$ in which is stored the list of positions where the factor, associated to the leaf, appears in x.

The searching phase consists in looking into the buckets of the text factors $y_j y_{j+1} \cdots y_{j+\ell-1}$ for all $j = k(m - \ell + 1) - 1$ with the integer k in the interval $[1, \lfloor (n - \ell)/m \rfloor]$.

The complete algorithm is shown in Fig. 7, it uses a function ADD-NODE given in Fig. 8.

4.2 Complexity

The construction of the trie $T(x)$ is linear in time and space in m providing that the alphabet size is constant, which is a reasonable assumption (see [7]). It is done using suffix links.

The time complexity of the searching phase of this algorithm is $O(mn)$. The expected search phase time is $O(\ell(n/(m - \ell)))$.

```
ALPHASKIPSEARCH(x, m, y, n)
▷ Initialization
root ← CREATE-NODE()
suffix[root] ← ∅
height[root] ← 0
▷ Preprocessing phase
node ← root
for i ← 0 to m − 1
    do if height[node] = ℓ
          then node ← suffix[node]
       childNode ← child[node, x_i]
       if childNode = ∅
          then childNode ← ADD-NODE(node, x_i)
       if height[childNode] = ℓ
          then z[node] ← z[node] ∪ {i − ℓ − 1}
       node ← childNode
▷ Searching phase
j ← m − ℓ
while j < n − ℓ
    do node ← root
       for k ← 0 to ℓ − 1
           do node ← child[node, y_{j+k}]
       if node ≠ ∅
          then for all i in z[node]
                   do if y_{j−i} ··· y_{j−i+m−1} = x
                         then REPORT(j − i)
       j ← j + m − ℓ + 1
```

Fig. 7. ALPHASKIPSEARCH algorithm.

5 Experiments

We tested our three algorithms against three other, namely: the Boyer-Moore algorithm (BM) [1], the Tuned Boyer-Moore algorithm (TBM) [4] and the Reverse Factor algorithm (RF) [6] and [2]. We add to these algorithms a fast skip loop (*ufast* in the terminology of [4]) which consists in checking only if there is a match between the rightmost symbol of the pattern and its corresponding aligned symbol in the text before checking for a full match with all the other symbols. For the BM and TBM algorithms we also use Raita's trick [9] (already introduced in [5]) which consists in storing the first symbol of the pattern in a variable and checking if the leftmost symbol of the window match this variable before checking for a full match with all the other symbols.

```
ADD-NODE(node, s)
childNode ← CREATE-NODE()
child[node, s] ← childNode
height[childNode] ← height[node] + 1
suffixNode ← suffix[node]
if suffixNode = ∅
  then suffix[childNode] ← node
  else suffixChildNode ← child[suffixNode, s]
       if suffixChildNode = ∅
         then suffixChildNode ← ADD-NODE(suffixNode, s)
       suffix[childNode] ← suffixChildNode
return childNode
```

Fig. 8. Add a new node labelled by s along the suffix path.

All these algorithms have been implemented in C in a homogeneous way such as to keep their comparison significant. The texts used are composed of 500000 symbols and were randomly built. The target machine is a Hyunday SPARC HWS-S310 running SunOS 4.1.3. The compiler is gcc.

For each pattern length, we searched for hundred patterns randomly chosen in the texts. The time given in the Table 1 are the times in seconds for searching hundred patterns.

Running times give a good idea of the efficiency of an algorithm but are very dependent of the target machine. We also present the number of inspections per text symbols (Table 2) which is a more theoretical measure.

Table 1. Running times for an alphabet of size 2.

algo.\ m	10	20	40	80	160	320	640
BM	**38.05**	16.03	11.81	9.46	7.47	6.65	6.13
TBM	50.61	32.16	30.76	31.11	30.62	31.31	32.77
RF	42.55	**14.47**	**9.82**	**6.55**	**5.77**	6.91	10.77
SKIP	65.27	37.27	35.59	34.98	34.26	33.87	34.66
KMPSKIP	82.61	61.87	61.10	60.91	60.70	60.45	61.25
ALPHASKIP	54.63	19.01	11.69	8.07	5.87	**4.83**	**4.47**

Table 2. Number of inspections per symbol for an alphabet of size 2.

algo.\ m	10	20	40	80	160	320	640
BM	0.6121	0.4505	0.3291	0.2700	0.2104	0.1815	0.1598
TBM	1.2369	1.2954	1.2536	1.2793	1.2499	1.2708	1.3114
RF	**0.5127**	**0.2942**	**0.1696**	**0.0970**	**0.0560**	**0.0338**	0.0238
SKIP	1.1980	1.0999	1.0502	1.0255	1.0138	1.0083	1.0087
KMPSKIP	0.9230	0.8604	0.8199	0.8028	0.7890	0.7931	0.7929
ALPHASKIP	0.7165	0.3897	0.2103	0.1141	0.0630	0.0361	**0.0211**

The tests show that the algorithm ALPHASKIP is the most efficient for searching long patterns, both practically and theoretically, when dealing with small alphabets (it is also the case for an alphabet of size 4).

6 Concluding Remarks

We presented a new string matching algorithm with an expected number of inspections of text symbols in $O(\log_\sigma m(n/(m-\log_\sigma m)))$ where m is the length of the pattern, n is the length of the text and σ is the size of the alphabet. The algorithm uses a new shifting technique, based on collecting positions of short substrings of the pattern. The goal is to quickly discover whether and where the pattern x can occur in a window onto y. This algorithm performs well in both theory and practice.

This shifting technique can be extended for searching patterns with classes of symbols or for searching a finite set of patterns.

References

[1] R. S. Boyer and J. S. Moore. A fast string searching algorithm. *Comm. ACM*, 20(10):762–772, 1977.

[2] M. Crochemore, A. Czumaj, L. Gąsieniec, S. Jarominek, T. Lecroq, W. Plandowski, and W. Rytter. Speeding up two string matching algorithms. *Algorithmica*, 12(4/5):247–267, 1994.

[3] M. Crochemore and W. Rytter. *Text algorithms*. Oxford University Press, 1994.

[4] A. Hume and D. M. Sunday. Fast string searching. *Software–Practice & Experience*, 21(11):1221–1248, 1991.

[5] D. E. Knuth, J. H. Morris, Jr, and V. R. Pratt. Fast pattern matching in strings. *SIAM J. Comput.*, 6(1):323–350, 1977.

[6] T. Lecroq. A variation on the Boyer-Moore algorithm. *Theoret. Comput. Sci.*, 92(1):119–144, 1992.

[7] T. Lecroq. Experiments on string matching in memory structures. *Software–Practice & Experience*, 28(5):562–568, 1998.

[8] J. H. Morris, Jr and V. R. Pratt. A linear pattern-matching algorithm. Report 40, University of California, Berkeley, 1970.

[9] T. Raita. Tuning the Boyer-Moore-Horspool string searching algorithm. *Software–Practice & Experience*, 22(10):879–884, 1992.

[10] G. A. Stephen. *String searching algorithms*. World Scientific Press, 1994.

Approximate Word Sequence Matching over Sparse Suffix Trees

Knut Magne Risvik

Department of Computer and Information Science
the Norwegian University of Science and Technology
N-7034 Trondheim
Norway

kmr@idi.ntnu.no

Abstract. In this paper, we discuss word sequence matching, and we adapt the common edit distance metric for approximate string matching to searching for words and sequences of words. We furthermore create a variant of the Sparse Suffix Tree([3]) and adapt algorithms for approximate word and word sequence matching over the Sparse Suffix Tree variant. The algorithms have been implemented and tested in WWW information retrieval environment, and performance data is presented.

1 Introduction

Approximate string matching is a thoroughly discussed and analyzed area. However, not many metrics exists for approximate matching of word sequences or phrases.

Several indixes have been proposed for approximate string matching. However, approximate string matching in natural language documents has other limitations than other string processing applications. A natural language has a limited count of words, but when the number of occurrences of each word increases, performance problems occur.

The outline of this paper is as follows:

In section 1, we briefly mention some preliminaries. In section 2 we discuss metrics for approximate matching of words and word sequences. In section 3, we introduce the Sparse Suffix Tree variant, namely the Word spaced Sparse Suffix Tree. In Section 4, we discuss the adapted and designed algorithms for approximate matching over the Word spaced Sparse Suffix Tree. In section 5, we present the results from testing the performance of an implementation of the discussed algorithms and data structures. Section 6 summaries and concludes the paper.

2 Preliminaries

In this paper, we discuss the use of indexes for collections of text documents. We assume that the text can be divided naturally into **words.**

We will use the notion of a **word sequence:**

Definition (Word sequence). A Word sequence is a sequence of separated, consecutive words. A word sequence $S = s_1, s_2, \ldots, s_n$ consists of the single words (or strings) $s_1, s_2, \ldots s_n$. ∎

3 Approximate Matching and Distance Metrics

Approximate matching is the process of matching a query approximately according to some criteria. Two different approaches are common for approximate matching:

- Defining the criterions for proximity.
- Use queries in a form that defines the set of correct matches. Specifying the set of matches could be done in a compact form, for instance a regular expression.

By defining the criterions for proximity, we usually utilize a distance function to calculate the distance between a query and a candidate match. We will first define the problem of approximate word matching and approximate matching of word sequences.

The problem of approximate word matching can be stated as follows:

Given a Text $T = t_1t_2 \ldots t_n$ and a word $W = k_1k_2 \ldots k_m$. Find all occurrences of W in T according to a proximity metric.

A common metric is the edit distance, briefly described in 3.2.1

3.1 Approximate Word Sequence Matching

Moving one level up from approximate or exact word matching, we might be interested in matching a pattern that consists of a sequence of words, like a phrase in natural language.

We could further define the notion of proximity for sequences as well as we defined the proximity for single word matching. The problem of matching a sequence pattern approximately can then be stated as follows:

Given a text T consisting of the n words $w_1w_2 \ldots wt_n$ where each of the words is a string of characters.

A sequence pattern P consists of the m words $p_1, p_2, \ldots, p_m$. The sequence pattern P is said to have an approximate occurrence in T if the sequence $p_1, p_2, \ldots, p_m$ is close enough to a sequence $w_i, w_{i+1}, \ldots w_j$, for some i,j such that $1 \leq i \leq j \leq n$.

3.2 Word Matching Metrics

Formally, a distance metric, D, is a function of objects satisfying the following axioms[1]:

Non-negative property	$D(x,y) \geq 0 \ \forall x,y$	(1)
Zero property	$D(x,y) = 0 \Leftrightarrow x = y$	(2)
Symmetry	$D(x,y) = D(y,x) \ \forall x,y$	(3)
Triangle Inequality	$D(x,z) \leq D(x,y) + D(y,z) \ \forall x,y$	(4)

3.2.1 Edit distance

The Levenstein[4] distance, or edit distance is defined as the minimum number of edit operations we need to perform to transform one string to the other. An edit operation is given by any rewriting rule:

- $(a \rightarrow \varepsilon)$, deletion
- $(\varepsilon \rightarrow a)$, insertion
- $(a \rightarrow b)$, change

Let p and m be two words of size i and j, respectively. Then $D(i,j)$ denotes the edit distance between the ith prefix of p and the jth prefix of m. The edit distance can then be recursively defined as:

$$D(i,0) = D(0,i) = i$$

$$D(i,j) = \min \begin{cases} D(i-1,j)+1 \\ D(i,j-1)+1 \\ D(i-1,j-1)+\partial(i,j) \end{cases} \tag{5}$$

where

$$\partial(i,j) = 0 \text{ if } p_i = m_j \text{ else } 1$$

The edit distance can be computed in $O(|x|\ |y|)$ by dynamic programming (improved versions by Ukkonen[6] among others). Common variations are non-uniform cost functions and the transpose edit operation $(ab \rightarrow ba)$

3.3 Sequence Matching Metrics

The operations allowed in the edit distance metric (deletion, insertion and change) intuitively applies to words as well as characters. Common errors in matching phrases are missing, extra or changed words. We will adapt the edit distance metric discussed above to the approximate word sequence matching problem.

3.3.1 Sequence Edit Distance

When discussing word matching metrics, we used the Levenstein[4] distance, or the edit distance to allow for proximity in all the three different similarity dimensions. We will now use the edit distance metric as a metric for approximate word sequence matching.

Definition (Edit operations for sequences). For transforming one sequence S of words into another sequence P of words, the edit operations allowed on the words in the sequences may be written according to the following rewrite rules:

- $(a \rightarrow \varepsilon)$, deletion of word a from the sequence.
- $(\varepsilon \rightarrow a)$, insertion of word a into the sequence.
- $(a \rightarrow b)$, change of word a into word b.
- $(ab \rightarrow ba)$, transpose of the adjacent words a and b. ■

The edit operations are identical to those used in the edit distance metric for word matching. The only differences are that we apply the edit operations with words as the atoms of our operations, instead of characters as used in the word matching metric.

The cost function $c_{edit}(x \rightarrow y)$ is a constant defined as

$$c_{edit}(x \rightarrow y) = \begin{cases} 1 & \text{delete} \\ 1 & \text{insert} \\ 1 & \text{transpose} \\ \partial(x, y) & \text{change} \end{cases} \tag{6}$$

where $\partial(x,y)$ is defined as

$$\partial(x, y) = \begin{cases} 0 & x = y \\ 1 & \text{else} \end{cases} \tag{7}$$

Using the edit operations defined above we define the edit distance for sequences.

Definition (Edit distance for sequences). The edit distance metric for sequences defines the distance $D_{seq}(S,P)$ between the sequence $S = s_1,s_2,\ldots,s_n$ and the sequence $P = p_1,p_2,\ldots,p_m$ as the minimum sum of the costs, $c(x \rightarrow y)$, for a sequence of edit operations transforming the sequence S into the sequence P. ■

The edit distance for sequences as defined above is probably not fair. In most cases, shorter words tend to have lower influence on the semantic interpretation of the phrase. Of course, there are short words in the English language, like "not", that totally reverse the semantic interpretation of the phrase.

However, we are not interested in the exact semantic interpretation of the phrase or sequence, we are more interested in the context of the semantic interpretation. "not exactly the nine o'clock news" and "exactly the nine o'clock news" have opposite semantic interpretation, but the context is the same (something about the news at nine o'clock).

Assuming that shorter words tend to be less important to the semantic context than longer words, we enhance the edit distance metric for sequences to weight the cost of the edit operations by the size of the words operated upon.

Definition (Word size dependent edit distance for sequences). The word size dependent edit distance for sequences is defined as the minimal sum of costs for the editing operations needed to transform one sequence into the other. The cost functions are dependent of the word size of its operands. ■

We suggest a possible definition of the cost functions in (8), where l denotes the average length of a word in the two sequences being compared. The cost of each edit operation is weighted by a size proportional to change in total length of the sequence, or by the ratio of the current word length and the average word length in the sequences considered.

$$\begin{aligned} c_{insert}(\varepsilon \rightarrow a) &= \frac{|a|}{l} \\ c_{delete}(a \rightarrow \varepsilon) &= \frac{|a|}{l} \\ c_{transpose}(ab \rightarrow ba) &= 1 \\ c_{change}(a \rightarrow b) &= \frac{\max(||a|-|b||, l)}{l} \end{aligned} \qquad \textbf{(8)}$$

Now, our edit distance metric reflects our assumption of some relation between the word length and how important the word is to the semantic context of the word sequence.

The edit distance metric could be further enhanced by allowing the use of proximity at the character level when the change edit operation $(a \rightarrow b)$ is used. Replacing a word a by another word b should be related to the similarity between these two words. The new cost function for the change edit operation can be defined as:

$$c_{change}(a \rightarrow b) = \partial_{approx}(a,b)\frac{\max(||a|-|b||, \bar{l})}{\bar{l}} \qquad \textbf{(9)}$$

where:

$$\partial_{approx}(a,b) = D(a,b), \qquad \textbf{(10)}$$

Where $D(a,b)$ is some normalized distance measuring function for words (0 means full similarity, 1 means no similarity).

4 The Word spaced Sparse Suffix Tree

A large number of data structures have been proposed to create an index for the text documents used as a base for searching. These data structures allow for sublinear searching algorithms.

Kärkkäinen and Ukkonen [3] introduce the Sparse Suffix Tree, based on ideas by Morrison [5]:

Definition (Sparse Suffix Tree). A sparse suffix tree, SST(T) of the text T contains only a subset of the suffixes present in a complete suffix tree, ST(T). ■

A general sparse suffix tree SST(T) of text T contains some arbitrary subset of the suffixes present in ST(T). These suffixes are denoted the SST-*suffixes*, and the starting point in the corresponding text for these suffixes are called the *suffix points*. Finding all the occurrences of a pattern P starting at a suffix point in the SST, can be done efficiently. However, for an arbitrary set of suffixes, finding patterns that do not start at suffix points in the SST, cannot be done efficiently.

To enable full sublinear pattern matching over the sparse suffix tree, we need to control which subset of suffixes that is present in the SST. Kärkkäinen and Ukkonen([3]) uses a strategy called an evenly spaced sparse suffix tree. However, when searching for entire words, we can create a non-evenly spaced Sparse Suffix Tree, by storing the suffixes starting at word boundaries only.

Definition (Word spaced Sparse Suffix Tree). A Word spaced Sparse Suffix Tree, $SST_{WS}(T)$ of a text T is a Sparse Suffix Tree containing only the suffixes starting at a word separator character in the text. ■

Two examples of Word spaced Sparse Suffix Trees are shown in Figure 1. Parts of the suffixes have been omitted to enhance the readability.

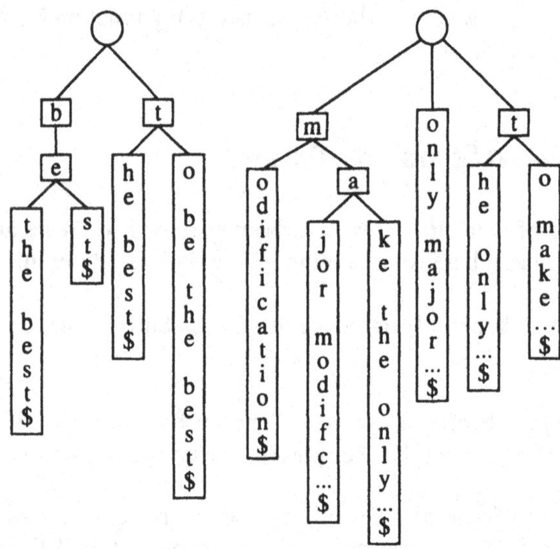

Figure 1 – Word spaced Sparse Suffix Trees for *T* = "*to be the best*" (left), and *T* = "*to make the only major modification*" (right).

Searching for patterns in the word spaced suffix tree, is reduced to traversal of the tree, branching at each level based on the current pattern character.

Building the word spaced sparse suffix tree for a text *T* of size *n*, could be done in $O(n)$ time.

4.1 Reducing the Suffix Length

We divide the text naturally into words, which are stored independently in the tree. Since our atom in searching is the word, we will also terminate each suffix at the end of the word. This reduces the sparse suffix tree to a PATRICIA-trie [5]. However, we will still maintain some sequence information(see section 4.2), so the structure is not identical to a PATRICIA trie.

Reducing the suffix length, requires that we change our representation of the leaf node. Pointers to the original text are replaced by the suffix string itself. We illustrate this suffix cutting in Figure 2 for the same two strings as used previously.

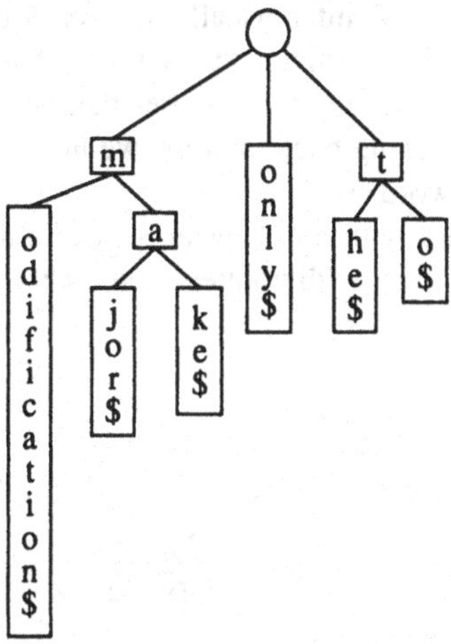

Figure 2 – Word spaced Sparse Suffix Trees with suffixes cut off at word boundaries. *T* = "*to be the best*" at left, and *T* = "*to make the only major modification*" at the right.

Using this representation of the text, we will not need to store an entire copy the original text in memory.

Now, one suffix may occur in many documents, and in many places of each document. Therefore, we add a list of the documents where the word occurs to the leaf node.

4.2 Sequence Information

We have abandoned the use of pointers to the original text in the leaf nodes of our suffix tree, reducing the structure to a PATRICIA-tree.

We will use a different approach to store the sequence information. Instead of using the implicit sequence information found in the original text, we will explicitly store sequence information in the suffix tree. This is done by using pointers between leaf nodes that represent consecutive words in the original text. Since we have a list of all the occurrences of the word represented by a particular leaf node, we must add a pointer to the next consecutive leaf

A leaf nodes contains only the suffix of the word it represents, so when traversing the sequence pointers in the occurrence list only the suffixes of each of the consecutive words are revealed. Two remedies can be applied to gain access to the entire word, represented by the leaf node.

1. Adding backpointers for all leaf nodes, enabling the possibility to backtrack the suffix tree to reveal the entire word.
2. Storing the entire word in the leaf node, not only the suffix.

Since the space usage of a pointer usually is 4 or 8 times the space usage of a character (pointer is 4 or 8 bytes, character is 1 or 2 bytes), the overhead of using backpointers to the parent nodes is much higher than storing the entire word in the leaf node. In addition, when using backpointers, we need $O(m)$ time to find the entire word, where m denotes the word size.

Annotating the leaf nodes with an occurrence list as described above, gives us a structure using an occurrence list with pointers to the next consecutive word and to its occurrence. See Figure 3.

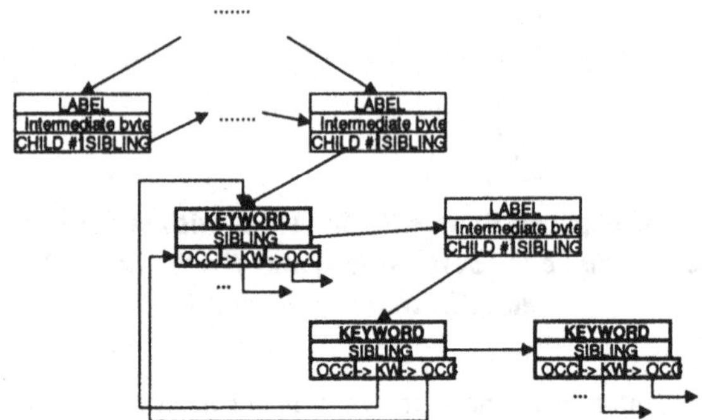

Figure 3 – Explicitly stored word sequence information

4.3 Organizing the occurrence list

The organization of the occurrence lists has a great impact on the performance of a sequence matching algorithm. Some words will have a very large number of occurrences (like "to", "be", "and", etc.), and scanning the entire list should be avoided. Using an occurrence list sorted on the next word in the sequence, will improve searching performance for the exact case. However, approximate matching will not benefit from this organization of the occurrence list.

A different approach uses a PATRICIA-trie [5] as the structure for organizing the occurrence list. The PATRICIA-trie enables us to access the list of all consecutive words matching p_2 in $O(l)$ time, where l denotes the length of p_2. To simplify finding all occurrence of a singular word, we could store an extra unsorted list of the occurrences in the leaf node. Now, using the PATRICIA-trie to organize the list of occurrences, we get a completely defined tree structure for storing words from a text, maintaining the sequence information. A typical leaf node, with both the PATRICIA trie for the organized occurrence list as well as the extra unsorted list of occurrences is shown in Figure 4.

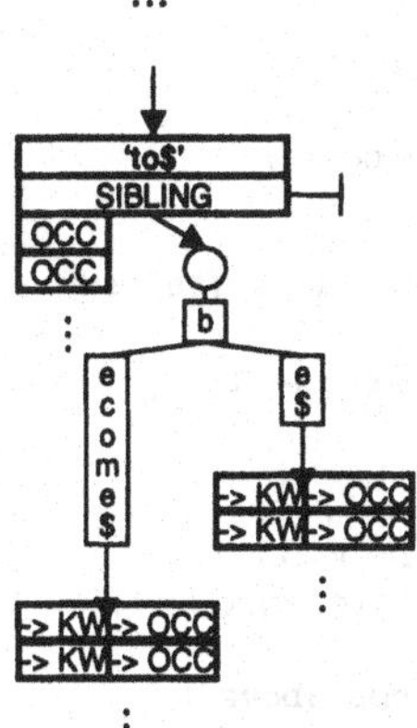

Figure 4 – Leaf node structure using both a PATRICIA trie and a simplified occurrence list.

5 Algorithms

Some algorithms have been adapted to the Word spaced Sparse Suffix Tree to allow approximate searching. In addition, a straightforward approximate word sequence matching algorithm has been designed. These algorithms are presented in this section.

5.1 Approximate Word Matching

Approximate word matching in Word spaced Sparse Suffix Tree is done by combining the calculation of the edit distance matrix and a traversal of the suffix tree. This algorithm is due to Chang and Merettal[6]:

```
FindApproximate(root,p,k)
 node ← root;
 i ← 1;

 nodes ← Children(node);
 for all v ∈ nodes do
   if IsLeaf(v) then
     for j ← i to length(Suffix(node)) do
       wj ← Suffix(node)j-i;
       if wj = '$' then        output w1..j;
         return;
       if EditDist(i) = ∞ then
         break;
   else
     i ← i+1;
```

```
    if EditDist(i) = ∞ then
      break;
    nodes ← Children(v) ∪ nodes;
// end for

EditDistance(j)      // Calculates jth row
for i ← 1 to length(P) do
   if p_i = w_j then ∂ ← 0 else ∂ ← 1;
   c_1 = D[i-1,j]+c_Ins(m_j);
   c_2 = D[i,j-1]+c_Del(p_i);
   c_3 = D[i-1,j-1]+c_change(p_i,m_j);
   c_4 = D[i-2,j-2]+c_transpose(p_i,m_j-1);
   D[i,j] ← c_fraction(j/l)· min(c_1,c_2,c_3,c_4);
 if D[i,j] > k
   return ∞; // All distances above k
 return D[i,j]
```

The expected worst case running time of this algorithm is $O(k\ |\Sigma|^k)$ according to Chang and Merettal[6].

5.2 Approximate Word Sequence Matching

Approximate word sequence matching requires us to calculate the word sequence edit distance for all possible matches. However, we can limit the number of possible matches by starting calculation of the edit distance only on the possible words. The cost of deleting a word from the sequences determines the number of possible start words. If the accumulated cost of deleting the i first words in a query sequence raises above the given error threshold, the candidate sequence starting with the ith word in the query cannot possibly be a match.

Therefore, for a query sequence of i words, at most i possible start words will be tried. Since we don't have backpointers in the sequence structure of the tree, we will not get all possible matches. Adding backpointers would solve this problem.

The algorithm is shown below:

```
ApproxSequenceMatch_ED(root,P(=p_1,p_2,...,p_m,k)
m ← |P|
matches ← ∅
startError ← 0
startIndex ← 1

while startError ≤ k OR startIndex ≤ m
do
startNode ← FindExact(p_startIndex);
  list ←
     UnorderedOccurenceList(startNode);
for all v ∈ lists do
   if ApproxMatchRest(v,P,k,startError)
```

```
    then
      matches ← matches ∪ v;

    startError ← startError + C_del(p_startIndex);
    startIndex ← startIndex + 1;
```

and the *ApproxMatchRest* function is defined as:

```
ApproxMatchRest(u,P,k,startError)
 error ← startError;
 lastError ← startError;
 column ← 0;
node ← u;
 for v ← p_2 to p_|P| do
   node ← NextOccurrence(node);
   word ← Keyword(node);
   lastError ← error;
   error ← startError + EditDistance(column);
   if error > k AND lastError > k then
     return false;

 return true;
```

The *FindExact* function used to find the leaf node matching the first word in the sequence is a simple traversal of the tree, and its running time is $O(|p_1|)$ where p_1 denotes the first word in the query sequence P. Calculating the edit distance can be done in $|P|^2$ time using straightforward dynamic programming, or in $O(k)$ time (where k denotes the error threshold) using improved versions of the calculation algorithm([7]).

Letting $\Sigma n_{occ}(p_i)$ denote the total sum of the number of occurrences of each word p_i in the word sequence, the expected worst case running time is $O(k\Sigma n_{occ}(p_i))$.

6 Experimental Results

An experimental system was implemented to test the proposed methods and metrics. Several tests were performed to evaluate the performance and usability of the proposed methods.

A collection of 10000 HTML documents was used in the tests. The total size of the document collection is 63,7 MB, and it contains about 420,000 distinct words. The total tree size was 113.4 MB.

The tests were run on a Pentium II system running at 266 MHz and with 384 MB of RAM. The total tree structure was resident in primary memory. The tests were repeated 100 times each.

6.1 Approximate Word Matching

The performance of the approximate word matching over the Word spaced Sparse Suffix Tree was tested by searching for several words. The error threshold k was not set to a constant value, rather it was set to be 20% of the query word length. The chart in Figure 5 shows the results of these tests.

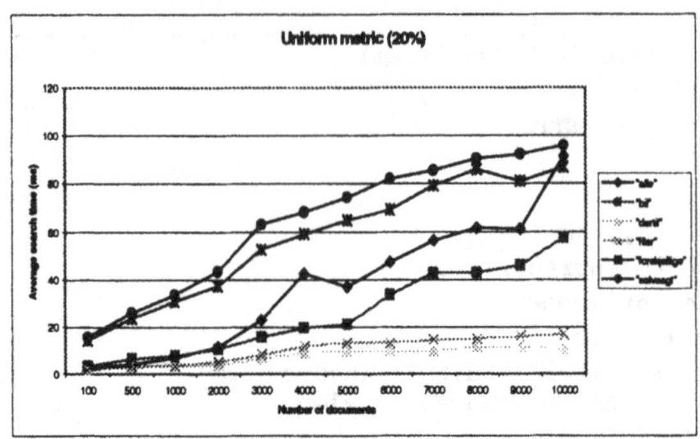

Figure 5 – Performance of approximate word matching over Word spaced Sparse Suffix Tree.

The average search time varies approximately linearly with the number of hits. This linear variation is due to reporting the hits. The average search time is about 50 ms, allowing up to 20 searches per second on a single CPU computer.

6.2 Approximate Sequence Matching

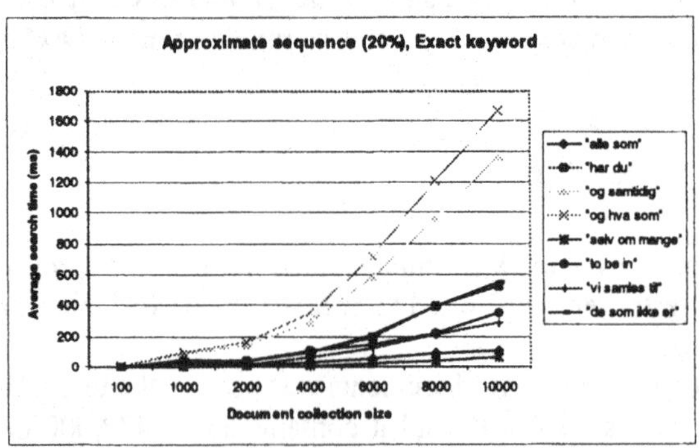

Figure 6 – Average search times for approximate word sequence matching.

The chart in Figure 6 shows how the average search time varies as the document collection increases. The main factor of the running time, the $\Sigma n_{occ}(p_1)$, expresses the number of matches of the first word in the query sequence. The search times vary from 50 ms to 1200 ms, thus enabling from 1-20 approximate word sequence searches per second on a single CPU computer.

Normalizing the chart with the $\Sigma n_{occ}(p_l)$ factors shows that the normalized approximate search time has a linear dependency on the number of documents(). This linearity is most likely due to reporting the list of occurrences to the end user.

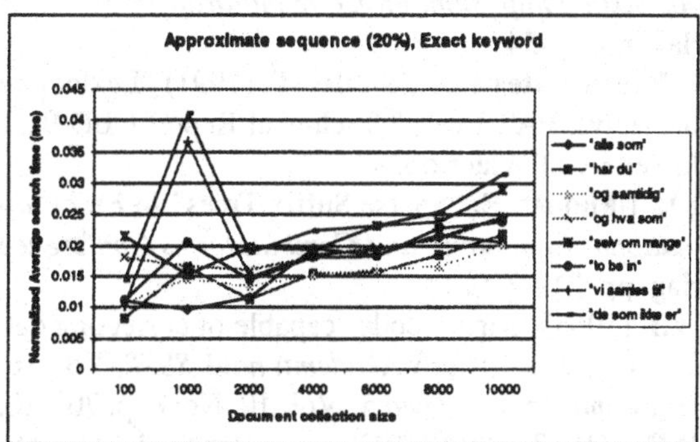

Figure 7 – Average search times for approximate word sequence matching, normalized with the $\Sigma n_{occ}(p_l)$ factor.

7 Conclusions

We have defined word sequence matching in texts. The edit distance metric used for approximate string matching is adapted to approximate word sequence matching.

The sparse suffix tree defined by Kärkkainen and Ukkonen[3] is extended to a non-evenly spaced version with explicit sequence information. This structure allows for approximate word and approximate word sequence matching.

The proposed structure and metrics have been implemented in a WWW information retrieval test system. The test system has shown good performance.

The use of approximate word sequence matching, could be useful in information retrieval applications where sequence matching has been limited to boolean operators and exact phrase matching.

8 References

[1] **Cobbs A. L.** (1995) "Fast Approximate Matching using Suffix Trees," In *Proceedings of Sixth Symposium on Combinatorial Pattern Matching (CPM'95)* Springer Verlag, pp. 41-54.

[2] **Gonnet G.H, Baeza-Yates R.A., Snider T.** (1991) "Lexicographical indices for text: Inverted files vs. PAT trees.," Technical Report OED-91-10, Center for the new OED, University of Waterloo.

[3] **Kärkkäinen J., Ukkonen E.** "Sparse Suffix Trees" In *Proceedings of the Second Annual International Computing and Combinatorics Conference (COCOON '96)*, Springer Verlag, pp. 219-230.

[4] **Levenstein, V.I.** (1965) "Binary codes capable of correcting deletions, insertions, and reversals," (Russian) *Doklady Akademii nauk SSSR,* Vol. 163, No. 4, p. 845-8 (also *Cybernetics and Control Theory,* Vol. 10, No. 8, p. 707-10, 1966).

[5] **Morrison D.R.** (1968) "PATRICIA – Practical Algorithm To Retrieve Information Coded in Alphanumeric," *Journal of the ACM*, 15, pp. 514-534.

[6] **Shang H., Merrettal T.H.** (1996) "Tries for Approximate String Matching," *IEEE Transactions on Knowledge and Data Engineering,* Vol 5, No. 4, p. 540-547.

[7] **Ukkonen E.** (1985) "Finding Approximate Patterns in Strings," *Journal of Algorithms,* vol. 6, pp. 132-137.

[8] **Weiner P.** (1973) "Linear pattern matching algorithms," In *Proceedings of the IEEE 14th Annual Symposium on Switching and Automata Theory,* pp. 1-11.

Efficient Parallel Algorithm for the Editing Distance between Ordered Trees

Kaizhong Zhang *

Department of Computer Science
University of Western Ontario
London, Ont. N6A 5B7
CANADA
kzhang@csd.uwo.ca

Abstract. Ordered labeled trees are trees whose nodes are labeled and in which the left-to-right order among siblings is significant. The tree editing problem for input ordered labeled trees T_1 and T_2 is defined as transforming T_1 into T_2 by performing a series of weighted edit operations on T_1 with overall minimum cost. An edit operation can be the deletion, the insertion, and the substitution. Previous results on this problem are only for some special cases and the time complexity depends on the actual distance, though for the more restricted version of degree-2 edit distance problem there are efficient solutions. In this extended abstract, we show polylogrithmic time algorithm for this problem.

1 Introduction

Rooted ordered labeled trees can be used to represent many phenomena. Comparing such trees by some meaningful distance metric is therefore useful. Edit distance is one of such metrics with applications in biology [5, 6]. Several sequential algorithms have been developed [8, 12]. Parallel algorithms developed so far [7] are only for some special case where the edit operation has unit cost. Also whether the problem is in class NC was not known since the complexity of the algorithms [7] depend on the actual distance between trees. Degree-1 and degree-2 edit distance [4, 2, 9, 11] between trees are the more restricted version of this problem. [10] present an efficient parallel algorithm for these restricted versions. Note that tree edit distances are the generalization of sequence edit distance for which many efficient parallel algorithms are well known [1, 3]. In this extended abstract, we show polylogrithmic time algorithms for this general ordered tree edit problem. Our result shows that this problem is in class NC.

2 Preliminaries

Given a rooted tree T, we use $V(T)$ to represent the set of its vertices, and use $r(T)$ to represent its root. For $v \in V(T)$, we use $t[v]$ to represent node v or

* Research supported by the Natural Sciences and Engineering Research Council of Canada under Grant No. OGP0046373.

the label of node v if there is no confusion. If $v \in V(T)$, then $T[v]$ denotes the subtree rooted at v and $F[v]$ denotes the forest resulting from removing v from $T[v]$.

We use $size(T[v])$ to denote the size of $T[v]$. For $v \in V(T)$, by $d(v)$ we denote its degree, by $C(v)$ the sequence of its children from left to right.

2.1 Edit operations

We consider three kinds of operations: relabel: change the label of a node; delete: delete a tree node and then make its children become the children of its parent; insert: inverse of delete. There is no restriction on insertion and deletion. These edit operations were proposed by Tai [8].

Following [8, 12], we represent an edit operation as $a \to b$, where a is either λ or a label of a node in tree T_1 and b is either λ or a label of a node in tree T_2. We call $a \to b$ a relabel operation if $a \neq \lambda$ and $b \neq \lambda$; a delete operation if $b = \lambda$; and an insert operation if $a = \lambda$.

Let S be a sequence $s_1, ..., s_k$ of edit operations. An S-derivation from tree A to tree B is a sequence of trees $A_0, ..., A_k$ such that $A = A_0$, $B = A_k$, and $A_{i-1} \to A_i$ via s_i for $1 \leq i \leq k$. Let γ be a cost function which assigns to each edit operation $a \to b$ a nonnegative real number $\gamma(a \to b)$.

We constrain γ to be a distance metric. That is, i) $\gamma(a \to b) \geq 0$, $\gamma(a \to a) = 0$; ii) $\gamma(a \to b) = \gamma(b \to a)$; and iii) $\gamma(a \to c) \leq \gamma(a \to b) + \gamma(b \to c)$.

We extend γ to the sequence of edit operations S by letting $\gamma(S) = \sum_{i=1}^{|S|} \gamma(s_i)$.

2.2 Edit distance and edit distance mapping

Edit distance. The *edit distance* between two trees is defined by considering the minimum cost edit operations sequence that transforms one tree to the other [12]. Formally the edit distance between T_1 and T_2 is defined as:

$D_e(T_1, T_2) = \min_S \{\gamma(S) \mid S \text{ is an edit operation sequence taking } T_1 \text{ to } T_2\}$.

By this definition, it is easy to see that D is a distance metric.

Edit distance mappings. The edit operations give rise to a mapping which is a graphical specification of what edit operations apply to each node in the two trees.

We define a triple (M, T_1, T_2) to be an edit distance mapping [12] from T_1 to T_2, where M is a binary relation on $V(T_1) \times V(T_2)$ such that any pair of (v_1, w_1) and (v_2, w_2) in M satisfying:

(a) $v_1 = v_2$ iff $w_1 = w_2$ (one-to-one)
(b) v_1 is an ancestor of v_2 iff w_1 is an ancestor of w_2 (ancestor order preserved).
(c) v_1 is to the left of v_2 iff w_1 is to the left of w_2 (sibling order preserved).

We will use M instead of (M, T_1, T_2) if there is no confusion. Let M be a edit distance mapping from T_1 to T_2. We can define the cost of M:

$$\gamma(M) = \sum_{(v,w)\in M} \gamma(v \to w) + \sum_{v \notin M} \gamma(v \to \lambda) + \sum_{w \notin M} \gamma(\lambda \to w)$$

The relation between an edit distance mapping and a sequence of edit operations is as follows:

Lemma 1. *Given S, a sequence $s_1, \ldots, s_k$ of edit operations from T_1 to T_2, there exists an edit distance mapping M from T_1 to T_2 such that $\gamma(M) \leq \gamma(S)$. Conversely, for any mapping M, there exists a sequence of edit operations such that $\gamma(S) = \gamma(M)$.*

Based on the lemma, the following theorem states the relation between the edit distance and the edit distance mappings.

Theorem 2. $D_e(T_1, T_2) = \min_M \{\gamma(M) \mid M$ *is an edit mapping from* T_1 *to* $T_2\}$

2.3 Algorithms for the edit distance between trees

A simple fast algorithm for computing the edit distance between trees is presented in [12]. The time complexity of the algorithm is $O(|T_1||T_2| \min\{depth(T_1), leaves(T_1)\} \min\{depth(T_2), leaves(T_2)\})$ and the space complexity is $O(|T_1||T_2|)$.

Let $T[i..j]$ be the ordered sub-forest of T induced by the nodes whose postorder left-to-right numbers are from i to j. Let $l(i)$ be the postorder number of the leftmost leaf descendant of node $t[i]$. The main formula in [12] is as follows.

$$D_e(T_1[1..i], T_2[1..j]) = \min \begin{cases} D_e(T_1[1..i-1], T_2[1..j]) + \gamma(t_1[i] \to \lambda) \\ D_e(T_1[1..i], T_2[1..j-1]) + \gamma(\lambda \to t_2[j]) \\ D_e(T_1[1..l(i)-1], T_2[1..l(j)-1]) + D_e(T_1[i], T_2[j]) \end{cases}$$

This formula implies a directed acyclic computation graph. For each node in this graph, there are three incoming edges corresponding to the three cases in the above formula. If we only consider the first two cases, this graph is a grid. For each node, the first edge is a vertical edge from the node immediately above the current node; the second edge is a horizontal edge from the node immediately to the left of the current node; the third edge is from a node in the first quadrant of the current nodes whose distance to the current node depends on the sizes of the two subtrees involved. We will use this kind of graph in the next section.

Parallel algorithms for unit cost tree edit problem are presented in [7]. The time complexity is $O(k\log(k)\log(\min\{|T_1|, |T_2|\}))$, where is k is the actual distance between T_1 and T_2.

3 Parallel algorithm for edit distance between trees

3.1 Basic idea

We can derive a parallel algorithm directly from the algorithm presented in [12]. The problem with this approach is that the time complexity will depend on the depth of the trees which in general is not logarithmic of the size of the trees.

First our algorithm will transform the computation depending on the tree depth into a computation depending on some factor which is logarithmic of the tree size.

We start by dividing the tree into a path from the root to a leaf and the resulting forests when we remove this path from the tree. Let $l_1 \in V(T_1)$ be a leaf of T_1 and P_1 be the set of nodes that forms the path from l_1 to $r(T_1)$. We use $P_1[i]$ to represent the node on P_1 which is the ith node from l_1. By this definition, $P_1[1]$ is the leaf. Sometimes, we also use $root(P_1)$ to denote the last node on P_1. P_2, $P_2[j]$, and $root(P_2)$ of T_2 can be defined similarly.

We now define the following sub-problem: Given T_1 and T_2 and P_1 and P_2, assume that we have all the edit distances $D_e(T_1[u], T_2[v])$ where either u is not on path P_1 or v is not on path P_2, we want to compute all the edit distances $D_e(T_1[u], T_2[v])$ where u is on path P_1 and v is on path P_2. We call this problem tree path distance problem. The main result of this extended abstract is how to parallelize the computation of this problem.

If we can solve this tree path distance problem, then we can arbitrarily pick up a path from a leaf to the root for each tree, and put the distance computation of the nodes on the paths to be the last computation. If we repeat this process, we will divide the trees into paths. This means that we can reduce the edit distance problem to the tree path distance problem.

In summary, we want to partition the tree into paths of different levels. For any two paths on the same level, their corresponding subtrees are non-overlapping implying that their corresponding tree path distance computation can be done in parallel. In addition the computation of subtree distances for any path on one level dose not depend on the computation of subtree distances of paths on upper levels.

3.2 New algorithm strategy

We first divide the trees into paths and group these paths into at most log($leaves$) levels. For paths on the same level, the computations are independent. For paths on different levels, the computations of the upper levels may depend on the computations on the lower levels. This is described in Step 1.

> Step 1.
> Divide the trees into paths and then group the paths into at most log($leaves$) levels. This can be done as follows. We choose the leaf in the middle and put the path from this leaf to the root in the top level. By deleting this path from the tree, we create two forests (on left and right sides of the path) with about the same number of leaves. For each forest, choose the leaf in the middle and consider the path from this leaf to its current root. We put these two paths into the next level. We repeat this process until we exhaust all the leaves.

Let us consider level i of the paths from T_1 and level j of the paths from T_2. For any path in level i of T_1 and any path in level j of T_2, we compute the tree distance for the nodes on these paths. In fact this is exactly the tree path distance problem. This leads to the following Step 2.

Step 2.
for $l_1 = 1$ to $log(leaves(T_1))$
for $l_2 = 1$ to $log(leaves(T_2))$
for any path P_1 in level l_1 and any path P_2 in level l_2
Solve the tree path distance problem for P_1 and P_2.

For the correctness of the strategy, it is easy to show the following invariants: for any two paths P_1 and P_2, immediately before the computation, all subtree distances for subtrees in $T_1[root(P_1)]$ and $T_2[root(P_2)]$, except $D_e(T_1[P_1[i]], T_2[P_2[j]])$ are available; immediately after the computation, all the subtree distances are available.

3.3 Parallel algorithm

Since it is easy to parallelize the step 1 (divide the trees into log($leaves$) levels of paths) using tree contraction, the main task of parallelizing the algorithm is to parallelize the computation for any two paths, i.e. parallelize the computation of tree path distance problem.

Let P_1 and P_2 be two paths on T_1 and T_2 respectively, we now consider how to compute $D_e(T_1[P_1[i]], T_2[P_2[j]])$ for all i and j. The assumption here is that, for any u and v, if u is not in P_1 or v is not in P_2 then $D_e(T_1[u], T_2[v])$ is available. Figure 1 shows two trees and in each tree there is a path specified.

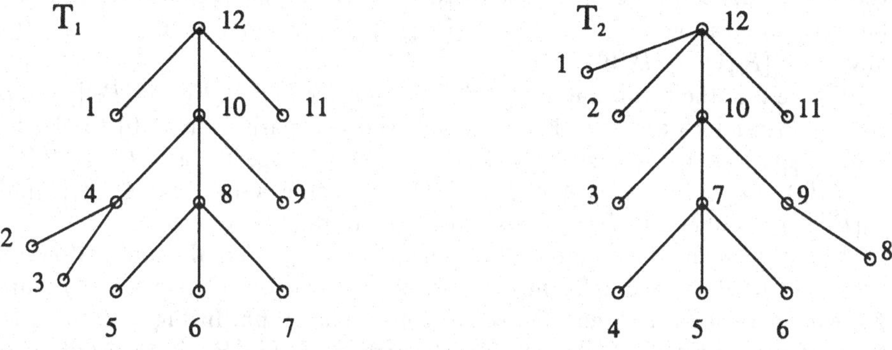

Fig. 1. In T_1, P_1 is the path from node 6 to the root. In T_2, P_2 is the path from node 5 to the root

Lemma 3.

$$D_e(T_1[P_1[i]], T_2[P_2[j]]) = \min \begin{cases} c_0, \\ D_e(T_1[P_1[i-1]], T_2[P_2[j]]) + c_1 \\ D_e(T_1[P_1[i]], T_2[P_2[j-1]]) + c_2 \\ \gamma(P_1[i] \rightarrow P_2[j]) + D_e(F_1[P_1[i]], F_2[P_2[j]]) \end{cases}$$

Proof: Consider the best mapping between $D_e(P_1[i], P_2[j])$, there are three cases:

1) $P_1[i]$ is not in the best mapping,
2) $P_2[j]$ is not in the best mapping,
3) $P_1[i]$ and $P_2[j]$ are both in the best mapping.

Case 1): Since $P_1[i]$ is not in the best mapping, $P_2[j]$ must be in the best mapping. And $P_2[j]$ must mapped to one of the subtrees rooted at $P_1[i]$.

If $P_2[j]$ is mapped to subtree rooted at $P_1[i-1]$, then $D_e(P_1[i], P_2[j]) = D_e(P_1[i-1], P_2[j]) + c_1$ where c_1 is the cost of deleting all the other subtrees and the node $P_1[i]$.

If $P_2[j]$ is mapped to other subtrees, then we can compute $D_e(P_1[i], P_2[j])$ using available distance values.

Case 2) is similar to case 1).

Case 3): $P_1[i]$ and $P_2[j]$ are mapped to each other, then $D_e(P_1[i], P_2[j]) = \gamma(P_1[i] \to P_2[j]) + D_e(F_1[P_1[i]], F_2[P_2[j]])$. □

From the Lemma, it is clear that we need to know how to compute the forest to forest distance $D_e(F_1[P_1[i]], F_2[P_2[j]])$.

Consider a best mapping between $F_1[P_1[i]]$ and $F_2[P_2[j]]$, we have two scenarios.

Scenario 1 There are no i_1 and j_1 such that $P_1[i_1]$ and $P_2[j_1]$ are mapped to each other in the best mapping between $F_1[P_1[i]]$ and $F_2[P_2[j]]$. In this case, since we have all $D_e(T_1[u], T_2[v])$ (for either u not on P_1 or v not on P_2), we have all the subtree distances we need. The method in [12] implies a computation graph for the computation of $D_e(F_1[P_1[i]], F_2[P_2[j]]))$. In this graph there is no edge for any $D_e(T_1[P_1[k]], T_2[P_2[l]])$.

Figure 2 is the computation graph for $D_e(F_1[root(P_1)], F_2[root(P_2)])$ where T_1, T_2, P_1, and P_2 are as in Figure 1. The shortest path from $(0,0)$ to $(11,11)$ is $D_e(F_1[root(P_1)], F_2[root(P_2)])$. And there is no edge for any $D_e(T_1[P_1[k]], T_2[P_2[l]])$ where $1 \le k \le 4$ and $1 \le l \le 4$. For example, for $D_e(T_1[P_1[3]], T_2[P_2[2]]) = D_e(T_1[10], T_2[7])$ there is no edge from $(1,3)$ to $(10,7)$.

Note that we need to compute $D_e(F_1[P_1[i]], F_2[P_2[j]])$ for all i and j. However we do not need to have different computation graphs for different i and j pairs. For each pair of path P_1 and P_2, we only need one graph. In Figure 2, in order to compute $D_e(F_1[10], F_2[7])$ (which is the same as $D_e(F_1[P_1[3]], F_2[P_2[2]])$), we only need to compute the shortest path from $(1,3)$ to $(9,6)$.

Scenario 2 There are i_1 and j_1 such that $P_1[i_1]$ is mapped to $P_2[j_1]$ in the best mapping. Without lose of generality we can assume that for any i_2 and j_2, where $i_1 < i_2 < i$ and $j_1 < j_2 < j$, $P_1[i_2]$ and $P_2[j_2]$ are not in the mapping.

In this situation, the path from $P_1[i_1]$ to $P_1[i]$ divides the tree $T_1[i]$ into four parts:

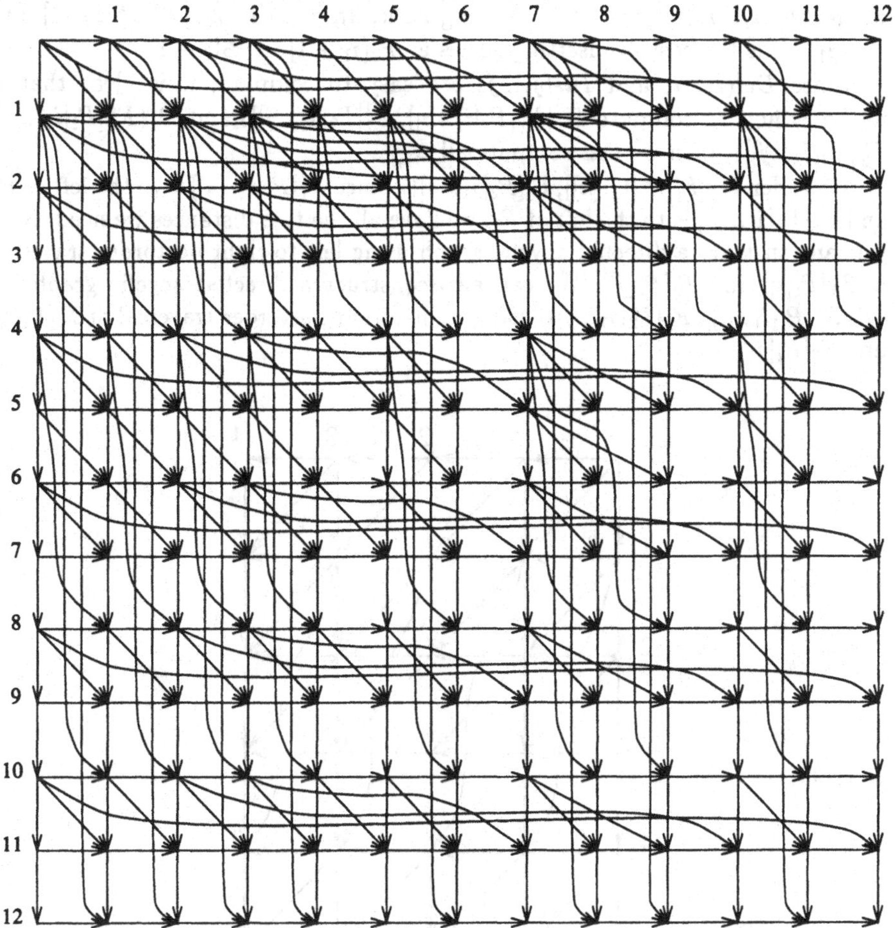

Fig. 2. Computational graph for scenario 1

1. the path from $P_1[i_1+1]$ to $P_1[i]$,
2. the tree $T_1[P_1[i_1]]$,
3. the forest to the left of the path ($F_1[l(P_1[i]) \ldots l(P_1[i_1])-1]$ using left-to-right post order notation where $l(\ )$ is the left-most leaf descendant), and
4. the forest to the right of the path.

Let the forest to the left of the path from $P_1[i_1]$ to $P_1[i]$ be $F_1^l[P_1[i_1 \ldots i]]$ and the forest to the right of the path from $P_1[i_1]$ to $P_1[i]$ be $F_1^r[P_1[i_1 \ldots i]]$. $F_2^l[P_2[j_1 \ldots j]]$ and $F_2^r[P_2[j_1 \ldots j]]$ are defined similarly.

With these definitions, it is easy to see that we have the following.

$$D_e(F_1[P_1[i]], F_2[P_2[j]]) = D_e(T_1[P_1[i_1]], T_2[P_2[j_1]]) + c_{i_1,j_1}$$

$$c_{i_1,j_1} = \begin{array}{l} D_e(F_1^l[P_1[i_1 \ldots i]], F_2^l[P_2[j_1 \ldots j]]) + D_e(F_1^r[P_1[i_1 \ldots i]], F_2^r[P_2[j_1 \ldots j]]) \\ +Del(i_1 \ldots i) + Ins(j_1 \ldots j) \end{array}$$

where $Del(i_1...i)$ is the cost of deleting nodes from $P_1[i_1+1]$ to $P_1[i-1]$ and $Ins(j_1...j)$ is the cost of inserting nodes as $P_2[j_1+1]$ to $P_2[j-1]$.

Since $Del(i_1...i)$ and $Ins(j_1...j)$ are easy to compute, it is clear that if we know how to compute $D_e(F_1^l[P_1[i_1...i]], F_2^l[P_2[j_1...j]])$ and $D_e(F_1^r[P_1[i_1...i]], F_2^r[P_2[j_1...j]])$, then we can compute all c_{i_1,j_1}.

Consider $D_e(F_1^l[P_1[i_1...i]], F_2^l[P_2[j_1...j]])$, since $F_1^l[P_1[i_1...i]]$ is to the left of P_1 and $F_2^l[P_2[j_1...j]]$ is to the left of P_2, we have all the tree distances we need. Now we can construct a directed acyclic graph using left-to-right postorder traversal of $T_1[P_1[i]]$ and $T_2[P_2[j]]$. We can also construct a directed acyclic graph for $D_e(F_1^r[P_1[i_1...i]], F_2^r[P_2[j_1...j]])$ using right-to-left postorder traversal of $T_1[P_1[i]]$ and $T_2[P_2[j]]$.

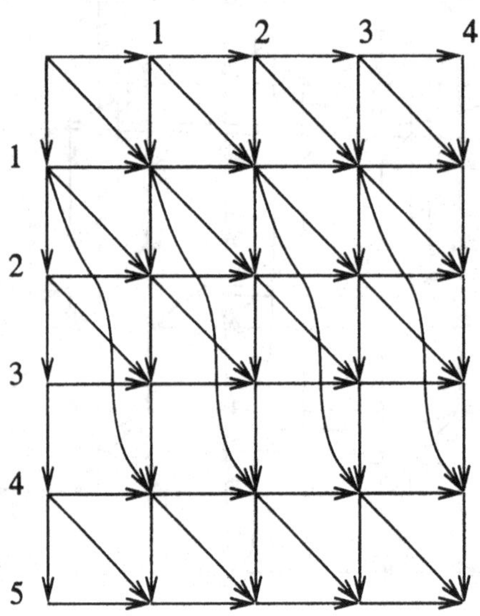

Fig. 3. Computational graph for scenario 2: left hand side of P_1 and P_2

Figure 3 shows the computation graph for $D_e(F_1^l[P_1[1...4]], F_2^l[P_1[1...4]])$ where trees and paths are as in Figure 1. The shortest path from $(0,0)$ to $(5,4)$ is $D_e(F_1^l[P_1[1...4]], F_2^l[P_2[1...4]])$.

Note that we need to compute $D_e(F_1^l[P_1[i_1...i]], F_2^l[P_2[j_1...j]])$ for all (i_1, j_1) and (i, j). Again, for a given pair of paths P_1 and P_2, we only need one graph for all the computations of the left hand side of P_1 and P_2. All the computations are the shortest paths in this graph. For example, in Figure 1 $D_e(F_1^l[P_1[2...4]], F_2^l[P_2[1...3]])$ is the shortest path from $(0,2)$ to $(4,4)$.

Combining both scenarios In both scenarios, we can consider the problem as problem of finding shortest paths in directed acyclic graph. For scenario 1, we need one graph and for scenario 2, we need two graphs.

This means that we can use parallel all pair shortest path algorithm for these computation.

Combining Lemma 3 and the analysis for scenario 1 and 2, the following lemma shows the formula for the computation of $D_e(T_1[P_1[i]], T_1[P_2[j]])$.

Lemma 4.

$$D_e(T_1[P_1[i], T_1[P_2[j]) = \min \begin{cases} c_0, \\ D_e(T_1[P_1[i-1]], T_1[P_2[j]]) + c_1 \\ D_e(T_1[P_1[i]], T_1[P_2[j-1]]) + c_2 \\ D_e(T_1[P_1[i_1]], T_1[P_2[j_1]]) + c_{i_1,j_1} \end{cases}$$

where $1 \leq i_1 < i$ *and* $1 \leq j_1 < j$.

Proof: Immediately from Lemma 3 and scenario 1 and 2. Note that in this formula c_0 is from Case 1), Case 2) of Lemma 3, and scenario 1. c_1 is from Case 1) of Lemma 3. c_2 is from Case 2) of Lemma 3. c_{i_1,j_1} is from scenario 2. □

Lemma 4 implies a computation graph G. This is a directed acyclic graph with a cost associated with not only each edge but also each node. Figure 4 is an example.

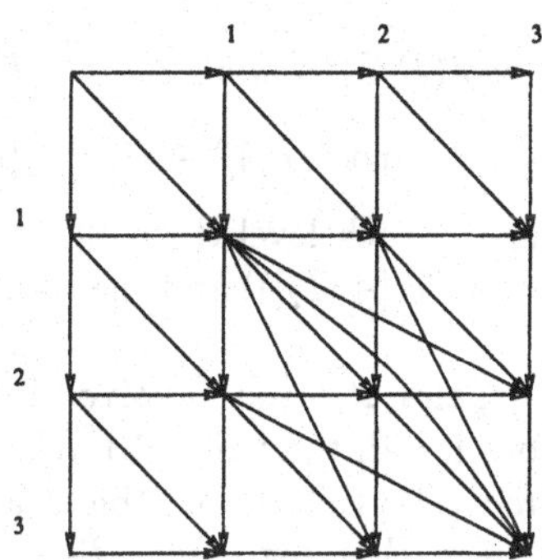

Fig. 4. Computation graph for $D_e(T_1[P_1[i]], T_2[P_2[j]])$.

From Lemma 4, for any node b in the graph, the value we want to compute is $\min_{a \in G}\{cost(a) + shortest\_path(a, b)\}$ Therefore, this again can be considered as a shortest path problem of a directed acyclic graph. We can first apply the parallel shortest path algorithm to this graph and the for each node we perform parallel minimization.

Algorithm. Algorithm:
1) preprocessing to divide trees into paths.
2) for $k = 2$ to $(\log(leaves_1) + \log(leaves_2))$
    for any path P_1 in T_1 and any path P_2 in T_2
        such that $level(P_1) + level(P_2) = k$ do,

2.1) compute $c_0, c_1, c_2, c_{i_1,j_1}$ for all $P_1[i]$ and $P_2[j]$.
2.2) compute all $D_e(T_1[P_1[i]], T_2[P_2[j]])$.

We use parallel all pair shortest paths algorithm to compute step 2.1. We use parallel all pair shortest paths algorithm and then minimization to compute step 2.2.

Complexity.

Theorem 5. *The time complexity of the algorithm is*

$$O(log^2(|T_1| + |T_2|)\log(leaves(T_1)) + leaves(T_2))).$$

Proof:

- For each pair of paths, time is bounded by the complexity of parallel all pair shortest path algorithm. Therefore the time complexity is

$$O(log^2(size(T_1[root(P_1)]) \times size(T_2[root(P_2)])))$$
$$= O(log^2(size(T_1[root(P_1)]) + size(T_2[root(P_2)]))).$$

The total work is

$$O(size(T_1[root(P_1)])^3 \times size(T_2[root(P_2)])^3 \times$$
$$log(size(T_1[root(P_1)]) + size(T_2[root(P_2)]))).$$

- For each pair of levels, since in each level the subtrees are non-overlapping, the time complexity is $O(log^2(|T_1|+|T_2|))$ and the total work is $O((|T_1||T_2|)^3 log(|T_1| + |T_2|))$.
- In one iteration, the total pairs of levels involved is $\min\{log(leaves(T_1)), \log(leaves(T_2))\})$. Therefore for all the pairs of levels in the same iteration, the time complexity is $O(log^2(|T_1|+|T_2|))$ and the total work is $O((|T_1||T_2|)^3 log(|T_1| + |T_2|) \min\{log(leaves(T_1)), \log(leaves(T_2))\})$.
- Since the total number of iteration is $\log(leaves(T_1)) + \log(leaves(T_2))$. The time complexity of the algorithm is

$$O(log^2(|T_1| + |T_2|)\log(leaves(T_1) + leaves(T_2))).$$

The total work of the algorithm is

$$O((|T_1||T_2|)^3 \log(|T_1| + |T_2|)\log^2(leaves(T_1) + leaves(T_2))).$$

□

4 Conclusion

We present fast polylogrithmic parallel algorithms for computing the edit distance between ordered labeled trees. It remains open if one can reduce the total work to within a polylogrithmic factor of the best sequential algorithm in [12].

References

1. A. Apostolico, M.J. Atallah, L.L. Larmore and S. Mcfaddin, 'Efficient parallel algorithms for string editing and related problems', *SIAM J. Comput.*, vol. 19, no. 5, pp. 968-988, 1990
2. S. Y. Lu, "A tree-to-tree distance and its application to cluster analysis", *IEEE Trans. PAMI*, vol. 1, pp.219-224, 1979
3. G. M. Landau and U. Vishkin, 'Fast parallel and serial approximate string matching', *J. Algorithms*, vol. 10, pp.157-169, 1989
4. S. M. Selkow, 'The tree-to-tree editing problem', *Information Processing Letters*, no. 6, 184-186, 1977
5. B. Shapiro, An algorithm for comparing multiple RNA secondary structures, *Comput. Appl. Biosci.*, pp. 387-393, 1988
6. B. Shapiro and K. Zhang, 'Comparing multiple RNA secondary structures using tree comparisons' *Comput. Appl. Biosci.* vol. 6, no. 4, pp.309-318, 1990
7. D. Shasha and K. Zhang, 'Fast algorithms for the unit cost edit distance between trees', *J. of Algorithms*, vol. 11, pp. 581-621, 1990
8. K. C. Tai, 'The tree-to-tree correction problem', *J. ACM*, vol. 26, pp.422-433, 1979
9. K. Zhang, 'A new editing based distance between unordered labeled trees', In A. Apostolico, M. Crochemore, Z. Galil, and U. Manber, editors, *Combinatorial Pattern Matching, Lecture Notes in Computer Science, 684*, pp. 254-265. Springer-Verlag, 1993; journal version is to appear in *Algorithmica.*
10. K. Zhang, 'Efficient parallel algorithms for tree editing problems', Proceedings of the Seventh Symposium on Combinatorial Pattern Matching,
11. K. Zhang, J. Wang and D. Shasha, 'On the editing distance between undirected acyclic graphs', Proceedings of the Sixth Symposium on Combinatorial Pattern Matching, Helsinki, Finland, July 1995. Springer-Verlag's Lecture Notes in Computer Science 937, pp 395-407.
12. K. Zhang and D. Shasha, 'Simple fast algorithms for the editing distance between trees and related problems', *SIAM J. Computing* vol. 18, no. 6, pp.1245-1262, 1989

Reporting Exact and Approximate Regular Expression Matches

Eugene W. Myers[1,*], Paulo Oliva[2,**], and Katia Guimarães[2,***]

[1] Dept. of Computer Science, University of Arizona Tucson, AZ 85721
[2] Dept. of Informatics, Federal University of Pernambuco, Recife, Brazil

Abstract. While much work has been done on determining if a document or a line of a document contains an exact or approximate match to a regular expression, less effort has been expended in formulating and determining what to report as "the match" once such a "hit" is detected. For exact regular expression pattern matching, we give algorithms for finding a longest match, all symbols involved in some match, and finding optimal submatches to tagged parts of a pattern. For approximate regular expression matching, we develop notions of what constitutes a significant match, give algorithms for them, and also for finding a longest match and all symbols in a match.

1 Introduction

Much attention has been given to the problem of efficiently determining if a string contains a match to a pattern. But the problem of selecting which substring or substrings of the target to report as matching, is an important practical issue, that probably many have had to face, but for which there is a dearth of published work. This paper serves as a systematic exploration of a range of possibilities. We consider here the problem of reporting matches to a regular expression R of size P in a text A of length N, in both the cases of exact and approximate matching. The individual problems we encounter vary in difficulty. We treat them all in order to give a cohesive treatment.

We first make the observation that for the problems being considered here, reporting the desired or interesting thing is more important than doing so extremely efficiently. Generally, one can use a very fast algorithm such as that in **egrep** or **agrep**, to find the lines or regions of a text that contain a match. Thus the N in our context is typically on the order of the length of a line, and not that of a document. That is, we can afford to spend more time on each line containing a match, working on delivering a meaningful match rather than one that is an artifact of the scanning/filtration algorithm.

* E-mail: gene@cs.arizona.edu. Partially supported by NLM grant LM-04960

** E-mail: pbo@di.ufpe.br. Partially supported by CNPq-PIBIC grant 111371/97-0

*** E-mail: katia@di.ufpe.br. Partially supported by CNPq grant 352775/96-3 and a member of Project PRONEX 107/97 (MCT/FINEP)

The widely accepted standard, e.g. `Perl` [WS91], `Tcl/Tk` [Ous94], and the IEEE Posix standard [IEE92], for exact regular expression pattern matching is to report the left-most longest match, i.e. the matching substring whose left end is leftmost, and if there are several with such a left end, then the longest of those. The motivation for this definition appears to stem more from the limitations of searching with a finite-state automata then it does from any conceptual principle: a traditional implementation of a finite automaton easily admits the reporting of left-most and right-most, longest and shortest matches. In a recent paper, Clarke and Cormack, argue that shortest matches have superior search properties when looking at patterns that involve matching several regular expressions [CC97]. On the other hand, we know of no reported work on reporting approximate matches to regular expressions, save that there are connections to work on finding locally optimal alignments [SW81,Sel84].

Our goal is to report matches in a meaningful way. For example, suppose one is given the regular expression **babb*bab** and one is requested to find matches with 2 differences or less. One might like to see a report such as:

where the symbols in grey are those that are in some substring that approximately matches the pattern, and the substring in the heavily bordered box is, in a sense to be defined later, the "best" match. This motivates us to consider, for the case of exact matching, the problems of finding the longest match, all symbols of the text involved in a match, and on parsing a match so as to deliver a consistent set of matching substrings that optimize some criterion. For the case of approximate matching, we first examine the problem of defining and delivering an essential or significant match, an issue that does not arise in the exact matching case. We then conclude by solving the longest-match and all-matches problems for the case of approximate matches as well.

2 Reporting Exact Matches

2.1 Finding the Longest Exact Match

Using traditional scan-based regular expression matching methods based on finite automaton, one can easily find the left-most or right-most longest or shortest match to the pattern in a given string. But often this is not the most interesting or specific match. For example, given the pattern **bb***, the left-most and right-most matches in **aabaabbbbbbbaabaa** are uninteresting compared to the long central match. This motivates the problem of finding the *longest* substring matching a regular expression, as it is the rarest match if symbols occur within the target with equal probability.

Interestingly, this problem can be solved in $O(PN)$ time using a specialization of the *approximate* regular expression matching algorithm of Miller and Myers

[MM88] that accommodates any additive alignment scoring scheme δ. Quickly we review this result and then proceed to the specialization. First, recall that any regular expression R can be converted to a state-labeled ϵ-NFA F that has at most $O(P)$ states and transitions and a single final/accepting state. Figure 2 in Appendix A gives such a conversion especially suited to our purpose. Let θ and ϕ be the unique start and final states of F that by construction are labeled with ϵ. Further let $\lambda_s \in \Sigma \cup \{\epsilon\}$ denote the label of state s.

Miller and Myers first observation was that the cost of the best alignment between $A[1..i]$ and a word in $L_F(s)$, where $L_F(s)$ is the set of words accepted at state s, is the minimum-fixed point to the system of equations given by the recurrence:

$$C(i,s) = \max \left\{ \begin{array}{c} \max\limits_{t\to s}\{C(i-1,t)+\delta(a_i,\lambda_s)\} \\ \max\limits_{t\to s}\{C(i,t)+\delta(\epsilon,\lambda_s)\} \\ C(i-1,s)+\delta(a_i,\epsilon) \end{array} \right\} \tag{1}$$

subject to the boundary condition $C(0,\theta)=0$. Modifying the boundary condition to $C(i,\theta)=0$ *for all* i, results in $C(i,s)$ being the score of the best match between *a suffix of* $A[1..i]$ and a word in $L_F(s)$. So in this instance, $C(i,\phi)$ is the score of the best match ending at i to a word recognized by R.

The second observation of Miller and Myers is that the graph of F is a reducible graph, so computing the desired fixed-point for a given i can be achieved by evaluating the relevant terms in two passes over the s parameter in topological order of the acyclic graph obtained by removing the *back-edges* from F (see Fig. 2). This gives the following algorithm outline:

```
C(0,θ) ← 0
for s ≠ θ in topological order of F (less backedges) do
    C(0,s) ← R.H.S. of (1) (exclude back-edge terms)
for i ← 1,2,...n do
 {  C(i,θ) ← 0
    for s ≠ θ in topological order of F (less backedges) do
        C(i,s) ← R.H.S. of (1) (exclude back-edge terms)
    for s ≠ θ in topological order of F do
        C(i,s) ← max{C(i,s), R.H.S. of (1) (include back-edge terms)}
 }
```

Turning now to the longest exact match problem, consider the scoring function δ_{long} defined as follows: $\delta_{long}(a,b) = 1$ if $a = b \neq \epsilon$, $\delta_{long}(a,b) = 0$ if $a = b = \epsilon$, and $\delta_{long}(a,b) = -\infty$ otherwise. Effectively, every pair of aligned equal symbols scores 1, every insertion of an ϵ-state of F has score 0, and every other mis-alignment is not allowed. It then follows that $C(i,s) \neq -\infty$ *iff* there is an exact match between a suffix of $A[1..i]$ and a word in $L_F(s)$, and furthermore, that $C(i,s) = len \neq -\infty$ *iff* the longest suffix of $A[1..i]$ matching a word in $L_F(s)$ is of length len. Thus $C(i,\phi)$, under the cost function δ_{long} is the length of the longest match to R ending at i, or $-\infty$ if there is no such match.

In practice, one may model the $-\infty$ of δ_{long} with the value $-(N+1)$ provided that integers of value $[-P(N+1), N]$ can be modeled on the given machine. In that case $C(i,s) \geq 0$ is equivalent to the condition that $C(i,s) \neq -\infty$ above.

Furthermore, the special form of δ_{long} allows one to specialize (1) for a non-asymptotic but practical gain in efficiency. First one must carefully construct F so that it is guaranteed to contain no cycles all of whose states are labeled with ϵ. Such an $O(P)$ time and space construction is novel and given in Figure 2 of Appendix A. Let F' be the subautomata consisting of the vertices of F and just those transitions directed into states labeled with ϵ. Because F contains no ϵ-cycles, it follows that F' is acyclic (*including* back edges). One then arrives at the following simple, one-pass algorithm:

```
Compute C(0, s) for all s as before
for i ← 1, 2, ... n do
{  C(i, θ) ← 0
   for s ≠ θ in topological order of F' do
      if λ_s = ε then
         C(i, s) ← max_{t→s} {C(i, t)}
      else if λ_s = a_i then
         C(i, s) ← max_{t→s} {C(i − 1, t) + 1}
      else
         C(i, s) ← −(N + 1)
}
```

The last consideration is how to actually report the longest matching substring. Running the algorithm above, one finds a left end of a longest possible match at a value of i, call it r, which maximizes $\max_i C(i, \phi)$. By then running the same algorithm on the reverse of R and $A[1..r]$ with the boundary condition that only $C(0, \theta) = 0$, one finds the left end of the longest match whose right end is r. The overall time is $O(PN)$ and only $O(P)$ space is required.

2.2 Finding All Matching Positions

Next consider finding *all* the places where the pattern R matches the text, i.e., determine for all i, if symbol a_i is in a match to R or not. Solving this problem requires extending the well-known $O(PN)$ state-set simulation algorithm for matching regular expressions [Tho68,Sed83]. Recall that the algorithm computes in increasing sequence of text position i, what we call here the *forward state-set* $S_f(i) = \{s : \text{a suffix of } A[1..i] \text{ is in } L_F(s)\}$. That is, $S_f(i)$ is the set of all states that F could be in after scanning the first i symbols of A. Connecting this with our algorithm for longest matches above, it is easy to see that $S_f(i) = \{s : C(i, s) \neq \infty\}$. While this demonstrates that $S_f(i)$ can be computed from $S_f(i-1)$ in $O(P)$ worst-case time, the traditional reaching algorithm of [Tho68] does so in $O(|S_f(i-1)| + |S_f(i)|)$ time as one is free to discover the states in $S_f(i)$ in *any* order. This is superior in practice as it is frequently the case that the average size of the state sets is $O(1)$.

Let $s \xrightarrow{w} t$ be a predicate denoting that there is a path in F between states s and t whose state-label sequence spells w. With this notation, we may give the following, automata-centric definition: $S_f(i) = \{s : \exists j \leq i+1, \theta \xrightarrow{A[j..i]} s\}$. To solve our problem we will also need to be able compute what we call the *reverse* state-set at position i, $S_r(i) = \{s : \exists j \geq i-1, s \xrightarrow{A[i..j]} \phi\}$. These reverse state sets are easily computed by symmetry to the forward case: run the same algorithm, but on the reverse of both F and A. The following lemma follows directly from the definitions of S_f and S_r:

Lemma 1: For all i, a_i is in a match to R *iff* there exists $s \in S_f(i) \cap S_r(i)$ such that $\lambda_s \neq \epsilon$.

The one potentially interesting computational issue is how to deliver both $S_f(i)$ *and* $S_r(i)$ for all choices of i. If one computes and saves the state sets for all values of i then the resulting algorithm takes $O(PN)$ time *and space* in the worst case. Using less space is difficult because $S_f(i)$ is easily computable from $S_f(i-1)$ but not from $S_f(i+1)$. Similarly $S_r(i)$ is easily computable from $S_r(i+1)$ but not from $S_r(i-1)$. Thus the "grain" of the computations for S_f and S_r oppose each other. If space is a problem in a particular context, then one can employ the "going-against the grain" algorithm of Myers and Jain [MJ96], to compute $S_r(1), S_r(2), S_r(3), \ldots S_r(N)$ in the given order using $O(tPN)$ time and $O(PN^{1/t})$ space for any choice of $t \geq 1$. Choosing $t = \log N$ gives an $O(PN \log N)$ time and $O(P \log N)$ space worst-case guarantee.

2.3 Finding All Match Parses

In numerous contexts one desires not only a substring of the text that matches the pattern, but also the precise way that subexpressions of R are matched. For example, it is common to find a notation in line-based editing commands for tagging subexpressions of a pattern so that whatever matched the tagged part may be used in forming a string to replace the overall match. Consider then matching a regular expression R in which any number of subexpressions may be enclosed in curly braces. Each subexpression so enclosed is said to be *tagged* and when we match R we desire not only the match to R but also to each tagged subexpression. For example, consider `{a*{b*}}(a+b)*` and its match to all of **aabbabab**. The match to the subexpression `a*b*` could be either **aabb** or ϵ. If the former, then the match to the tagged subexpression `b*` must be **bb**, otherwise it must be ϵ. That is, the substrings returned for the tagged subexpressions must be *consistent* with a given parse. The two possible *tag-matches* for the example are the *ordered* pairs $(aabb, bb)$, and (ϵ, ϵ). Another subtlety is that in some cases a match to one or more of the subexpressions may not occur in the match to the entire pattern. For example, `(a{ba}b|bb)*` matches **bb** without matching the designated sub-expression. Basically this can happen whenever a tagged subexpression is part of an alternation ('|') construct, a Kleene closure ('*') construct, or an option ('?') construct. We will call expressions where tags do not occur *within* such constructs, *unambiguous*, and make the distinction because such cases can be handled with greater efficiency. Finally, note that by using curly braces to denote tags we are assuming that tagged subexpressions either nest or are disjoint.

Assume that the substring B of the text that R is to match has been selected, so that we can hereafter consider matches of R to the entirety of B. Rather than develop a particular method for selecting a parse to R, let's consider the problem of determining the graph $\mathcal{M}_{<B,R>}$ of all possible paths through R that match all of B. Let $S(i) = S_f(i) \cap S_r(i) \cap S_\Sigma$ and let $E(i) = S_f(i) \cap S_r(i+1) \cap S_\epsilon$ where $S_\Sigma = \{s : \lambda_s \in \Sigma\}$ and $S_\epsilon = \{s : \lambda_s = \epsilon\}$. Let the vertices in $\mathcal{M}_{<B,R>}$ be the set of ordered pairs $\{(0,s) : s \in E(0)\} \cup \{(i,s) : i \in [0,N] \text{ and } s \in E(i) \cup S(i)\}$. There is an edge $(i,s) \to (j,t)$ in the graph if and only if $s \to t$ is a transition in F and either (1) $j = i+1$, or (2) $j = i$ and $t \in S_\epsilon$. It follows easily that there are at most $O(PN)$ vertices and edges in the graph and that it can be computed in time linear in its size. Note that $\mathcal{M}_{<B,R>}$ is acyclic and every path from vertex $(0,\theta)$ to vertex (N,ϕ) models a match between B and R. Also, given the graph and a selected path from $(0,\theta)$ to (N,ϕ), it is a simple matter to deliver the tagged submatches.

Now we develop criteria for selecting a path through the graph $\mathcal{M}_{<B,R>}$. Suppose that the k subexpressions $\alpha_1, \alpha_2, \cdots \alpha_k$ have been tagged. For a consistent match to these subexpressions, say $(a_1, a_2, \cdots a_k)$, let $\Sigma|a_i|$ be the *extent* of the match and let the k-tuple of integers, $(|a_1|, |a_2|, \cdots, |a_k|)$, be its *footprint*. Two possible selection criteria for a path through the graph are (1) to find the one whose submatches give the largest extent, or (2) to find the one that has the lexicographically largest footprint. Below we will solve for both of these problems by giving an algorithm that, more generally, works with respect to any ranking $\succeq$ of fingerprints satisfying the following monotonicity property: for all fingerprints F, G, H, $F \succeq G$ implies $F \oplus H \succeq G \oplus H$ where $\oplus$ is component-by-component (vector) addition.

In the case where k properly nested but otherwise arbitrary subexpressions are tagged, one must keep track of the highest-ranking fingerprint to each vertex v of $\mathcal{M}_{<B,R>}$ that involves a particular subset of tagged subexpressions, C, whose right-ends have been seen, and a particular subset of tagged subexpressions, I, disjoint from C, whose left-ends have been seen, but not yet their right ends. We say the tagged subexpressions in C are *complete*, and those in I are *in-progress*. Formally, we keep track of $Best_{C,I}(v)$ for every choice of C and I and every vertex v, computing these quantities in a topological order of $\mathcal{M}_{<B,R>}$. In the algorithm outline of Figure 1 below we do not worry about whether a particular (C,I) is legitimate for a given vertex v, but simply use $-\infty$ to fill illegitimate components of a fingerprint. That is, if $x \in C$ but there does not exist a path to v whose projection onto F passes through x's subautomaton, then the x component of the candidate will be $-\infty$. Similarly, if $x \in I$ but there does not exist a path to v whose projection onto F enters x's subautomaton but does not leave it, then the x component of the candidate will be $-\infty$. For each *Best* value the algorithm retains a trace value *Trace* recording which predecessor vertex gave rise to the best value, so that one can trace back a desired path at the end of the computation. If the time to compare fingerprints under $\succeq$ is $O(c)$ then the algorithm of Figure 1 takes $O(c4^kPN)$ time and space in the worst case as all 2-partitions, (C,I) of all subsets of $[1,k]$ are considered. Note

```
for v ≡ (i, s) a vertex in M<B,R> in topological order do
{ for I ∈ 2^[1,k] and C ∈ 2^[1,k]−I do
    { Best_(C,I)(v) ←< −∞, −∞, ..., −∞ >
      for w → v in M<B,R> do
          if Best_(C,I)(w) ⪰ Best_(C,I)(v) then
              (Best_(C,I)(v), Trace_(C,I)(v)) ← (Best_(C,I)(w), w)
    }
  for x a tagged subexpr. starting at s do
      for I ∈ 2^[1,k]−x and C ∈ 2^[1,k]−I do
        { f ← Best_(C,I)(v)
          f[x] ← 0
          (Best_(C−x,I+x)(v), Trace_(C−x,I+x)(v)) ← (f, Trace_(C,I)(v))
        }
  if λ_s ≠ ε then
      for I ∈ 2^[1,k], C ∈ 2^[1,k]−I, and x ∈ I do
          Best_(C,I)(v)[x] ← Best_(C,I)(v)[x] + 1
  for x a tagged subexpr. ending at s do
      for I ∈ x + 2^[1,k] and C ∈ 2^[1,k]−I do
        { if Best_(C,I)(v) ⪰ Best_(C+x,I−x)(v) then
              (Best_(C+x,I−x)(v), Trace_(C+x,I−x)(v))
                  ← (Best_(C,I)(v), Trace_(C,I)(v))
          Best_(C,I)(v)[x] ← −∞
        }
}
```

Fig. 1. Optimal R.E. Parsing Algorithm.

that c is $O(k)$ if we seek the lexicographically largest footprint, and c is $O(1)$ if we seek the footprint with the largest extent.

The algorithm above can be significantly improved by observing that for a given regular expression and a given choice of k tagged subexpressions within it, the set of (C, I) pairs that are legitimate at some vertex in $\mathcal{M}_{B,R}$ is usually much less than 4^k. For example, for the expression **x(x{1}x{2}x|x{3}x)*x{4}x**, of the 81 2-partitions of subsets of $\{1,2,3,4\}$, only the following 22 pairs are legitimate at some vertex: $(\emptyset,\emptyset)$, $(\emptyset,\{1\})$, $(\{1\},\emptyset)$, $(\{1\},\{2\})$, $(\{1,2\},\emptyset)$, $(\emptyset,\{3\})$, $(\{3\},\emptyset)$, $(\{2\},\{1\})$, $(\{3\},\{1\})$, $(\{1,3\},\emptyset)$, $(\{1,3\},\{2\})$, $(\{1,2,3\},\emptyset)$, $(\{1,2\},\{3\})$, $(\{2,3\},\{1\})$, $(\emptyset,\{4\})$, $(\{1,2\},\{4\})$, $(\{3\},\{4\})$, $(\{1,2,3\},\{4\})$, $(\{4\},\emptyset)$, $(\{1,2,4\},\emptyset)$, $(\{3,4\},\emptyset)$, and $(\{1,2,3,4\},\emptyset)$. Moreover the number of legitimate pairs at each vertex of the graph is even smaller, and is maximal at vertices whose state is final for the automaton F. In our current example, a maximum of 4 pairs need to be computed at each vertex as the legal pairs at the final vertex of the automaton is $(\{4\},\emptyset)$, $(\{1,2,4\},\emptyset)$, $(\{3,4\},\emptyset)$, and $(\{1,2,3,4\},\emptyset)$.

Lemma 2 gives recurrences bounding the number of subset pairs required for a particular regular expression and tags. Let T_R denote the number of legitimate (C, I) pairs required for expression R, excluding the initial pair $(\emptyset,\emptyset)$. Simultaneously, we will need to compute recurrences for M_R, the number of legitimate $(C,\emptyset)$ pairs found at the final state of R (including the pair $(\emptyset,\emptyset)$), and C_R, which is 1 or 0 depending on whether there is or is not, respectively, a path through R's automaton not involving a tagged subexpression.

Lemma 2:

$$\begin{aligned}
C_a &= (\textbf{if } a \textbf{ is tagged then } 0 \textbf{ else } 1) \\
M_a &= 1 \\
T_a &= (\textbf{if } a \textbf{ is tagged then } 2 \textbf{ else } 0) \\[6pt]
C_{RS} &= (\textbf{if } RS \textbf{ is tagged then } 0 \textbf{ else } \max(C_R, C_S)) \\
M_{RS} &= M_R M_S \\
T_{RS} &= T_R + M_R T_S + (\textbf{if } RS \textbf{ is tagged then } (1 + M_{RS})) \\[6pt]
C_{R|S} &= (\textbf{if } R|S \textbf{ is tagged then } 0 \textbf{ else } \max(C_R, C_S)) \\
M_{R|S} &= M_R + M_S - \min(C_R, C_S) \\
T_{R|S} &= T_R + T_S + (\textbf{if } R|S \textbf{ is tagged then } (1 + M_{R|S})) \\[6pt]
C_{R*} &= (\textbf{if } R* \textbf{ is tagged then } 0 \textbf{ else } 1) \\
M_{R*} &= 2^{M_R - C_R} \\
T_{R*} &\leq 2^{M_R - C_R}(\tfrac{3}{4}T_R - (M_R - C_R) + 1) - 1 \\
&\quad +(\textbf{if } R* \textbf{ is tagged then } (1 + M_{R*})) \\[6pt]
C_{R+} &= (\textbf{if } R+ \textbf{ is tagged then } 0 \textbf{ else } C_R) \\
M_{R+} &= 2^{M_R - C_R} - (1 - C_R) \\
T_{R+} &\leq 2^{M_R - C_R}(\tfrac{3}{4}T_R - (M_R - C_R) + 1) - 1 \\
&\quad +(\textbf{if } R+ \textbf{ is tagged then } (1 + M_{R+})) \\[6pt]
C_{R?} &= (\textbf{if } R? \textbf{ is tagged then } 0 \textbf{ else } 1) \\
M_{R?} &= M_R + (1 - C_R) \\
T_{R?} &= T_R + (\textbf{if } R? \textbf{ is tagged then } (1 + M_{R?}))
\end{aligned}$$

While the lemma gives recurrences for bounding the size of legitimate sets, it is a simple step to extend them to recurrences for enumerating the legitimate sets at each vertex with a given state of F, in a prepass over F. The prepass takes $O(T_R)$ time and it can be shown that the maximum at any state is given by M_R. It is thus possible to modify the coarse algorithm above to only compute the legitimate pairs at each vertex, giving an $O(cM_RPN + T_R)$ time and space algorithm. A simple corollary is that if the tags are unambiguous, then there are at most $O(1)$ legitimate pairs at each vertex and T_R is $O(k)$. So the refined algorithm takes only $O(cPN)$ time in this case.

3 Reporting Approximate Matches

3.1 Finding the Most Significant Approximate Match

We now consider some problems in *approximate* regular expression pattern matching. A k-match to a regular expression R is a string whose minimal distance from a string exactly matching R is k, where distance is the standard unit-cost difference metric. The most significant issue in this context is what constitutes a match. For example, consider the expression **babb*bab** and suppose we are searching for all matches with 2-or-less differences to it, i.e., a *2-match*. When run against the text ...**aba|babbbab|aaa**... there is a 0-match to R shown between bars, but there are also 12 induced 2-matches in the vicinity that can be obtained with insertions and deletions at either end of the 0-match, i.e., **bababbbab**, **ababbba**, **ababbbab**, **ababbbaba**, **babbb**, **babbba**, **babbbaba**, **babbbabaa**, **abbba**, **abbbab**, **abbbaba**, and **bbbab**. Indeed, wherever there is a 0-match to the pattern there will *always* be another 12 2-matches *induced* by it. In addition, just by fortuitous circumstance, there can be additional overlapping matches, e.g., **abababbbab** in the example. Here we propose two schemes, first one for filtering out the induced matches, and then one for filtering the fortuitous matches.

It will be convenient in the ensuing treatment to think about alignments in terms of paths through an *edit graph* between A and the pattern R. Basically the edit graph $\mathcal{G}_{<A,R>}$ is just the dependency graph of the recurrence (1) with each edge weighted according to the δ-part of its recurrence term. Specifically, there is a vertex for each term (i,s), an insertion edge from (i,t) to (i,s) weighted $\delta(\epsilon,\lambda_s)$ for every edge $t \rightarrow s$ in F, a deletion edge from $(i-1,s)$ to (i,s) weighted $\delta(a_i,\epsilon)$, and a substitution edge from $(i-1,t)$ to (i,s) weighted $\delta(a_i,\lambda_s)$ for every edge $t \rightarrow s$ in F. By construction every path from (i,s) to (j,t) models an alignment between $a_{i+1}a_{i+2}\ldots a_j$ and a string spelled on the projection path from s to t in F excluding the first symbol on s. Moreover, the weight of the path is the score of the alignment it models. Thus in general the value $C(i,s)$ is the score of the least cost path to (i,s) from a θ-vertex of $\mathcal{G}_{<A,R>}$ (i.e. a vertex (j,θ) for some j).

Filtering Non-Essential Matches In our first approach, we consider an alignment *essential* if (1) it begins and ends with aligned symbols (they need not be equal), and (2) the alignment has the lowest possible score of all alignments between the two strings involved. Note immediately, that none of the induced matches in the example above constitute essential matches. Also note that condition (2) is important: for example, there is a 2-alignment between **babb*bab** and **ababbbab** that begins and ends with aligned symbols, but there is also a 1-alignment that begins with an insertion. It is not difficult to prove that wherever there is a non-essential match, there is also at least one essential match. Thus every matching region will be reported when one restricts attention to just the essential matches. One must be careful to add a sentinel character at each end of the string A being searched in order that matches involving its ends be found.

Let a state, s, of F be termed θ-reachable if there is a path from θ to s all of whose states are labeled ε including s. Further let Θ be the set of θ-reachable states. With this definition we may then develop the recurrences below for $B(i,s)$ and $E(i,s)$ which are the best score of a path in $\mathcal{G}_{<A,R>}$ from a θ-vertex to (i,s) that (1) begins with aligned symbols, or (2) begins and ends with aligned symbols, respectively. Essentially the recurrences are exactly that for $C(i,s)$ save that certain edges in $\mathcal{G}_{<A,R>}$ are not permitted.

Lemma 3:

$$B(i,s) = \min \left\{ \begin{array}{c} \min\limits_{t \to s}\{B(i-1,t) + \delta(a_i, \lambda_s)\}, \\ \min\limits_{t \to s}\{B(i,t) + \delta(\epsilon, \lambda_s)\}, \\ B(i-1,s) + \delta(a_i, \epsilon), \\ \delta(a_i, \lambda_s) \text{ if } t \to s \in \Theta \times \bar{\Theta} \\ \infty \end{array} \right\}$$

$$E(i,s) = \min \left\{ \begin{array}{c} \min\limits_{t \to s:\lambda_s \neq \epsilon}\{B(i-1,t) + \delta(a_i, \lambda_s)\}, \\ \min\limits_{t \to s:\lambda_s = \epsilon} E(i,s) \\ \infty \end{array} \right\}$$

It follows that $E(i, \phi)$ is the score of a best alignment beginning and ending with aligned symbols between a suffix of $A[1..i]$ and a word in R. We can thus report as the left end of an essential k-match only those i for which $E(i, \phi) \leq k$ *and* $E(i, \phi) = C(i, \phi)$ is true.

Filtering Fortuitous Matches While our first attempt at defining "true" matches removes the potential $O(k^2)$ induced matches, it does not distinguish or eliminate *fortuitous* matches which occur because by chance there is another way to complete the beginning or tail portion of a "match", e.g. **bababbbab** in our running example. For a given finite alphabet Σ, one can computationally approximate the limit:

$$r_\Sigma = \lim_{n \to \infty} E[\mathit{diff}(A,P)/|P| : A,\ P \text{ chosen uniformly from } \Sigma^n]$$

where $\mathit{diff}(A,P)$ is the score of the best alignment between A and P. Intuitively this is the number of differences per unit alignment length one expects to see in a "random" match. Any match with a lower *difference ratio* can be considered to be significant. Utilizing this, in our search for a good definition of an interesting match, let a *significant* alignment between two sequences be one for which the difference ratio of every prefix and suffix of the match is less than r_Σ. Intuitively, every "extension" of the match is significant. In the early 1980's Sellers [Sel84] explored algorithms for finding such matches in the context of molecular biology. This work appears to have been forgotten in the wake of the current popularity of the Smith-Waterman algorithm [SW81].

Sellers' basic idea is as follows. Suppose scoring is with respect to a general additive scoring scheme δ, and suppose one wants to detect only matches for which $\delta(A,P)/|P| \leq r$. Sellers observed that this is equivalent to finding matches for which $\delta_r(A,P) \geq 0$ where δ_r is a maximization scoring scheme derived from δ as follows: $\delta_r(a,b) = -\delta(a,b)$ if $b = \epsilon$, and $\delta_r(a,b) = r - \delta(a,b)$ if $b \neq \epsilon$. That this holds follows easily from the fact that δ_r has been constructed so that $\delta_r(A,P) = r|P| - \delta(A,P)$.

It now follows that what we seek are paths in the edit graph of A versus R that begin at a θ-vertex, end at a ϕ-vertex, and are both *prefix and suffix positive* under the weighting supplied by δ_r. A prefix positive path is one for which the score on every prefix of the path is positive. A suffix positive path is similarly defined. A basic exercise gives the following recurrence for $\mathit{PreP}(i,s)$ which is true if and only if there is a prefix positive path to (i,s).

Lemma 4:

$$PreP(i,s) = \begin{cases} or \begin{cases} or_{t\to s}\{PreP(i-1,t) \text{ and } (C(i-1,t)+\delta_r(a_i,\lambda_s) > 0)\}, \\ or_{t\to s}\{PreP(i,t) \text{ and } (C(i,t)+\delta_r(\epsilon,\lambda_s) > 0)\}, \\ PreP(i-1,s) \text{ and } (C(i-1,s)+\delta_r(a_i,\epsilon) > 0) \end{cases} & \text{if } s \neq \theta \\ true & \text{if } s = \theta \end{cases}$$

Given the vertices on prefix positive paths, one can quickly infer the edges on such paths (i.e., $v \to w$ is on a prefix positive path if *PreP*(v) is true and $C(v) + \delta_r(v \to w) > 0$). The suffix positive vertices and edges can similarly be found by developing recurrences for *SufP*(i, s) over the reverse of R and A. By taking the set of vertices and edges that are on both prefix and suffix positive paths, one arrives at the subgraph $\mathcal{S}_{<A,R>}$ of $\mathcal{G}_{<R,A>}$ modeling all paths that are prefix and suffix positive, or equivalently all the significant matches, according to our definition of significant. One can compute $\mathcal{S}_{<A,R>}$ in $O(PN)$ time and space.

Now one may wish to report matches that are both essential and significant. While it is the case that a significant essential match occurs where ever a significant non-essential match is found, it is not true that an essential match is necessarily significant, or for that matter, that a k-match is significant. Consider then first computing the subgraph, $\mathcal{E}_{<R,A>}$ of all vertices and edges on essential matches. Intersecting this subgraph with $\mathcal{S}_{<R,A>}$ leads to a subgraph whose connected components may not include a θ- or ϕ-vertex, corresponding intuitively to a region where there is an essential match that is not significant. For such components, we suggest that one might either find the least cost extensions that reach a θ- and ϕ-vertex, or that one recompute $\mathcal{S}$ with increasing values of r until the intersection does admit an end-to-end path. This second approach yields an interesting subproblem in parametric dynamic programming that we leave open.

The computation of $\mathcal{E} \cap \mathcal{S}$ can be efficiently organized as we noted earlier in Section 2.2, where the grain or direction of two recurrences oppose each other, as do the recurrences for *PreP* and *SufP* here. With the method of [MJ96] we can compute $SufP_R(0)$, $SufP_R(1)$, ... $SufP_R(N)$ in the order given using $O(tPN)$ time and $O(PN^{1/t})$ space for any choice of $t \geq 1$, where $SufP_R(i)$ is the set of values $\{SufP(i,s) : s \in F\}$. Given this we may then simultaneously compute the intersection of C, B, E, and $PreP$ with $SufP$ in a single forward scan, in time and space dominated by the terms for delivering *SufP* against its grain. In particular, this gives us an $O(PN \log N)$ time, $O(P \log N)$ space algorithm for delivering the significant portion of an essential match of R to substrings of A.

3.2 Reporting Longest Matches and All Matches

We conclude, by sketching solutions for the longest-match and all-matches problems for the case of approximate matching. To compute the longest essential match, we need only augment the computation of $\mathcal{E}_{<R,A>}$ with a trace-back record of a minimum inducing predecessor that has the longest match achieving that minimum. Note that what we are doing is simply delivering the longest match achieving the minimum. If rather we want the longest match that is within the threshold k then this requires that we keep track of the longest solution with each of the scores in $[0, k]$. The additional complexity for doing so is $O(kPN)$ time and $O(kP)$ space. One can then further combine this with our result for finding the significant part of essential matches, resulting in an $O(PN(k+\log N))$ time and $O(P(k+\log N))$ space algorithm for finding the longest significant part.

Computing all match positions is straightforward. Simply compute $C(i, s)$ in the forward direction over A and also compute $C^r(i, s)$ in the reverse direction of A and R. Report all positions i for which there exists a state t for which $C(i,t)+C^r(i,t) \leq k$. Again this requires the simultaneous delivery of recurrences opposing each other, and can be solved with the by now, well understood, complexities.

References

[CC97] C.A. Clarke and G.V. Cormack. On the use of regular expressons for searching text. *ACM Trans. on Prog. Languages and Systems*, 19(3):413–426, 1997.

[IEE92] IEEE. *Portable Operating System Interface (POSIX)*. IEEE Std 1003.2, Inst. of EE Engineers, New York, 1992.

[MJ96] E. Myers and M. Jain. Going against the grain. In Carleton University Press, editor, *Proc. 3rd South American Workshop on String Processing*, International Informatics Series #4, pages 203–213, 1996.

[MM88] E. Myers and W. Miller. Approximate matching of regular expressions. *Bulletin of Mathematical Biology*, 51(1):5–37, 1988.

[Ous94] J.K. Ousterhout. *Tcl and the TK Toolkit*. Addison-Wesley, Reading, Mass., 1994.

[Sed83] R. Sedgewick. *Algorithms*. Addison-Wesley, Reading, Mass., 1983.

[Sel84] P.H. Sellers. Pattern recognition in genetic sequences by mismatch density. *Bulletin of Mathematical Biology*, 46:501–514, 1984.

[SW81] T.F. Smith and M.S. Waterman. Identification of common molecular sequence. *J. of Molecular Biology*, 147:195–197, 1981.

[Tho68] K. Thompson. Regular expression search algorithm. *Comm. of ACM*, 11(6):419–422, 1968.

[WS91] L. Wall and R.L. Schwartz. *Programming Perl*. O'Reilly and Associates, Sebastopol, Calif., 1991.

Appendix A: ε-Cycle Free Automata

We quickly show here the inductive construction of an ε-NFA for a regular expression that has $O(P)$ states and vertices and does not contain an ε-cycle. To this end, we need the predicate $Nil(r)$ that is true if and only if ε is a word in the language specified by regular expression r. The following recurrence for Nil follows easily by induction:

$$\begin{aligned} Nil(a) &\equiv (a \notin \Sigma) \\ Nil(r^*) &\equiv \mathbf{true} \\ Nil(r^+) &\equiv Nil(r) \\ Nil(r?) &\equiv \mathbf{true} \\ Nil(rs) &\equiv Nil(r) \text{ and } Nil(s) \\ Nil(r|s) &\equiv Nil(r) \text{ or } Nil(s) \end{aligned}$$

Given the Nil predicate for each subexpression of a regular expression r, we construct the ε-cycle free automata F for it as shown in Figure 2. The dashed edges labeled "*if Nil(?)*" are to be placed in the construction only if the predicate is true. The induction of the construction is that the machine built for expression r is one that accepts all words in r *except* for ε, if it happens to be matched by r. In the very last step of constructing F, we add a path accepting ϵ if r accepts ϵ.

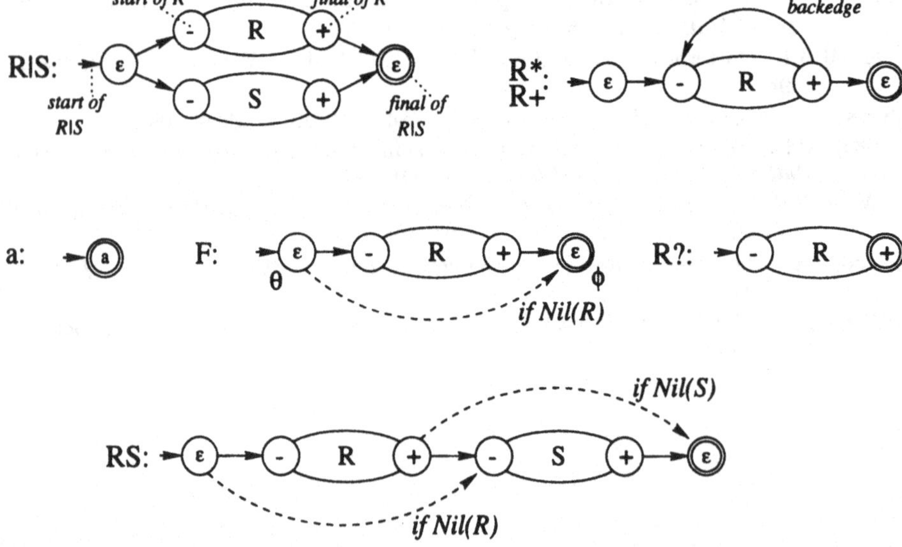

Fig. 2. Inductive RE to ϵ-cycle free NFA Construction.

An Approximate Oracle for Distance in Metric Spaces*

Yanling Yang[1] Kaizhong Zhang[2] Xiong Wang[3]
Jason T. L. Wang[4] Dennis Shasha[5]

[1] Department of Mathematics, Beijing Institute of Light Industry, 11 Fucheng Road, Beijing 100037, China
[2] Department of Computer Science, The University of Western Ontario, London, Ontario, Canada N6A 5B7 (kzhang@csd.uwo.ca)
[3] Department of CIS, New Jersey Institute of Technology, Newark, NJ 07102, USA (xiong@cis.njit.edu)
[4] Department of CIS, New Jersey Institute of Technology, Newark, NJ 07102, USA (jason@cis.njit.edu)
[5] Courant Institute of Mathematical Sciences, New York University, 251 Mercer Street, New York, NY 10012, USA (shasha@cs.nyu.edu)

Abstract. In this paper we present a new data structure for estimating distances in a pseudo-metric space. Given are a database of objects and a distance function for the objects, which is a pseudo-metric. We map the objects to vectors in a pseudo-Euclidean space with a reasonably low dimension while preserving the distance between two objects approximately. Such a data structure can be used as an approximate oracle to process a broad class of pattern-matching based queries. Experimental results on both synthetic and real data show the good performance of the oracle in distance estimation.

1 Introduction

Consider a database of objects $\mathcal{D} = \{p_0, p_1, \ldots, p_k\}$ and a function d where for any $p_i, p_j \in \mathcal{D}$, $d(p_i, p_j)$ (or $d_{i,j}$ for short) represents the distance between p_i and p_j. This paper presents a data structure for distance estimation, assuming only that the pairwise distances between the objects in $\mathcal{D}$ are given and the distance function d is a pseudo-metric. That is, for any $0 \le i, j, l \le k$, $d_{i,i} = 0$, $d_{i,j} \ge 0$, $d_{i,j} = d_{j,i}$ (symmetry) and $d_{i,l} \le d_{i,j} + d_{j,l}$ (triangle inequality) [8]. The proposed data structure contributes to the processing of various pattern-matching based queries, including nearest neighbor search [11], which finds the objects closest to a given target, ϵ-range search, which finds the objects within distance ϵ of the target, and so on. Such retrieval operations arise in many applications including vision [5], data mining [3], computational biology [13], document processing [11] and multimedia information management [2].

* Work supported in part by the Natural Sciences and Engineering Research Council of Canada under Grant No. OGP0046373, and by the U.S. NSF grants IRI-9224601, IRI-9224602, IRI-9531548 and IRI-9531554.

Data structures for distance calculations and their applications to pattern-matching based query processing have been studied in the past. A common assumption is that calculating the distance between two objects is the dominating cost, which should be minimized. Common techniques include using the triangle inequality to prune the search space or mapping the objects to a Euclidean space where the cost incurred by computing Euclidean distances is negligible. For example, in [3], Faloutsos and Lin proposed the *FastMap* approach to solving the ϵ-range search problem. The authors mapped all objects (including the target) to vectors in a Euclidean space and used the Euclidean distances between vectors to approximate the target-object distances. The approximation can find all the qualifying data objects by examining the vectors within some distance of the target vector in the Euclidean space. However if the dimension of the Euclidean space is chosen inappropriately, many unqualified data objects appear to qualify according to the data structure. In contrast to *FastMap*, we map objects to a pseudo-Euclidean space [7]. This technique yields fewer false positives than *FastMap*.

In [2], Berchtold *et al.* described a parallel method for nearest-neighbor search in high-dimensional feature space. The FQ tree proposed by Baeza-Yates *et al.* [1] and the approximate distance map described in [10] approached the nearest neighbor search problem by exploiting the triangle inequality to prune the search space. There are other related techniques [4]. However, none of the work considered mapping objects to a high precision pseudo-Euclidean space.

The rest of the paper is organized as follows. Section 2 describes how to map data objects to vectors in a pseudo-Euclidean space with a reasonably low dimension that preserves the distance function approximately. In practice, such a mapping can be done in the off-line phase. Section 3 shows how to project a given target, possibly arriving in the on-line phase, onto the same vector space. Section 4 describes applications of our approach and reports some experimental results on the performance of the proposed data structure.

2 Mapping Data Objects to a Vector Space

We are given a database of $k+1$ objects $\mathcal{D}$, a distance function d, which is a pseudo-metric, and pairwise distances $d_{i,j}$, for all $0 \leq i,j \leq k$. Thus, ($\mathcal{D}$, d) is a pseudo-metric space [8]. We first describe how to map the $k+1$ objects to a k-dimensional pseudo-Euclidean space, R^k. Then Sect. 2.2 establishes an orthogonal basis for R^k. Section 2.3 considers a lower dimension space R^n, $n \leq k$, by ignoring those dimensions dim_j where after mapping all the objects of $\mathcal{D}$ to R^k, the differences among the jth components of the corresponding vectors are small. To further reduce the dimension, Sect. 2.4 considers an orthonormal basis and Sect. 2.5 establishes an m-dimensional pseudo-Euclidean space R^m, $m \ll k$. The objects corresponding to the dimensions of R^m are chosen as *reference objects.* These reference objects will be compared with the target in the very beginning of the on-line phase, so that the calculated distances can be used for projecting the target onto R^m.

2.1 Pseudo-Euclidean Space R^k

Our notation is mainly based on [6, 9]. We define a mapping α as follows: $\alpha : \mathcal{D} \to R^k$ such that $\alpha(p_0) = a_0 = (0, \ldots, 0)$, $\alpha(p_i) = a_i = (0, \ldots, 1_{(i)}, \ldots, 0)$, $1 \le i \le k$. Let $M(\psi_{<a>}) = (m_{i,j})_{1 \le i,j \le k}$, where $m_{i,j} = (d_{i,0}^2 + d_{j,0}^2 - d_{i,j}^2)/2$, $1 \le i, j \le k$. We define another mapping ψ as follows: $\psi : R^k \times R^k \to R$ such that $\psi(x, y) = x^T M(\psi_{<a>}) y$, where x^T is the transpose of vector x. Notice that $\psi(a_i, a_j) = m_{i,j}$. ψ is a *symmetric bilinear form* of R^k. $M(\psi_{<a>})$ is the matrix of ψ w. r. t. the basis $\{a_i\}_{1 \le i \le k}$. The vector space R^k equipped with the symmetric bilinear form ψ is called a *pseudo-Euclidean space* [7]. For any two vectors $x, y \in R^k$, $\psi(x, y)$ is called the *inner product* of x and y, and $\|x - y\|^2 = \psi(x - y, x - y)$ is called the *squared distance* between x and y.

2.2 ψ-Orthogonal Basis $\{e_i\}$

Since the matrix $M(\psi_{<a>})$ is real symmetric, there is an orthogonal matrix $Q = (q_{i,j})_{1 \le i,j \le k}$ and a diagonal matrix $D = diag(\lambda_i)_{1 \le i \le k}$ such that

$$Q^T M(\psi_{<a>}) Q = D \tag{1}$$

where Q^T is the transpose of Q, $\lambda_i's$ are eigenvalues of $M(\psi_{<a>})$ arranged in some order, and columns of Q are the corresponding eigenvectors [6]. Let $(e_1, \ldots, e_k) = (a_1, \ldots, a_k)Q$ or equivalently

$$(a_1, \ldots, a_k) = (e_1, \ldots, e_k) Q^T \tag{2}$$

Then $\{e_i\}_{1 \le i \le k}$ is another basis of R^k. Note that the coordinate of e_j w. r. t. $\{a_i\}_{1 \le i \le k}$ is the jth column of matrix Q, and the coordinate of a_j w. r. t. $\{e_i\}_{1 \le i \le k}$ is the jth row of Q.

Since there will often be three different bases of a space in our discussion, we introduce a new notation, which is not common, but convenient. Let $x = (x^1, \ldots, x^k)$ be a vector and $\{a_i\}_{1 \le i \le k}$ be a basis of R^k. The coordinate of x w. r. t. $\{a_i\}_{1 \le i \le k}$ is denoted by $x_{<a>} = (x^i_{<a>})_{1 \le i \le k}$. Using this notation, the relation between $\{a_j\}$ and $\{e_j\}$ may be written as $e_{j<a>} = (q_{1,j}, \ldots, q_{k,j})$, $1 \le j \le k$, and $a_{j<e>} = (q_{j,1}, \ldots, q_{j,k})$, $1 \le j \le k$, where $e_{j<a>}$ is the coordinate of e_j w. r. t. $\{a_i\}_{1 \le i \le k}$, and $a_{j<e>}$ is the coordinate of a_j w. r. t. $\{e_i\}_{1 \le i \le k}$. Let x be a vector in R^k. Then

$$(x^1_{<a>}, \ldots, x^k_{<a>}) = (x^1_{<e>}, \ldots, x^k_{<e>}) Q^T \tag{3}$$

Therefore the matrix of the bilinear form ψ w. r. t. $\{e_i\}_{1 \le i \le k}$ is $M(\psi_{<e>}) = Q^T M(\psi_{<a>}) Q = D$. That is, the basis $\{e_i\}_{1 \le i \le k}$ is *ψ-orthogonal*. Let x, y be two vectors in R^k. Then $\psi(x, y) = x^T_{<a>} M(\psi_{<a>}) y_{<a>} = x^T_{<e>} Q^T M(\psi_{<a>}) Q y_{<e>} = x^T_{<e>} D y_{<e>} = \Sigma_{i=1}^k \lambda_i x^i_{<e>} y^i_{<e>}$. $\|x - y\|^2 = \Sigma_{i=1}^k \lambda_i (x^i_{<e>} - y^i_{<e>})^2$. Especially, we have $\psi(a_i, a_j) = \Sigma_{l=1}^k \lambda_l q_{i,l} q_{j,l}$.

Remark. If the matrix $M(\psi_{<a>})$ has negative eigenvalues, the squared distance between two vectors in the pseudo-Euclidean space may be negative. That's why we never say the "distance" between vectors in a pseudo-Euclidean space. Furthermore, the fact that the squared distance between two vectors vanishes does not imply that these two vectors are the same. These situations cannot happen in a Euclidean space.

2.3 Pseudo-Euclidean Space R^n, $n \leq k$

Assume that the eigenvalues of the matrix $M(\psi_{<a>})$ are ordered as follows: first n^+ positive eigenvalues, then n^- negative ones and finally zeroes. $n = n^+ + n^-$. Then $R^k = V \oplus R^0$ where $\oplus$ denotes the direct sum of two subspaces: $V = R^{(n^+ + n^-)}$ is the subspace generated by $\{e_i\}_{1 \leq i \leq n}$, and R^0 is the subspace generated by $\{e_i\}_{n+1 \leq i \leq k}$ [9]. Let $\phi = \psi|_{V \times V}$. Then ϕ is a non-degenerate bilinear form over $V \times V$. The set of vectors $\{e_i\}_{1 \leq i \leq n}$ is a ϕ-orthogonal basis of subspace V.

Let x be a vector in R^k. We define the ψ-orthogonal projection $\Pi : R^k \rightarrow R^n$ such that $\Pi(x^1_{<e>}, \ldots, x^n_{<e>}, x^{n+1}_{<e>}, \ldots, x^k_{<e>}) = (x^1_{<e>}, \ldots, x^n_{<e>})$. Let v_j denote $\Pi(a_j)$. Let $Q_{[kn]}$ be the $k \times n$ matrix consisting of the first n columns of the orthogonal matrix Q, namely $Q_{[kn]} = (q_{i,j})_{1 \leq i \leq k, 1 \leq j \leq n}$. Then from the definition of Π and (2), we have

$$(v_1, \ldots, v_k) = (e_1, \ldots, e_n) Q^T_{[kn]} \tag{4}$$

i.e. the coordinate of v_j w. r. t. $\{e_i\}_{1 \leq i \leq n}$ includes the first n elements of the jth row of the matrix Q, namely $v_{j<e>} = (q_{j,1}, \ldots, q_{j,n})$.

All the discussions about the inner product can now be summarized as follows: $\phi(v_i, v_j) = \Sigma_{l=1}^n \lambda_l q_{i,l} q_{j,l} = \Sigma_{l=1}^k \lambda_l q_{i,l} q_{j,l} = \psi(a_i, a_j) = (d^2_{i,0} + d^2_{j,0} - d^2_{i,j})/2$. Thus, the vector representation of the pseudo-metric space $(\mathcal{D}, d)$ is the mapping $\beta : \mathcal{D} \rightarrow R^{(n^+, n^-)}$ satisfying $\beta(p_0) = \Pi(\alpha(p_0)) = \Pi(a_0) = (0, \ldots, 0)_n$, and $\beta(p_j) = \Pi(\alpha(p_j)) = \Pi(a_j) = v_j$, $1 \leq j \leq k$.

Definition 1. A vector representation ρ of the pseudo-metric space $(\mathcal{D}, d)$ is an *isometric representation* if for any $p_i, p_j \in \mathcal{D}$, $\|\rho(p_i) - \rho(p_j)\|^2 = d^2_{i,j}$.

From the above discussions, we have

Theorem 2. *The mapping β is an isometric representation of the pseudo-metric space $(\mathcal{D}, d)$ in the pseudo-Euclidean space $R^{(n^+ + n^-)}$. That is, for any pair of indices i, j, $0 \leq i, j \leq k$, $\|v_i - v_j\|^2 = \phi(v_i - v_j, v_i - v_j) = d^2_{i,j}$.*

Theorem 2 describes the relation between the distance d in the pseudo-metric space and the squared distance in the corresponding pseudo-Euclidean space, stating the fact that the mapping β preserves d.

2.4 ϕ-Orthonormal Basis $\{\tilde{e}_i\}$

Define $sign(\lambda_i)$ to be 1 if $\lambda_i > 0$, 0 if $\lambda_i = 0$, and -1 if $\lambda_i < 0$. Let $J = diag(sign(\lambda_i))_{1 \le i \le k}$ and $\tilde{D} = diag(d_i)_{1 \le i \le k}$, where $d_i = |\lambda_i|$ if $\lambda_i \ne 0$ and 1 otherwise. Let $\tilde{Q} = Q \times \tilde{D}^{-1/2}$. Then, $\tilde{Q}^T M(\psi_{<a>})\tilde{Q} = \tilde{D}^{-1/2} Q^T M(\psi_{<a>}) Q \tilde{D}^{-1/2}$ $= \tilde{D}^{-1/2} diag(\lambda_i) \tilde{D}^{-1/2} = J$. This means that the first n columns of the matrix $\tilde{Q}$ are ψ-orthonormal vectors. Let $\tilde{e}_i = e_i/\sqrt{d_i}$, $1 \le i \le k$, or equivalently

$$(\tilde{e}_1, \ldots, \tilde{e}_k) = (e_1, \ldots, e_k)\tilde{D}^{-1/2} \tag{5}$$

Then, the set of vectors $\{\tilde{e}_i\}_{1 \le i \le n}$ is a *ϕ-orthonormal basis* of R^n. From (2) and (5), we have

$$(a_1, \ldots, a_k) = (\tilde{e}_1, \ldots, \tilde{e}_k)\tilde{D}^{1/2} Q^T \tag{6}$$

From (4) and (5), we have

$$(v_1, \ldots, v_k) = (\tilde{e}_1, \ldots, \tilde{e}_n)\tilde{D}^{1/2}_{[n]} Q^T_{[kn]} \tag{7}$$

where $\tilde{D}_{[n]}$ is the nth leading principal submatrix of the matrix $\tilde{D}$, i.e. $\tilde{D}_{[n]}$ $= diag(|\lambda_i|)_{1 \le i \le n}$. The coordinate of v_j w. r. t. the basis $\{\tilde{e}_i\}_{1 \le i \le n}$ includes the first n elements of the jth row of the matrix $\tilde{T} = Q \times \tilde{D}^{1/2}$, i.e. $v_{j<\tilde{e}>}$ $= (\sqrt{|\lambda_1|} q_{j,1}, \ldots, \sqrt{|\lambda_n|} q_{j,n})$. Let x, y be two vectors in R^n. Then $\psi(x, y) =$ $\Sigma_{i=1}^n sign(\lambda_i) x^i_{<\tilde{e}>} y^i_{<\tilde{e}>}$, and $\|x - y\|^2 = \Sigma_{i=1}^n sign(\lambda_i)(x^i_{<\tilde{e}>} - y^i_{<\tilde{e}>})^2$.

2.5 Pseudo-Euclidean Space R^m, $m < n$

In practice, the number of objects in $\mathcal{D}$, i.e. $k+1$, may be rather large. The dimension of R^n could still be large. From (3), we know that the eigenvalues represent the extension of variances of the objects in $\mathcal{D}$ in the corresponding dimension. To avoid dealing with a space of very high dimensionality, we ignore the dimensions along which the eigenvalues are small. Specifically, suppose the eigenvalues are sorted in descending order by their absolute values. Let $\{\lambda_i\}_{1 \le i \le m}$ be the first m eigenvalues, $m < n$, $m = m^- + m^+$, $m^- \le n^-$ and $m^+ \le n^+$. The mapping $\gamma : \mathcal{D} \to R^{(m^+ + m^-)}$ is the projection of the exact vector representation β onto the subspace spanned by the first m vectors in the ϕ-orthonormal basis. The first m elements of the ith row of the corresponding $\tilde{T}$ would give the coordinates of $\gamma(p_i)$ for the reduced vector representation, i.e. $\gamma(p_i) = (\sqrt{|\lambda_1|} q_{i,1}, \ldots, \sqrt{|\lambda_m|} q_{i,m})$.

Let x,y be two vectors in R^n. Then $\varphi(x, y) = \Sigma_{i=1}^m sign(\lambda_i) x^i_{<\tilde{e}>} y^i_{<\tilde{e}>}$ is the approximate representation of $\psi(x, y)$ for the corresponding vectors in R^m, and

$$\varphi(x - y, x - y) = \Sigma_{i=1}^m sign(\lambda_i)(x^i_{<\tilde{e}>} - y^i_{<\tilde{e}>})^2 \tag{8}$$

is the approximate representation of $\|x - y\|^2$. Note that $\|x - y\|^2 - \varphi(x - y, x - y) = \Sigma_{i=m+1}^n \lambda_i (x^i_{<\tilde{e}>} - y^i_{<\tilde{e}>})^2$. Let w_i be v_i projected onto R^m, i.e. w_i $= (\sqrt{|\lambda_1|} q_{i,1}, \ldots, \sqrt{|\lambda_m|} q_{i,m})$. We have $\|w_i - w_j\|^2 - \varphi(w_i - w_j, w_i - w_j) =$ $\Sigma_{l=m+1}^n \lambda_l (q_{i,l} - q_{j,l})^2$.

Proposition 3. *Let* $\Delta_{i,j} = \|w_i - w_j\|^2 - \varphi(w_i - w_j, w_i - w_j)$. *Then* $|\Delta_{i,j}| \le 4|\lambda_{m+1}|$.

3 Projection of a Target Object

3.1 Projection of an Embeddable Target

Now, suppose we are given a target p_*, and want to find the object in $\mathcal{D}$ that is closest to p_*. We project p_* onto R^m based on the distances between p_* and the reference objects ref_j, $0 \leq j \leq m$. To begin with, add p_* into $\mathcal{D}$. Let the distances between p_* and p_j be given as: $d_{*,j} = \pi(p_*, p_j)$, $0 \leq j \leq k$. Assume that the new object p_* is isometrically represented by a vector $u_* \in R^k$, i.e. $\|u_* - v_j\|^2 = d^2_{*,j}$, $0 \leq j \leq k$, or equivalently

$$\phi(u_*, v_j) = (d^2_{*,0} + d^2_{j,0} - d^2_{*,j})/2, 1 \leq j \leq k \tag{9}$$

$\phi(u_*, u_*) = d_{*,0}$. Let $Q_{[km]}$ be the matrix consisting of the first m columns of the matrix Q, namely $Q_{[km]} = (q_{i,j})_{1\leq i\leq k, 1\leq j\leq m}$. Let $\tilde{D}_{[m]}$ be the mth leading principal submatrix of the matrix $\tilde{D}$, i.e., $\tilde{D}_{[m]} = diag(|\lambda_i|)_{1\leq i\leq m}$. Then from (4) and (7),

$$(w_1, \ldots, w_k) = (e_1, \ldots, e_m)Q^T_{[km]} = (\tilde{e}_1, \ldots, \tilde{e}_m)\tilde{D}^{1/2}_{[m]}Q^T_{[km]} \tag{10}$$

Let r_* be the ϕ-orthogonal projection of u_* onto R^m. r_* can be represented as a linear combination of the set of vectors $\{w_i\}_{1\leq i\leq m}$, $r_* = \Sigma^m_{i=1} r^i_* w_i$. Taking the inner product of r_* and w_j, $1 \leq j \leq m$, we obtain $\phi(r_*, w_j) = \Sigma^m_{i=1} r^i_* \phi(w_i, w_j)$. Owing to the ϕ-orthogonality, $\phi(r_*, w_j) = \phi(u_*, w_j), 1 \leq j \leq m$. Hence

$$\Sigma^m_{i=1} r^i_* \phi(w_i, w_j) = \phi(u_*, w_j), 1 \leq j \leq m \tag{11}$$

Let the Gram matrix $G(w_1, \ldots, w_m) = (\phi(w_i, w_j))_{1\leq i,j\leq m}$, $b = (\phi(u_*, w_j))_{1\leq j\leq m}$. Then (11) can be re-written as $G(w_1, \ldots, w_m)r_* = b$. Since $G(w_1, \ldots, w_m)$ is non-singular, i.e., its determinant is not zero, $r_* = [G(w_1, \ldots, w_m)]^{-1}b$.

Note that this equation gives the coordinate of r_* w. r. t. the basis $\{w_i\}_{1\leq i\leq m}$. To obtain the coordinate w. r. t. $\{e_i\}$ or $\{\tilde{e}_i\}$, we need the matrices of coordinate transformation. Let $Q_{[mm]}$ be the mth leading principal submatrix of the orthogonal matrix Q. Then from (10),

$$(w_1, \ldots, w_m) = (e_1, \ldots, e_m)Q^T_{[mm]} = (\tilde{e}_1, \ldots, \tilde{e}_m)\tilde{D}^{1/2}_{[m]}Q^T_{[mm]} \tag{12}$$

So, $r_{*<e>} = Q^T_{[mm]}[G(w_1, \ldots, w_m)]^{-1}b$ and

$$r_{*<\tilde{e}>} = \tilde{D}^{1/2}_{[m]}Q^T_{[mm]}[G(w_1, \ldots, w_m)]^{-1}b \tag{13}$$

where $b = (\phi(u_*, w_j))_{1\leq j\leq m}$.

These equations can be simplified. From (12), we know that the coordinate of w_i w. r. t. $\{e_j\}_{1\leq j\leq m}$ is the ith row of $Q_{[mm]}$, i.e., $w_{i<e>} = (q_{i,1}, \ldots, q_{i,m})$. According to the formula for the inner product, $\phi(w_i, w_j) = \Sigma^m_{l=1}\lambda_l q_{i,l} q_{j,l}$, $1 \leq i, j \leq m$. Therefore, $G(w_1, \ldots, w_m) = (\phi(w_i, w_j))_{1\leq i,j\leq m} = Q_{[mm]}D_{[m]}Q^T_{[mm]}$.

Substituting this into (13), $r_{*<\tilde{e}>} = \tilde{D}_{[m]}^{1/2} Q_{[mm]}^T (Q_{[mm]}^T)^{-1} D_{[m]}^{-1} Q_{[mm]}^{-1} b$ $= \tilde{D}_{[m]}^{1/2} D_{[m]}^{-1} Q_{[mm]}^{-1} b = J_{[m]} \tilde{D}_{[m]}^{-1/2} Q_{[mm]}^{-1} b$. Thus,

$$r_{*<\tilde{e}>} = J_{[m]} \tilde{D}_{[m]}^{-1/2} Q_{[mm]}^{-1} b \tag{14}$$

Note that, after computing m eigenvalues and eigenvectors, one obtains the matrices $Q_{[mm]}$ and $\tilde{D}_{[m]}$. However, in general we do not know how large $\phi(u_*, w_j)$ is. What we know is $\phi(u_*, v_j) = (d_{*,0}^2 + d_{j,0}^2 - d_{*,j}^2)/2$, $1 \le j \le k$. Thus we have to use $\phi(u_*, v_j)$ as an approximate value of $\phi(u_*, w_j)$ to compute r_*. In other words, the formulae we use in practice are: $\overline{r}_{*<e>} = Q_{[mm]}^T [G(w_1, \ldots, w_m)]^{-1} \overline{b}$ and

$$\overline{r}_{*<\tilde{e}>} = \tilde{D}_{[m]}^{1/2} Q_{[mm]}^T [G(w_1, \ldots, w_m)]^{-1} \overline{b} \tag{15}$$

where $\overline{b} = (\phi(u_*, v_j))_{1 \le j \le m}$. Following the way to simplify $r_{*<\tilde{e}>}$, (15) can be simplified as

$$\overline{r}_{*<\tilde{e}>} = J_{[m]} \tilde{D}_{[m]}^{-1/2} Q_{[mm]}^{-1} \overline{b} \tag{16}$$

One may ask how well this works? The following three propositions estimate the error between $r_{*<\tilde{e}>}$ and $\overline{r}_{*<\tilde{e}>}$ when $\phi(u_*, v_i)$ is used in place of $\phi(u_*, w_i)$. All these propositions are based on the assumption that there is an object p_h in $\mathcal{D}$ that is very close to p_*. That is, if $\Delta_j = d_{*,j} - d_{h,j}$, $0 \le j \le k$, then there exists a small positive real number ϵ such that

$$|\Delta_j| \le \epsilon, 0 \le j \le k \tag{17}$$

Proposition 4. $\|u_* - a_h\|_2 \le \overline{\epsilon}$, *where*

$$\overline{\epsilon} = \frac{\epsilon}{|\lambda_{min}|} [\Sigma_{j=1}^k (d_{h,0} + d_{h,j})^2]^{1/2}$$

λ_{min} *is the non-zero eigenvalue with the smallest absolute value in* $\{\lambda_i\}_{1 \le i \le n}$, *and* u_*, a_h *are the vector representations of* p_* *and* p_h, *respectively.*

Proposition 5. *For each* i, $1 \le i \le k$, $|\phi(u_*, v_i) - \phi(u_*, w_i)| \le |\lambda_{m+1}|(1 + \overline{\epsilon})$.

Proposition 6. $\|\Delta r_*\|_2 = \|\overline{r}_{*<\tilde{e}>} - r_{*<\tilde{e}>}\|_2 \le \sqrt{m/|\lambda_m|}|\lambda_{m+1}|(1+\overline{\epsilon})\|Q_{[mm]}^{-1}\|_2$ $\le \sqrt{m|\lambda_{m+1}|}(1 + \overline{\epsilon})\|Q_{[mm]}^{-1}\|_2$.

From these propositions, it can be seen that the error is negligible whenever m is not large and λ_{m+1} is small enough. It should be pointed out that the coordinates of $\{w_i\}_{1 \le i \le k}$ are derived from $Q_{[km]}$, which are projections of the eigenvectors, whereas the coordinate of the target is calculated using (15) or (16). Thus, the projection of the target may be different from any of the objects in $\mathcal{D}$, even if the target is entirely the same as one of the objects of $\mathcal{D}$. Under this circumstance, one should calculate the coordinates of the objects of $\mathcal{D}$ using the formula (15) during the off-line phase.

Proposition 7. *If* $\{\overline{w}_i\}_{1 \le i \le k}$ *are the coordinates calculated based on the formula (15), then* $\|\Delta w_i\|_2 = \|\overline{w}_i - w_i\|_2 \le \sqrt{m|\lambda_{m+1}|}\|Q_{[mm]}^{-1}\|_2$.

Note that this upper bound is just the first term of that for Δr_* in Proposition 6, which is reasonable, since $\epsilon = \overline{\epsilon} = 0$ in this case.

3.2 Projection of an Unembeddable Target

In many cases, the target will not be isometrically embedded into R^k. However, we still can derive a projection formula which is basically the same as (15). The problem with an unembeddable target is that (9) in Sect. 3.1 does not hold. As a consequence, the projection formula of (13) can not be established. To address this problem, we construct a $(k+1)$-dimensional space with the target p_* as the $(k+1)^{th}$ dimension. Then we project all the $(k+2)$ objects (i.e. the $k+1$ data objects in $\mathcal{D}$, plus the target) onto R^m. The projection of the $(k+2)^{th}$ object establishes the formula for the target. We then introduce a new mapping η to connect R^{k+1} with R^k, thus resulting in a formula very similar to the previous one for an embeddable target.

To begin with, let us first establish a $(k+1)$-dimensional space. Let $\mathcal{D}_* = \mathcal{D} \cup \{p_*\}$ and $\alpha_* : \mathcal{D}_* \rightarrow R^{k+1}$, such that (i) $\alpha_*(p_0) = a_{*0} = (0, \ldots, 0, 0)$, (ii) $\alpha_*(p_j) = a_{*j} = (0, \ldots, 1_{(j)}, \ldots, 0, 0)$, $1 \leq j \leq k$, (iii) $\alpha_*(p_*) = a_{*(k+1)} = (0, \ldots, 0, 1)$. Next we define a symmetric bilinear form ψ_* over $R^{k+1} \times R^{k+1}$, such that (i) $\psi_*(a_{*i}, a_{*j}) = (d^2_{i,0} + d^2_{j,0} - d^2_{i,j})/2$, $1 \leq i, j \leq k$, and (ii) $\psi_*(a_{*(k+1)}, a_{*j}) = (d^2_{*,0} + d^2_{j,0} - d^2_{*,j})/2$, $1 \leq j \leq k$. Then define the matrix of the bilinear form w. r. t. $\{a_{*j}\}_{1 \leq j \leq k+1}$: $M_*(\psi_{*<a*>}) = (\psi_*(a_{*i}, a_{*j}))_{1 \leq i,j \leq k+1}$. Comparing the definition of ψ_* with that of ψ in Sect. 2.1, one can see that for each pair of subscripts i, j, $1 \leq i, j \leq k$, $\psi_*(a_{*i}, a_{*j}) = \psi(a_i, a_j)$. Moreover, the matrix $M(\psi_{<a>})$ is simply the kth leading principal submatrix of $M_*(\psi_{*<a*>})$.

Analogously to how we dealt with ψ in Sections 2.2 through 2.5, we can compute the eigenvectors of the matrix $M_*(\psi_{*<a*>})$ to obtain a ψ_*-orthonormal basis, say $\{\tilde{e}_{*i}\}_{1 \leq i \leq k+1}$, of R^{k+1}. To derive a formula similar to (13) for an embeddable target, we need another ψ_*-orthonormal basis in R^{k+1}. We define a mapping $\eta : R^k \rightarrow R^{k+1}$ such that $\eta(x^1, \ldots, x^k) = (x^1, \ldots, x^k, 0)$, where $(x^1, \ldots, x^k)$ is the coordinate of a vector in R^k with respect to some basis of it.

Proposition 8. *Let $x_1, \ldots, x_l$ be vectors in R^k and let $c_1, \ldots, c_l$ be real numbers. If $\Sigma_{i=1}^{l} c_i x_i = 0$, then $\Sigma_{i=1}^{l} c_i \eta(x_i) = 0$.*

Proposition 9. *Let x and y be two vectors in R^k. Then $\psi_*(\eta(x), \eta(y)) = \psi(x, y)$.*

Consider the subspace R^n of R^k defined in Sect. 2. The mapping η associates R^n with a subspace, say R^n_*, in R^{k+1}. The space R^{k+1} can be represented as the direct sum of R^n_* and its ψ_*-orthogonal complement [9]. It follows that the union of a ψ_*-orthogonal basis of R^n_* and a ψ_*-orthogonal basis of its ψ_*-orthogonal complement will become a ψ_*-orthogonal basis of R^{k+1}. The subspace R^n is spanned by $\{\tilde{e}_i\}_{1 \leq i \leq n}$. According to Proposition 8, the set of vectors $\{\eta(\tilde{e}_i)\}_{1 \leq i \leq n}$ spans R^n_*. According to Proposition 9, $\{\eta(\tilde{e}_i)\}_{1 \leq i \leq n}$ is ψ_*-orthonormal, since $\{\tilde{e}_i\}_{1 \leq i \leq n}$ is ψ-orthonormal. Therefore, there is a ψ_*-orthonormal basis of R^{k+1} which includes $\{\eta(\tilde{e}_i)\}_{1 \leq i \leq n}$ as a subset. The coordinate of a vector in R^{k+1} with respect to the basis mentioned above may be obtained from its coordinate w. r. t. $\{\tilde{e}_{*i}\}_{1 \leq i \leq k+1}$, through multiplying the latter one by a certain non-singular matrix (i.e. through coordinate transformation).

Note that $a_{*j} = \eta(a_j)$, $1 \le j \le k$. According to Proposition 9 and (6), $(a_{*1}, \ldots, a_{*k}) = (\eta(\tilde{e}_1), \ldots, \eta(\tilde{e}_k))\tilde{D}^{1/2}Q^T$. Therefore the coordinate of the projection of a_{*j} w. r. t. $\{\eta(\tilde{e}_i)\}_{1\le i\le k}$ is simply the coordinate of a_j w. r. t. $\{\tilde{e}_i\}_{1\le i\le k}$. In parallel with the introduction of the subspace R^n_*, we can introduce a subspace R^m_* of R^{k+1} from R^m in R^k, and then consider the projection of the target p_* onto R^m_*. Let w_j be the ψ-orthogonal projection of a_j onto R^m. Then $w_{*j} = \eta(w_j)$ is the ψ_*-orthogonal projection of a_{*j} onto R^m_*. Since the set of projections $\{w_j\}_{1\le j\le m}$ spans R^m, according to Proposition 8, the set of projections $\{w_{*j}\}_{1\le j\le m}$ spans R^m_*. Furthermore, from (12), $(w_{*1}, \ldots, w_{*m}) = (\eta(\tilde{e}_1), \ldots, \eta(\tilde{e}_m))\tilde{D}^{1/2}_{[m]}Q^T_{[mm]}$.

According to Proposition 9, the Gram matrix of $\{w_{*j}\}$ is simply the Gram matrix of $\{w_j\}$. Summarizing these results, we know that the coordinate of projecting a_{*j} onto R^m_* w. r. t. $\{\eta(\tilde{e}_i)\}_{1\le i\le m}$ can be computed using the equation: $r_{*<\eta(\tilde{e})>} = \tilde{D}^{1/2}_{[m]}Q^T_{[mm]}[G(w_1, \ldots, w_m)]^{-1}b_*$, where the matrices $\tilde{D}_{[m]}$, $Q_{[mm]}$ and $G(w_1, \ldots, w_m)$ are the same as those in (15), and $b_* = (\psi_*(a_{*(k+1)}, w_{*j}))_{1\le j\le m}$. Again we don't know how large $\psi_*(a_{*(k+1)}, w_{*j})$ is. What we can do is to replace it by $\psi_*(a_{*(k+1)}, a_{*j})$, thus obtaining

$$\overline{r}_{*<\eta(\tilde{e})>} = \tilde{D}^{1/2}_{[m]}Q^T_{[mm]}[G(w_1, \ldots, w_m)]^{-1}\overline{b}_* \tag{18}$$

where $\overline{b}_* = (\psi_*(a_{*(k+1)}, a_{*j}))_{1\le j\le m}$. By comparing (15) with (18), we conclude that no matter whether or not the target is embeddable to R^k, one can always use the same formula to calculate the projection of the target, though the resulting coordinates are with respect to the same basis represented in different dimensional spaces (more precisely, with respect to $\{\tilde{e}_i\}_{1\le i\le m}$ and $\{\eta(\tilde{e}_i)\}_{1\le i\le m}$, respectively).

4 Experiments and Applications

We have implemented the proposed data structure and tested it on three datasets, containing 90 artificial objects, proteins and dictionary words respectively. The pairwise distances between the artificial objects are randomly generated over a uniform distribution between 0 and 350, and those for the proteins and dictionary words are calculated using the edit distance [12]. Let p, q be two objects in $\mathcal{D}$ and let x, y be their vectors in R^m. We define the *vector distance* between p and q, denoted $vecdist(p, q)$, to be $\sqrt{\varphi(x-y, x-y)}$ if $\varphi(x-y, x-y) \ge 0$ and $-\sqrt{-\varphi(x-y, x-y)}$ otherwise, where $\varphi(x-y, x-y)$ is the squared distance between x and y (cf. (8)). The measures used for evaluating the performance of the data structure are the *average absolute error* (Err_a), *standard deviation* (Dev_a), and *average relative error* (Err_r). Let $\delta(p, q) = |vecdist(p, q) - d(p, q)|$ and $\Delta = \sum_{p,q\in\mathcal{D}} \delta(p, q)$. There are (90, 2) = 4005 combinations of pairs of objects. $Err_a = \Delta/4005$, $Dev_a = \sqrt{(\sum_{p,q\in\mathcal{D}}(\delta(p,q) - Err_a)^2)/4005}$, and $Err_r = (\sum_{p,q\in\mathcal{D}} \delta(p, q)/\sum_{p,q\in\mathcal{D}} d(p, q)) \times 100\%$. Figure 1 graphs Err_a, Dev_a and Err_r

as a function of the dimension m of different pseudo-Euclidean spaces. The performance is data dependent, and as expected, the larger the m, the smaller the errors (i.e., the better performance the data structure has).

Our future work is concerned with the question: *if one wants correct results, then how can one use the data structure as an approximate oracle?* We have studied two applications. First, we partitioned the three datasets of 90 objects into clusters solely based on the oracle. (We omit the details here.) An object p is *mis-clustered* if, based on the oracle, p belongs to a cluster C, whereas based on the real distances between objects, p is not in C; or vice versa, i.e. when p really is in C, but the oracle indicates otherwise. Our experimental results showed that when the dimension m was 40, the number of mis-clustered objects was 9, 16 and 3 for artificial objects, proteins, and words, respectively (out of 90 in each case). This shows that the oracle offers an excellent first cut at clustering. In the second application, we solved the nearest neighbor search problem by using the oracle to approximate target-object distances in combination with the triangle inequality. It was estimated that the approach did fewer than half the comparisons needed in using the triangle inequality alone.

References

1. R. Baeza-Yates, W. Cunto, U. Manber, and S. Wu. Proximity matching using fixed-queries trees. In *Combinatorial Pattern Matching, Lecture Notes in Computer Science*, pages 198–212, June 1994.
2. S. Berchtold, C. Bohm, B. Braunmuller, D. A. Keim, and H.-P. Kriegel. Fast parallel similarity search in multimedia databases. In *Proceedings of the ACM SIGMOD International Conference on Management of Data*, pages 1–12, May 1997.
3. C. Faloutsos and K.-I. Lin. *Fastmap*: A fast algorithm for indexing, data-mining and visualization of traditional and multimedia datasets. In *Proceedings of the ACM SIGMOD International Conference on Management of Data*, pages 163–174, May 1995.
4. K. Fukunaga. *Introduction to Statistical Pattern Recognition.* Academic Press, Inc., San Diego, California, 1990.
5. L. Godfarb. A new approach to pattern recognition. In L. Kanal and A. Rosenfeld, editors, *Progress in Pattern Recognition*, volume 2, pages 241–402, North-Holland, Amsterdam, 1985.
6. G. H. Golub and C. F. Van Loan. *Matrix Computations.* The Johns Hopkins University Press, Baltimore, Maryland, 1996.
7. W. Greub. *Linear Algebra.* Springer-Verlag, Inc., New York, New York, 1975.
8. J. L. Kelley. *General Topology.* D. Van Nostrand Company, Inc., Princeton, New Jersey, 1955.
9. P. D. Lax. *Linear Algebra.* John Wiley & Sons, Inc., New York, New York, 1997.
10. D. Shasha and T. L. Wang. New techniques for best-match retrieval. *ACM Transactions on Information Systems*, 8(2):140–158, April 1990.
11. A. F. Smeaton and C. J. Van Rijsbergen. The nearest neighbor problem in information retrieval: An algorithm using upperbounds. *ACM SIGIR Forum*, 16:83–87, 1981.
12. R. A. Wagner and M. J. Fischer. The string-to-string correction problem. *Journal of the ACM*, 21(1):168–173, Jan. 1974.

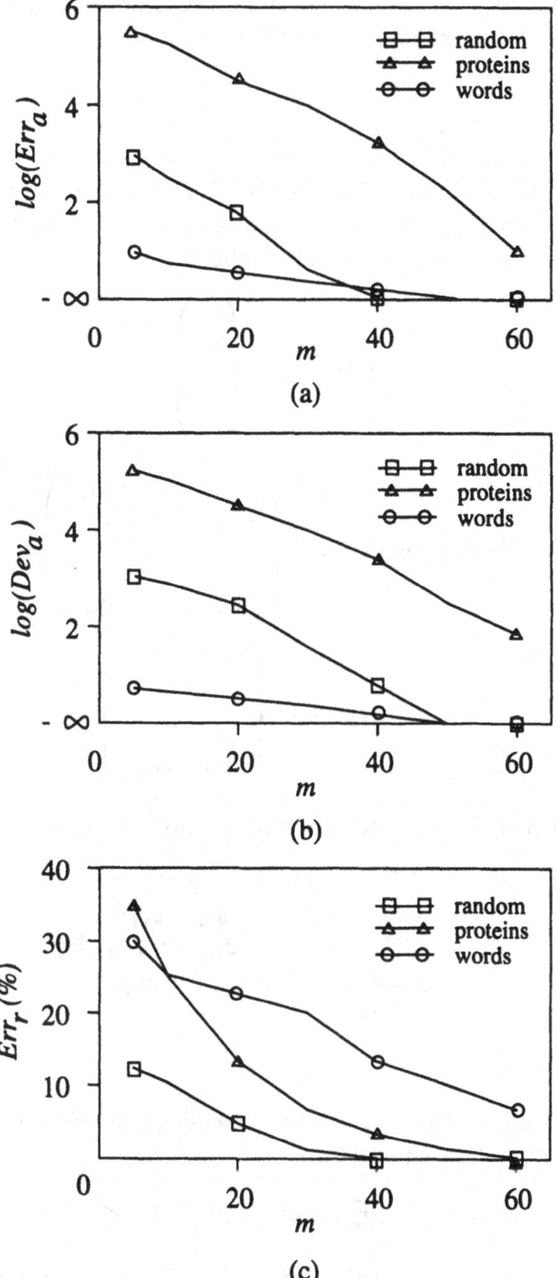

Fig. 1. Error measures of the data structure as a function of the dimension m of the pseudo-Euclidean spaces. In each case, the data set had 90 objects. The curves are very similar when the data sets were increased to 1000 objects.

13. J. T. L. Wang, B. A. Shapiro, and D. Shasha, editors. *Pattern Discovery in Biomolecular Data*. Oxford University Press, in press.

Appendix

The appendix contains proofs of the propositions presented in the text. We use $\{a_i\}_{1\leq i\leq n}$, or simply $\{a_i\}$ when the context is clear, to represent $\{a_1, a_2, \ldots, a_n\}$ where a_i's are vectors. Let c_i's be real numbers.

$$(c_i)_{1\leq i\leq n} = \begin{pmatrix} c_1 \\ c_2 \\ \vdots \\ c_n \end{pmatrix}$$

$$(c_{i,j})_{1\leq i\leq m, 1\leq j\leq n} = \begin{pmatrix} c_{1,1} & c_{1,2} & \cdots & c_{1,n} \\ c_{2,1} & c_{2,2} & \cdots & c_{2,n} \\ \vdots & & & \vdots \\ c_{m,1} & c_{m,2} & \cdots & c_{m,n} \end{pmatrix}$$

$$diag(c_i)_{1\leq i\leq n} = \begin{pmatrix} c_1 & 0 & \cdots & 0 \\ 0 & c_2 & \cdots & 0 \\ \vdots & & & \vdots \\ 0 & 0 & \cdots & c_n \end{pmatrix}$$

Proof of Proposition 3. The result follows by observing that

$$\begin{aligned} |\Delta_{i,j}| &\leq \Sigma_{l=m+1}^{n} |\lambda_l| (q_{i,l} - q_{j,l})^2 \\ &\leq |\lambda_{m+1}| \Sigma_{l=m+1}^{n} (q_{i,l} - q_{j,l})^2 \\ &\leq |\lambda_{m+1}| \Sigma_{l=m+1}^{n} 2(|q_{i,l}|^2 + |q_{j,l}|^2) \\ &\leq 2|\lambda_{m+1}| \Sigma_{l=1}^{n} (|q_{i,l}|^2 + |q_{j,l}|^2) \\ &= 4|\lambda_{m+1}| \end{aligned}$$

□

Proof of Proposition 4. Since p_* can be embedded isometrically into R^k, $\|u_* - a_j\|^2 = d_{*,j}^2$, $0 \leq j \leq k$, or equivalently

$$\psi(u_*, a_j) = (d_{*,0}^2 + d_{j,0}^2 - d_{*,j}^2)/2, 1 \leq j \leq k \tag{19}$$

Let $b_* = (b_*^j)_{1\leq j\leq k}$ where $b_*^j = (d_{*,0}^2 + d_{j,0}^2 - d_{*,j}^2)/2$, $1 \leq j \leq k$. Equation (19) can be written as $M(\psi_{<a>})u_{*<a>} = b_*$. Thus $(Q^T M(\psi_{<a>})Q)(Q^T u_{*<a>}) = Q^T b_*$. From (1) and (3), $Du_{*<e>} = Q^T b_*$. Since $\lambda_j = 0$, $n+1 \leq j \leq k$, $u_{*<e>}^j$, $n+1 \leq j \leq k$, can take any values. Let $\overline{D} = diag(\overline{\lambda_j})_{1\leq j\leq k}$ where

$$\overline{\lambda_j} = \begin{cases} \lambda_j & \text{if } j < n \\ \lambda_n & \text{if } n \leq j \leq k \end{cases}$$

We choose those u^j, $1 \leq j \leq k$, that satisfy $\overline{D}u_{*<e>} = Q^T b_*$. Thus $u_{*<e>} = \overline{D}^{-1} Q^T b_*$. Similarly, $a_{h<e>} = \overline{D}^{-1} Q^T b_h$. Thus $\|u_* - a_h\|_2 \leq \|\overline{D}^{-1}\|_2 \|Q^T\|_2 \|b_* - b_h\|_2$. Evaluating these norms, we get $\|\overline{D}^{-1}\|_2 \leq \frac{1}{|\lambda_{min}|}$, $\|Q^T\|_2 = 1$,

$$\begin{aligned} \|b_* - b_h\|_2 &= [\Sigma_{j=1}^k (b_*^j - b_h^j)^2]^{1/2} \\ &= [\Sigma_{j=1}^k \tfrac{1}{4}(d_{*,0}^2 - d_{*,j}^2 - d_{h,0}^2 + d_{h,j}^2)^2]^{1/2} \\ &= [\Sigma_{j=1}^k (d_{h,0}\Delta_0 - d_{h,j}\Delta_j + \tfrac{1}{2}\Delta_0^2 - \tfrac{1}{2}\Delta_j^2)^2]^{1/2} \end{aligned}$$

Omitting the infinitesimal of higher order and substituting inequality (17), we get $\|b_* - b_h\|_2 = \epsilon[\Sigma_{j=1}^k (d_{h,0} + d_{h,j})^2]^{1/2}$. Hence $\|u_* - a_h\|_2 \leq \frac{\epsilon}{|\lambda_{min}|}[\Sigma_{j=1}^k (d_{h,0} + d_{h,j})^2]^{1/2}$. □

Proof of Proposition 5. Let $u_{*<e>} = (u^j)_{1 \leq j \leq n}$. Since $v_{i<e>} = (q_{i,1}, \ldots, q_{i,n})$ (cf. (4)) and w_i is the projection of v_i to R^m, we obtain

$$\begin{aligned} |\phi(u_*, v_i) - \phi(u_*, w_i)| &= |\Sigma_{j=m+1}^n \lambda_j u^j q_{i,j}| \\ &= |\Sigma_{j=m+1}^n \lambda_j (q_{h,j} + \Delta v_j) q_{i,j}| \\ &\leq |\Sigma_{j=m+1}^n \lambda_j q_{h,j} q_{i,j}| + |\Sigma_{j=m+1}^n \lambda_j \Delta v_j q_{i,j}| \end{aligned}$$

where $\Delta v_j = u^j - q_{h,j}$, $1 \leq j \leq n$.

The first term on the right-hand side is easy to estimate. Since Q is orthogonal, $\Sigma_{j=1}^k q_{i,j}^2 = 1$, $1 \leq i \leq k$. Thus

$$\begin{aligned} |\Sigma_{j=m+1}^n \lambda_j q_{h,j} q_{i,j}| &\leq \Sigma_{j=m+1}^n |\lambda_j||q_{h,j}||q_{i,j}| \\ &\leq |\lambda_{m+1}|[\Sigma_{j=m+1}^n q_{h,j}^2]^{1/2}[\Sigma_{j=m+1}^n q_{i,j}^2]^{1/2} \\ &\leq |\lambda_{m+1}| \end{aligned}$$

Similarly,

$$\begin{aligned} |\Sigma_{j=m+1}^n \lambda_j \Delta v_j q_{i,j}| &\leq \Sigma_{j=m+1}^n |\lambda_j||\Delta v_j||q_{i,j}| \\ &\leq |\lambda_{m+1}|[\Sigma_{j=m+1}^n (\Delta v_j)^2]^{1/2}[\Sigma_{j=m+1}^n q_{i,j}^2]^{1/2} \\ &\leq |\lambda_{m+1}|[\Sigma_{j=m+1}^n (\Delta v_j)^2]^{1/2} \end{aligned}$$

By Proposition 4,

$$\begin{aligned} [\Sigma_{j=m+1}^n (\Delta v_j)^2]^{1/2} &\leq [\Sigma_{j=1}^k (\Delta v_j)^2]^{1/2} \\ &= \|u_* - a_h\|_2 \\ &\leq \overline{\epsilon} \end{aligned}$$

Hence $|\phi(u_*, v_i) - \phi(u_*, w_i)| \leq |\lambda_{m+1}|(1 + \overline{\epsilon})$. □

Proof of Proposition 6. Subtracting (14) from (16), $\Delta r_* = J_{[m]} \tilde{D}_{[m]}^{-1/2} Q_{[mm]}^{-1} (\overline{b} - b)$. Hence

$$\|\Delta r_*\|_2 \leq \|J_{[m]}\|_2 \|\tilde{D}_{[m]}^{-1/2}\|_2 \|Q_{[mm]}^{-1}\|_2 \|\overline{b} - b\|_2 \tag{20}$$

Evaluating these norms, we get $\|J_{[m]}\|_2 = 1$, $\|\tilde{D}_{[m]}^{-1/2}\|_2 = \frac{1}{\sqrt{|\lambda_m|}}$. By Proposition 5,

$$\begin{aligned}\|\bar{b} - b\|_2 &= [\Sigma_{i=1}^{m}(\phi(u_*, v_i) - \phi(u_*, w_i))^2]^{1/2} \\ &\le [\Sigma_{i=1}^{m}(|\lambda_{m+1}|(1+\bar{\epsilon}))^2]^{1/2} \\ &= \sqrt{m}|\lambda_{m+1}|(1+\bar{\epsilon})\end{aligned}$$

Substituting these into inequality (20), we get

$$\begin{aligned}\|\Delta r_*\|_2 &\le \sqrt{\tfrac{m}{|\lambda_m|}}|\lambda_{m+1}|(1+\bar{\epsilon})\|Q_{[mm]}^{-1}\|_2 \\ &\le \sqrt{m|\lambda_{m+1}|}(1+\bar{\epsilon})\|Q_{[mm]}^{-1}\|_2\end{aligned}$$

□

Proof of Proposition 7. By replacing u_* with v_i, $1 \le i \le k$, in (14) and (16), we get $w_{i<\bar{\epsilon}>} = J_{[m]}\tilde{D}_{[m]}^{-1/2}Q_{[mm]}^{-1}b_i$, $\overline{w}_{i<\bar{\epsilon}>} = J_{[m]}\tilde{D}_{[m]}^{-1/2}Q_{[mm]}^{-1}\overline{b_i}$, where $b_i = (\phi(v_i, w_j))_{1\le j\le m}$ and $\overline{b_i} = (\phi(v_i, v_j))_{1\le j\le m}$. Observe that

$$\begin{aligned}\|\overline{b_i} - b_i\|_2 &= [\Sigma_{j=1}^{m}(\phi(v_i, v_j) - \phi(v_i, w_j))^2]^{1/2} \\ &= [\Sigma_{j=1}^{m}(\Sigma_{l=m+1}^{n}\lambda_l q_{i,l}q_{j,l})^2]^{1/2} \\ &\le [\Sigma_{j=1}^{m}(|\lambda_{m+1}|\Sigma_{l=m+1}^{n}|q_{i,l}q_{j,l}|)^2]^{1/2} \\ &\le \sqrt{m}|\lambda_{m+1}|\end{aligned}$$

Thus $\|\overline{w}_i - w_i\|_2 = \|J_{[m]}\|_2\|\tilde{D}_{[m]}^{-1/2}\|_2\|Q_{[mm]}^{-1}\|_2\|\overline{b_i} - b_i\|_2 \le \sqrt{m|\lambda_{m+1}|}\|Q_{[mm]}^{-1}\|_2$.

□

Proof of Proposition 8. Let $x_i = (x_i^j)_{1\le j\le k}, 1 \le i \le l$. We have $\Sigma_{i=1}^{l}c_i x_i^j = 0, 1 \le j \le k$. Let $y_i = (y_i^j)_{1\le j\le k+1}, 1 \le i \le l$ and $\eta(x_i) = y_i, 1 \le i \le l$. By the definition of η, $y_i^j = x_i^j$, $1 \le j \le k$, $1 \le i \le l$ and $y_i^{k+1} = 0$, $1 \le i \le l$. Thus $\Sigma_{i=1}^{l}c_i y_i^j = 0, 1 \le j \le k+1$. Namely $\Sigma_{i=1}^{l}c_i y_i = 0$. □

Proof of Proposition 9. By the definition of η,

$$\eta(x) = \begin{pmatrix} x \\ 0 \end{pmatrix}$$

$$\eta(y) = \begin{pmatrix} y \\ 0 \end{pmatrix}$$

$$M_*(\psi_{*<a*>}) = \begin{pmatrix} M(\psi_{<a>}) & M_1 \\ M_1^T & \psi_*(a_{*(k+1)}, a_{*(k+1)}) \end{pmatrix}$$

where $M_1 = (\psi_*(a_{*(k+1)}, a_{*j}))_{1\le j\le k}$. Thus

$$\begin{aligned}\psi_*(\eta(x), \eta(y)) &= \begin{pmatrix} x^T & 0 \end{pmatrix}\begin{pmatrix} M(\psi_{<a>}) & M_1 \\ M_1^T & \psi_*(a_{*(k+1)}, a_{*(k+1)}) \end{pmatrix}\begin{pmatrix} y \\ 0 \end{pmatrix} \\ &= x^T M(\psi_{<a>})y = \psi(x, y)\end{aligned}$$

□

A Rotation Invariant Filter for Two-Dimensional String Matching[1]

Kimmo Fredriksson and Esko Ukkonen

Department of Computer Science, University of Helsinki
PO Box 26, FIN–00014 Helsinki, Finland
kfredrik@cs.Helsinki.FI, ukkonen@cs.Helsinki.FI

Abstract. We consider the problem of finding the occurrences of two–dimensional pattern $P[1..m, 1..m]$ in two–dimensional text $T[1..n, 1..n]$ when also rotations of P are allowed. A fast filtration–type algorithm is developed that finds in T the locations where a rotated P can occur. The corresponding rotations are also found. The algorithm first reads from P a linear string of length m in all $\Theta(m^2)$ orientations that are relevant. We also show that the number of different orientations which P can have is $\Theta(m^3)$. The text T is scanned with Aho–Corasick string matching automaton to find the occurrences of any of these $\Theta(m^2)$ linear strings of length m. Each such occurrence indicates a potential set of occurrences of whole P which are then checked. Some preliminary running times of a prototype implementation of the method are reported.

1 Introduction

The pattern matching problem with fixed orientation of the pattern is a well studied problem. The classical problem is to find the exact or approximate occurrences of pattern P from text T, when the orientation of P is known *a priori*. See e.g. [2, 6, 4, 8, 7, 11].

The idea of reducing the problem of finding fixed orientation occurrences of a two–dimensional pattern to the problem of finding one–dimensional string is not a new one [12, 13, 3]. However, in this paper we present a filtration method that reduces the problem of finding a rotated two–dimensional pattern to the well studied problem of finding the occurrences of a set of strings. The method is based on a careful study of the combinatorial structure of the problem.

Other methods for rotation invariant pattern matching are e.g. geometric hashing [9], template matching [5] and various methods of log–polar transformation and Fourier transform techniques; for a summary see e.g. [5].

[1] A work supported by the Academy of Finland under grant 8745.

An interesting variation of two dimensional pattern matching problem and techniques to solve it was presented in [10] by Landau and Vishkin. They did not consider the rotations of the pattern, but a pattern matching problem in which the pattern and the text are specified in terms of their "continuous" properties. The pattern matching problems arising from this point of view and the techniques used to solve the problem are somewhat similar to ours.

2 Problem Definition

Let *pattern* $P = P[1..m, 1..m]$ be a two-dimensional $m \times m$ array and *text* $T = T[1..n, 1..n]$ an $n \times n$ array over some finite alphabet Σ, such that $m < n$ (see Fig. 1). For simplicity we restrict the consideration on square arrays only.

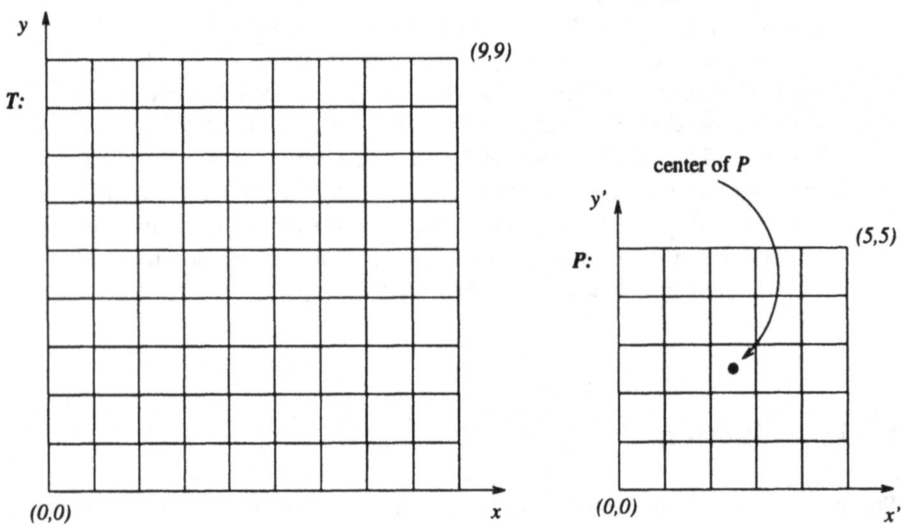

Fig. 1. The array of pixels for T and P ($n = 9$, $m = 5$).

To define a possibly rotated occurrence of P in T we need a geometric interpretation of P and T. To this end, we say that the array of *unit pixels* for T consists of n^2 unit squares called *pixels*, in the real plane $\mathbf{R}^2$ ((x, y)–plane). The four corners of the pixel for $T[1, 1]$ are points $(0,0), (0,1), (1,0)$ and $(1,1)$ in $\mathbf{R}^2$, and generally, the corners of the pixel for $T[i, j]$ are $(i-1, j-1), (i, j-1), (i-1, j)$ and (i, j). Hence the pixels for T form a regular $n \times n$ array covering the area between $(0,0), (n,0), (0,n)$ and (n,n). Each pixel has a *center* which is the geometric center point of the pixel, i.e., the center of the pixel for $T[i, j]$ is $(i - \frac{1}{2}, j - \frac{1}{2}) \in \mathbf{R}^2$.

The array of pixels for pattern P is defined similarly. In the sequel we identify the original array entries with the corresponding pixels and speak of, say, pixel

$T[i,j]$ whose value is in Σ. The values are called *colors*; hence each pixel $T[i,j]$ has a color in Σ.

The *center* of the whole pattern P is the center of the pixel which is in the middle of P. Precisely, assuming for simplicity that m is odd, the center of P is the center of pixel $P[\frac{m+1}{2}, \frac{m+1}{2}]$, that is, point $(\frac{m}{2}, \frac{m}{2}) \in \mathbf{R}^2$.

Consider now a rigid motion (translation and rotation) that moves P on top of T. We restrict the consideration on the special case such that the translation moves the center of P exactly on top of the center of some pixel of T. We call this the *center-to-center translation*.

Assume that the center of P is on top of the center of $T[i,j]$ after the translation. Then some rotation is applied that creates angle α between the x-axis of T and the x-axis of P (see Fig. 2). We say now that P is at *location* $((i,j),\alpha)$ on top of T.

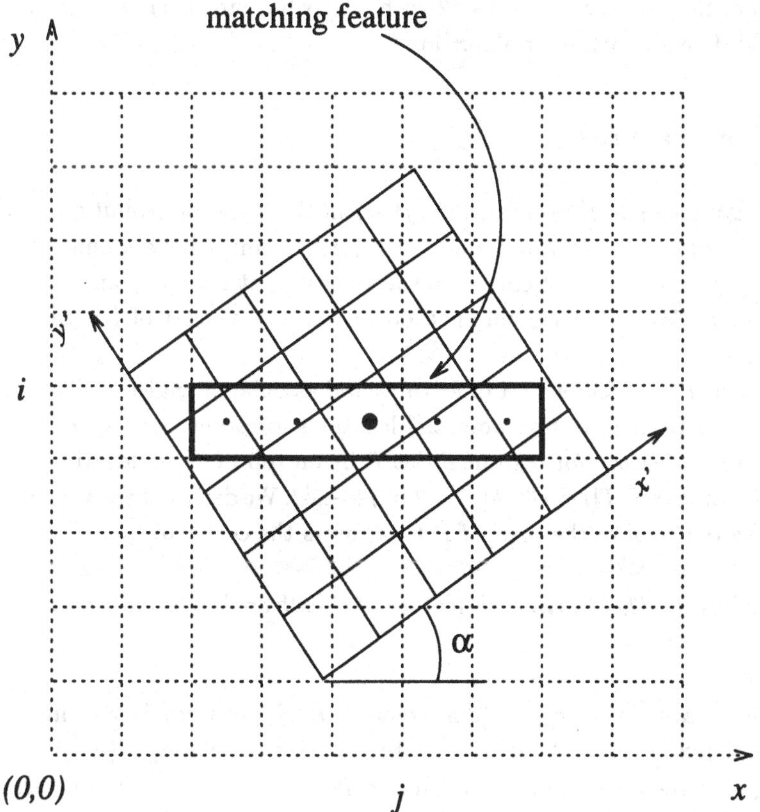

Fig. 2. An occurrence with a matching feature.

Pattern P is said to have an *occurrence* at location $((i,j),\alpha)$ if the pixel colors of P match with the pixel colors of T below them. More precisely, we will require that the colors match at the centers of the pixels of P. To this end, let us define the *matching function* M at location $((i,j),\alpha)$ as follows. For each pixel $P[r,s]$ of P, let $T[r',s']$ be a pixel of T such that the center of $P[r,s]$ belongs to the area covered by $T[r',s']$. Then $M(P[r,s]) = T[r',s']$.

We may assume that function M is uniquely defined; otherwise adjust α "infinitesimally" such that no center of P hits the pixel boundaries of T.

Definition 1. *Pattern P occurs at location $((i,j),\alpha)$ of T if the color of $P[r,s]$ is the same as the color of $M(T[r,s])$ for all pixels $P[r,s]$ of P.*

Our problem is to develop an efficient algorithm that, given T and P, finds all locations $((i,j),\alpha)$ such that P occurs in T at $((i,j),\alpha)$.

One easily observes that even if P occurs at some location of T, there can be pixels of T that are covered by P but not by any center of P. The occurrence test does not depend on the color of such a pixel of T. (The reader can convince himself of this by examining Fig. 2 with somewhat larger α). This phenomenon will slightly complicate our algorithm.

3 The Features

Let us assume that P occurs at $((i,j),\alpha)$ in the sense of Definition 1. Then it should be obvious that row i and column j of T must have around the center pixel $T[i,j]$ a sequence of colors that also occurs, at least in some approximate sense, in P. We will base our method on such sequences of length m, called *features*.

Pattern P will define a set of features and associated angles α. They are then searched for from T, to find potential locations for occurrences of P.

To read a feature for α from P, let P be on top of T, on location $((i,j),\alpha)$. Consider the pixels $T[i,j-\frac{m-1}{2}],\ldots,T[i,j+\frac{m-1}{2}]$. We denote them as $t_1,t_2,\ldots,t_m$. Let c_i be the color of the pixel of P that covers the center of t_i.

Definition 2. *The (horizontal)* feature *of P with angle α is the sequence $F(\alpha) = c_1c_2\cdots c_m$.*

Note that feature $c_1\cdots c_m$ contains colors from P, not from T. We only used the centers of pixels $t_1,\ldots,t_m$ of T as technical trick in defining $F(\alpha)$.

Fig. 2 gives an example of reading a feature. Here $m=5$, and the highlighted block contains the pixels $t_1,t_2,\ldots,t_5$. Their centers are shown as dots. The center of t_1 also belongs to pixel $P[4,1]$ of P. Therefore the first color c_1

of the feature is the color of $P[4,1]$. The colors of $P[4,2]$, $P[3,3]$, $P[2,4]$, and $P[2,5]$ form the rest of the feature.

If P really occurs at $((i,j),\alpha)$ in the sense of Definition 1, what is then the relation between $F(\alpha)$ and the colors of $T[i, j - \frac{m-1}{2}], \ldots, T[i, j + \frac{m-1}{2}]$? The answer is given by the following theorem.

Theorem 1. *For $k = 1, \ldots, m$, if c_k is not equal to the color of $t = T[i, j - \frac{m-1}{2} + k - 1]$, then it is equal to the color of some of the four pixels above, below, left, or right to pixel t in T.*

Proof. Assume that c_k is not equal to the color of t. Recall that c_k is the color of the pixel, say p, of P that covers the center of t. The center of p can not belong to the area of t; otherwise p and t would have the same color because P was assumed to occur at $((i,j),\alpha)$. However, the distance between the centers of p and t must be $< \sqrt{2}/2$. Hence the center of p can belong only to some pixel above, below, left, or right to t (but not to any pixel in some diagonal direction from t. Because P has an occurrence, one of these four pixels must have color c_k. □

Theorem 1 suggest the following filtration scheme for finding the occurrences of P from T.

1. Extract all possible features $F(\alpha)$ from P.
2. Find the occurrences of features $F(\alpha)$ from the rows of T. The occurrence can be approximate in the sense of Theorem 1, i.e., if a text pixel does not match, try the four pixels above, below, left, and right of it.
3. If $F(\alpha)$ occurs centered at $T[i,j]$, check if the whole P occurs at $((i,j),\alpha)$.

This needs further refinement in order to understand how the proper values of α can be found.

4 Counting the Matching Functions

The matching function M gives the correspondence between P and T that is used in the occurrence test at $((i,j),\alpha)$. The locations of the centers of the pixels of P with respect to T determine M. Consider what happens to M when angle α grows continuously, starting from $\alpha = 0$. Function M changes only at values α such that some pixel center of P hits some pixel boundary of T. Some elementary analysis shows that the set of possible angles α, $0 \le \alpha \le \pi/2$, is

$$A = \{\beta, \pi/2 - \beta \mid \beta = \arcsin \frac{h + \frac{1}{2}}{\sqrt{i^2 + j^2}} - \arcsin \frac{j}{\sqrt{i^2 + j^2}};$$
$$i = 1, 2, \ldots, \lfloor m/2 \rfloor; j = 0, 1, \ldots, \lfloor m/2 \rfloor; h = 0, 1, \ldots, \lfloor \sqrt{i^2 + j^2} \rfloor\}.$$

By symmetry, the set of possible angles α, $0 \le \alpha \le 2\pi$, is

$$\mathcal{A} = A \cup A + \pi/2 \cup A + \pi \cup A + 3\pi/2$$

Similarly, feature $F(\alpha)$ can change only at angles α such that some pixel boundary of P hits some center of pixels $t_1, t_2, \ldots, t_m$. The set of such angles is

$$\mathcal{B} = B \cup B + \pi/2 \cup B + \pi \cup B + 3\pi/2$$

where

$$B = \{\gamma = \arcsin \frac{h + \frac{1}{2}}{i} \mid i = 1, 2, \ldots, \lfloor m/2 \rfloor; h = 0, 1, \ldots, i - 1\}.$$

Obviously, the size of $\mathcal{A}$ is $O(m^3)$, the size of $\mathcal{B}$ is $O(m^2)$, and $\mathcal{B} \subseteq \mathcal{A}$.

Let $(\beta_1, \beta_2, \ldots)$ be the elements of $\mathcal{A}$ and $(\gamma_1, \gamma_2, \ldots)$ the elements of $\mathcal{B}$ in increasing order. By construction, $F(\alpha)$ stays unchanged for α such that $\gamma_i \le \alpha \le \gamma_{i+1}$. So $F(\alpha) = F(\gamma_i)$. However, there might be one or more angles β_k such that $\gamma_i < \beta_k < \gamma_{i+1}$ because $\mathcal{B}$ is a subset of $\mathcal{A}$. Let us denote as $K(\gamma_i)$ the set of such angles β_k. Feature $F(\gamma_i)$ represents all angles in $K(\gamma_i)$ or the corresponding functions M. If we find an occurrence of $F(\gamma_i)$, all the functions M that correspond to $K(\gamma_i)$ must be checked for an occurrence of P.

5 The Method

The method works in the following main steps.

1. Find and sort angle sets $\mathcal{A}$ and $\mathcal{B}$. This takes time $O(m^3 \log m)$.
2. For each α in $\mathcal{B}$, find the corresponding feature $F(\alpha)$. Associate with each $F(\alpha)$ the corresponding set of angles $K(\alpha) \subseteq \mathcal{A}$. The $O(m^2)$ different features $F(\alpha)$, each of length m, can be read from P in time $O(m^3)$. Using the sorted $\mathcal{A}$ the associated sets $K(\alpha)$ can also be found in time $O(m^3)$.
3. Build an Aho–Corasick string matching automaton [1] for features $F(\alpha)$. As the total length of the features is $O(m^3)$, the automaton can be constructed in time $O(m^3|\Sigma|)$ by standard methods.
4. Scan the rows of text T with the AC–automaton to find the occurrences of features $F(\alpha)$. The occurrences have to be found in the sense of Theorem 1, that is, if the text pixel does not match, then try the four neighboring text pixels. This can be implemented by maintaining a *set* of states of the AC–automaton that are reachable under different choices of matching pixels.
5. If the AC–automaton finds an occurrence of $F(\alpha)$ centered at (i, j) in T, check for each angle $\beta \in K(\alpha)$, whether or not P occurs at $((i, j), \beta)$.

In practise one can scan T in two directions, i.e., along the rows and the columns. A candidate location (i, j) is verified for an occurrence of P only if two features are found, centered at (i, j), and their associated angles are perpendicular. This effectively doubles the length of the features which considerably improves the filtration capability.

6 Experimental Results

The described algorithm was implemented in C and compiled using gcc version 2.8.0 on DEC Alpha workstation running Linux operating system. The performance was tested over different pattern sizes. Patterns were obtained from the input text in random positions and orientations. Summary of the experiments is shown in Table 1. The times reported are averages over ten runs. As expected, the preprocessing time grows fast when the feature length (the width of the pattern) increases.

feature length	preprocessing	scanning time	checking time	occurrences
3	0.003418	1.783105	0.031934	1.0
5	0.013086	1.216992	0.002637	1.0
7	0.032422	0.976758	0.002344	1.0
9	0.072168	1.013672	0.004297	1.0
11	0.132812	1.183496	0.004297	1.0
13	0.235059	1.493066	0.006836	1.0
15	0.382617	1.488770	0.007324	1.0
17	0.590625	1.617676	0.008789	1.0
19	0.855469	2.139160	0.010840	1.0
29	3.795020	3.797656	0.027246	1.0
39	11.247852	7.692480	0.074316	1.0
49	25.352930	12.933789	0.144336	1.0

Table 1. Experimental results for different feature lengths. Times included are for preprocessing, scanning the input text with AC–automaton and checking the potential occurrences. The times are given in seconds. The input text was a natural image ("Lena") of 512 × 512 pixels.

7 Conclusion

We have presented a filtering algorithm for exact two–dimensional pattern matching that allows rotating the pattern.

There are several directions for further study. It seems possible to remove the center–to–center assumption but the price is rather high because then the number of features becomes $O(m^5)$. If P is not very small, then it is not necessary to use features of maximal length m. A better average performance can probably be achieved with moderately short features. Generalizing the method for approximate matching is from the practical point of view the most important future goal. We are currently working on all these problems.

References

1. Alfred V. Aho and Margaret J. Corasick. Efficient string matching: an aid to bibliographic search. *Communications of the ACM*, 18(6):333–340, June 1975.
2. A. Amir, G. Benson, and M. Farach. Alphabet independent two dimensional matching. In N. Alon, editor, *Proceedings of the 24th Annual ACM Symposium on the Theory of Computing*, pages 59–68, Victoria, B.C., Canada, May 1992. ACM Press.
3. Ricardo Baeza-Yates and Mireille Regnier. Fast two-dimensional pattern matching. *Information Processing Letters*, 45(1):51–57, January 1993.
4. Robert S. Boyer and J. Strother Moore. A fast string searching algorithm. *Communications of the ACM*, 20(10):762–772, October 1977.
5. T. M. Caelli and Z. Q. Liu. On the minimum number of templates required for shift, rotation and size invariant pattern recognition. *Pattern Recognition*, 21:205–216, 1988.
6. Z. Galil and K. Park. Truly alphabet-independent two-dimensional pattern matching. In IEEE, editor, *Proceedings of the 33rd Annual Symposium on Foundations of Computer Science*, pages 247–257, Pittsburgh, PN, October 1992. IEEE Computer Society Press.
7. Juha Kärkkäinen and Esko Ukkonen. Two and higher dimensional pattern matching in optimal expected time. In Daniel D. Sleator, editor, *Proceedings of the 5th Annual ACM-SIAM Symposium on Discrete Algorithms*, pages 715–723, Arlington, VA, January 1994. ACM Press.
8. Marek Karpinski and Wojciech Rytter. Alphabet-independent optimal parallel search for three-dimensional patterns. Technical Report 85101-CS, University of Bonn, Department of Computer Science, November 1993.
9. Yehezkel Lamdan and Haim J. Wolfson. Geometric hashing: A general and efficient model-based recognition scheme. In *Second International Conference on Computer Vision: December 5–8, 1988, Innesbrook Resort, Tampa, Florida, USA*, pages 238–249. IEEE Computer Society Press, 1988.
10. G. M. Landau and U. Vishkin. Pattern matching in a digitized image. *Algorithmica*, 12(4/5):375–408, October 1994.
11. S. Ranka and T. Heywood. Two–dimensional pattern matching with k mismatches. *Pattern Recognition*, 24:31–40, 1991.
12. J. Tarhio. A sublinear algorithm for two-dimensional string matching. *PRL: Pattern Recognition Letters*, 17, 1996.
13. Rui Feng Zhu and Tadao Takaoka. A technique for two-dimensional pattern matching. *Communications of the ACM*, 32(9):1110–1120, September 1989.

Constructing Suffix Arrays for Multi-dimensional Matrices

Dong Kyue Kim   Yoo Ah Kim   Kunsoo Park *

Department of Computer Engineering
Seoul National University
Seoul 151-742, Korea

Abstract. We propose multi-dimensional index data structures that generalize suffix arrays to square matrices and cubic matrices. Giancarlo proposed a two-dimensional index data structure, the *Lsuffix tree*, that generalizes suffix trees to square matrices. However, the construction algorithm for Lsuffix trees maintains complicated data structures and uses a large amount of space. We present simple and practical construction algorithms for multi-dimensional suffix arrays by applying a new partitioning technique to lexicographic sorting. Our contributions are the following:

(1) We present the first algorithm for constructing two-dimensional suffix arrays directly. Our algorithm is ten times faster and five time space-efficient than Giancarlo's algorithm for Lsuffix trees.
(2) We present an efficient algorithm for three-dimensional suffix arrays, which is the first algorithm for constructing three-dimensional index data structures.

1 Introduction

The classical string matching for a pattern p and a text t finds all occurrences of p in t. In many applications [8] ranging from string matching to computational biology, the same text is queried many times with different patterns. Efficient solutions for this problem are based on constructing an index data structure of t that contains an occurrence of p as an *index* in t. Various kinds of index data structures for one-dimensional strings have been developed such as suffix trees [15, 22], suffix arrays [14], suffix automata [21], and suffix cactus [11].

The *suffix tree* is a compacted trie that represents all suffixes of a text string [15]. It was designed as a space-efficient alternative to Weiner's position tree [22]. A suffix tree for a text t of length n over an alphabet Σ can be built in $O(n \log |\Sigma|)$ time with $O(n)$ space. We can search a pattern p of length m in $O(m \log |\Sigma|)$ time using the suffix tree. Note that the construction time and the query time depend on the alphabet size. Although it was mainly designed for pattern matching purposes, the suffix tree is useful for many other applications of string processing [4, 8, 12]. The *suffix array* due to Manber and Myers [14] is

* E-mail: {dkkim, yakim}@algo.snu.ac.kr, kpark@theory.snu.ac.kr

basically a sorted list of all the suffixes of a text string and can be constructed in $O(n \log n)$ time. When the sorted list is coupled with information about *longest common prefixes* (*lcps*), string searches can be answered in $O(m + \log n)$ time using a simple augmentation to a classic binary search. In practice, suffix arrays use less space than suffix trees, but the construction takes more time [14].

Recently, many algorithms for two- and higher-dimensional matrices have been developed. For two-dimensional pattern matching that finds all occurrences of an $m \times m$ pattern P in an $n \times n$ text T, Amir, Benson and Farach [3] and Galil and Park [6] gave linear-time solutions. For index data structures in two dimensions, Giancarlo [7] proposed the *Lsuffix tree* that is a generalization of the suffix tree to square matrices. The Lsuffix tree can be built in $O(n^2 \log n)$ time using $O(n^2)$ space, but its construction algorithm maintains complicated data structures and uses a large amount of space. On the other hand, it seems that there is no obvious way to extend Manber and Myers's suffix array construction algorithm to two dimensions.

In this paper we propose *Isuffix arrays* and *Zsuffix arrays* that generalize the suffix array to $n \times n$ square matrices and $n \times n \times n$ cubic matrices, respectively. We first define linear representations of square matrices and cubic matrices. In order to sort the linearly represented suffixes, we develop a new partitioning technique based on Hopcroft's function partitioning technique [1, 9]. By applying the technique, we present a simple and practical algorithm for constructing Isuffix arrays. Our algorithm is independent of the alphabet size and can be easily extended to higher dimensions. Our contributions are the following:

(1) We present an $O(n^2 \log n)$ time construction algorithm for Isuffix arrays, which is the first algorithm for constructing two-dimensional suffix arrays directly. Our algorithm is ten times faster and five times space-efficient than Giancarlo's algorithm for Lsuffix trees, which is a remarkable improvement.
(2) We present an $O(n^3 \log n)$ time construction algorithm for Zsuffix arrays. This is the first algorithm for constructing three-dimensional index data structures.

2 Preliminaries

2.1 Linear representation of square matrices

Given an $n \times n$ matrix A, we denote by $A[i : k, j : l]$ the submatrix of A with corners (i, j), (k, j), (i, l), and (k, l). When $i = k$ or $j = l$, we omit one of the repeated indices. An entry of matrix A has a symbol from an alphabet Σ, on which a total order $\prec$ is defined. Consider a string x over alphabet Σ. The ith suffix (resp. prefix) of x is defined as the largest substring of x that starts (resp. ends) at position i. We generalize this definition of suffixes to higher dimensions: For $1 \leq i, j \leq n$, the *suffix* SA_{ij} *of matrix* A is the largest square submatrix of A that starts at position (i, j) in A. That is, $SA_{ij} = A[i : i + k, j : j + k]$ where $k = n - \max(i, j)$.

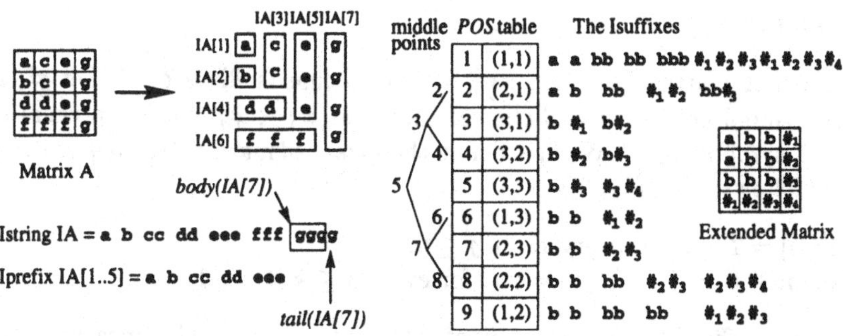

Fig. 1. Istring of a square matrix

Fig. 2. An Isuffix array of a square matrix

Giancarlo [7] proposed two constraints of a two-dimensional index data structure, i.e., completeness and common prefix constraints so that we can use it for pattern matching purposes when patterns are square matrices. The completeness constraint is that every square submatrix of A must be associated with a *prefix* of a suffix of A and the common prefix constraint is that the same square submatrices of A must be a common *prefix* of some suffixes of A, whatever the definition of *prefix* is. To satisfy these constraints, we adopt a linear representation of a square matrix. Let $I\Sigma = \bigcup_{i=1}^{\infty} \Sigma^i$, where the letter I represents linear shapes. We refer to the strings of $I\Sigma$ as *Icharacters* and we consider each of them as an *atomic* item. We refer to $I\Sigma$ as the *alphabet of Icharacters.* Two Icharacters are *equal* if and only if they are equal as strings over Σ. Moreover, given two Icharacters Iw and Iu of equal length, $Iw \prec Iu$ if and only if Iw as a string is lexicographically smaller than Iu as a string.

We describe a linearization method for a square matrix $A[1:n, 1:n]$. We linearize the matrix along its main diagonal [2, 7]. When we cut a matrix along the main diagonal, it is divided into an upper right half and a lower left half. Let $a(i) = A[i+1, 1:i]$ and $b(i) = A[1:i+1, i+1]$ for $1 \le i < n$, i.e., $a(i)$ is a row of the lower left half and $b(i)$ is a column of the upper right half. Then $a(i)$'s and $b(i)$'s can be seen as Icharacters.

The linearized string IA of matrix $A[1:n, 1:n]$, which is called the *Istring of matrix A*, is the concatenation of Icharacters $IA[1], \ldots, IA[2n-1]$ that are defined as follows: (See Fig. 1.)

(i) $IA[1] = A[1,1]$;

(ii) $IA[2i] = a(i)$, $1 \le i < n$;

(iii) $IA[2i+1] = b(i)$, $1 \le i < n$.

Since IA is composed of $2n-1$ Icharacters, the *Ilength* of Istring IA is $2n-1$. The kth *Iprefix* of an Istring IA, denoted by $IA[1..k]$, is the concatenation of Icharacters $IA[1], \ldots, IA[k]$. For each Icharacter $IA[l]$, $1 < l \le 2n-1$, $tail(IA[l])$ is the last character of $IA[l]$ and $body(IA[l])$ is the rest of $IA[l]$. See Fig. 1.

2.2 Isuffix arrays

Given a text matrix $T[1:n,1:n]$, we will define Isuffixes of T. Let $\#_i$ be a special symbol not in the alphabet Σ such that $\#_i \prec \#_j \prec a$ for integers $i<j$ and each symbol $a \in \Sigma$. We first define the *extended* matrix $A[1:n+1,1:n+1]$ of T as follows: (See Fig. 2.)

(i) $A[i,j]=T[i,j]$ for every $1 \le i,j \le n$;
(ii) $A[k,n+1]=A[n+1,k]=\#_k$, for every $1 \le k \le n+1$.

Consider a suffix SA_{ij}, $1 \le i,j \le n$, of extended matrix A. The linearized Istring of SA_{ij} is called an *Isuffix* of T and denoted by α_{ij}. Since special symbols were added, there cannot exist a pair of Isuffixes α_{ij} and α_{uv} such that $\alpha_{ij}=\alpha_{uv}$. The number of all Isuffixes of T is n^2.

Now we define two-dimensional suffix arrays, *Isuffix arrays*. The Isuffix array is a suffix array of all the Isuffixes of a given matrix, and consists of three tables POS, *Llcp*, and *Rlcp*. The three tables are basically the same as those of Manber and Myers. The basis of the Isuffix array is a lexicographically sorted table POS. We define a table $POS[1:n^2]$ of matrix T as follows: An element $POS[k]$ has the start position (i,j) if and only if α_{ij} is the kth smallest Isuffix in lexicographic order $\prec$. We will construct table POS by sorting all the Isuffixes α_{ij} for $1 \le i,j \le n$.

Given two Isuffixes β and γ, let $lcp(\beta,\gamma)$ be the length of the longest common prefix of β and γ when β and γ are regarded as one-dimensional strings. Consider all the possible triples (L,M,R) that can arise in a binary search on the interval $[1:n^2]$, where L, M, and R denote the left point, middle point, and right point of the interval that remains to be searched. There are exactly n^2-2 such triples, each with a unique midpoint $M \in [2:n^2-1]$ and we have $1 \le L < M < R \le n^2$ for each triple. Let (L_M,M,R_M) be the unique triple containing midpoint M. *Llcp* and *Rlcp* are tables of size n^2-2 such that $Llcp[M] = lcp(\alpha_{POS[L_M]},\alpha_{POS[M]})$ and $Rlcp[M]=lcp(\alpha_{POS[R_M]},\alpha_{POS[M]})$.

Example 1. In Fig. 2, we give an example of Isuffix arrays. Given a 3×3 text matrix T, we show a table POS that is a lexicographic sorted array of all Isuffixes of T. We also show all middle points that can arise in a binary search. If a middle point M is 7 then the unique triple is $(5,7,9)$. In this case, we have $Llcp[7]=lcp(\alpha_{3,3},\alpha_{2,3})=1$ and $Rlcp[7]=lcp(\alpha_{2,3},\alpha_{1,2})=2$.

3 Efficient partitioning technique

We will sort the Isuffixes of a matrix using partitioning techniques. Hopcroft [1, 9] first proposed the function partitioning technique that takes $O(N\log N)$ time. Paige, Tarjan and Bonic [17] gave a linear time solution for the single function partition problem. Crochemore [5] used the partitioning technique to find all squares of a string. Iliopoulos, Moore, and Park [10] also used this technique for finding all seeds of a string. Paige and Tarjan [16] gave an algorithm for

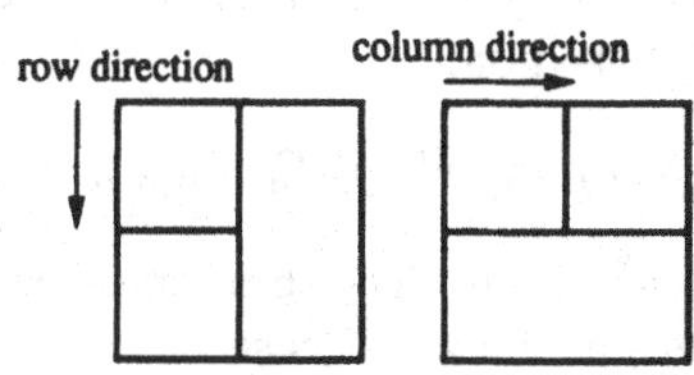

Fig. 3. Doubling the size of submatrices

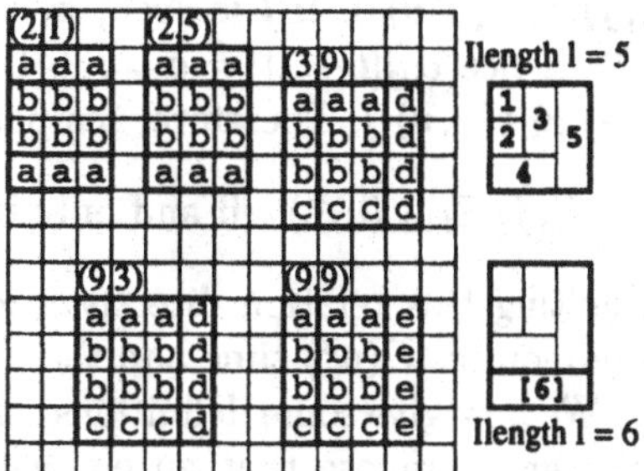

Fig. 4. Positions in equivalence classes

lexicographic sorting of *independent* strings using a naive partitioning. We apply Hopcroft's partitioning technique to lexicographic sorting of strings that have some geometric dependency, i.e., suffixes of one string.

3.1 Equivalence classes

We first define equivalence relations E_l for $1 \leq l \leq 2n+1$. E_l is defined on the set of start positions (i,j) of all Isuffixes α_{ij} of matrix T:

$$(i,j)\ E_l\ (u,v) \quad \text{if and only if} \quad \alpha_{ij}[1..l] = \alpha_{uv}[1..l].$$

That is, (i,j) and (u,v) are in an equivalence class of E_l if and only if the Ilength of the longest common Iprefix between two Isuffixes α_{ij} and α_{uv} is at least l.

To construct one-dimensional suffix arrays, Manber and Myers computed equivalence classes of E_l by doubling the value of l [14]. If we extend their method to two dimensions, we can double the size of submatrices first in the row (or column) direction and then in the other direction. See Fig. 3. However, both of the two ways violate the completeness constraint. For example, none of 3×3 submatrices can be a *prefix* of a suffix.

We will compute equivalence classes of E_l by increasing the value of l by 1. We first find equivalence classes of E_1 by sorting all symbols of matrix T because each symbol $T[i,j]$ corresponds to the first Icharacter of an Isuffix α_{ij}. The time complexity is $O(n^2 \log n)$ when the alphabet is general. Then, we compute equivalence classes of $E_2, E_3, \ldots$ successively by the partitioning technique until all classes are singleton sets.

Consider an equivalence class C of E_l. The *Istring of class* C, denoted by $Istr(C)$, is the common Iprefix of Ilength l of Isuffixes α_{ij}'s for all $(i,j) \in C$. We define the *reference* of $(i,j) \in C$, denoted by $r(i,j)$: if l is odd, $r(i,j) = (i+1,j)$; otherwise (i.e., l is even), $r(i,j) = (i,j+1)$.

At stage l of the partitioning we compute equivalence classes of E_{l+1} from equivalence classes of E_l. To do that, we do not compare $(l+1)$st Icharacters of two Isuffixes but use the position information only. Suppose that $(i,j)\ E_l\ (u,v)$. If l is odd, $(l+1)$st Icharacters of Isuffixes α_{ij} and α_{uv} are some subrows in T. In this case, $\alpha_{ij}[l+1] = \alpha_{uv}[l+1]$ if and only if $(i+1,j)\ E_l\ (u+1,v)$ (i.e.,

$r(i,j)\ E_l\ r(u,v)$). If l is even, $\alpha_{ij}[l+1]$ and $\alpha_{uv}[l+1]$ are some subcolumns in T. Hence $\alpha_{ij}[l+1] = \alpha_{uv}[l+1]$ if and only if $(i,j+1)\ E_l\ (u,v+1)$ (i.e., $r(i,j)\ E_l\ r(u,v)$). Therefore, the partitioning is based on:

$$(i,j)\ E_{l+1}\ (u,v) \quad \text{if and only if} \quad (i,j)\ E_l\ (u,v) \text{ and } r(i,j)\ E_l\ r(u,v).$$

Exploiting this relation directly leads to an $O(n^3)$ time algorithm, since each stage requires $O(n^2)$ time and there are $2n$ stages in the worst case.

We now give some definitions and facts that will be used in the construction algorithm. Suppose that an equivalence class C of E_l is split into subclasses $C_1, \ldots, C_r$ of E_{l+1} at stage l. For all (i,j) in a subclass C_k, $1 \le k \le r$, of C, there must exist one class X_k of E_l such that $r(i,j) \in X_k$. We say that X_k is the *reference class* of C associated with C_k. See Fig. 4. At stage 5, suppose that a class C of E_5 has its Istring $\alpha_{2,1}[1..5]$ and a subclass C_1 of C has its Istring $\alpha_{2,1}[1..6]$. Then the reference class X_1 of C associated with C_1 has its Istring $\alpha_{3,1}[1..5]$. The *relative order of class* C_k, $1 \le k \le r$, in $\{C_1, \cdots, C_r\}$, denoted by $order(C_k)$, is the lexicographic order of $Istr(C_k)$ in the set of Istrings $\{Istr(C_1), \cdots, Istr(C_r)\}$.

Fact 1. *Suppose that a class C is split into $C_1, \ldots, C_r$. Let X_k, $1 \le k \le r$, be the reference class of C associated with C_k. Then $order(C_k)$ is the relative order of X_k in $\{X_1, \cdots, X_r\}$.*

For each position (i,j) in matrix T, let q_{ij} be the index of the POS table such that $POS[q_{ij}] = (i,j)$. We define $rank(C)$ as the minimum of q_{ij}'s for all positions (i,j) in an equivalence class C. Then $rank(C) \le q_{ij} < rank(C) + |C|$ for every $(i,j) \in C$ because all positions in C take contiguous entries in the POS table. Notice that for each singleton class D that has one Isuffix α_{uv}, $POS[rank(D)] = (u,v)$.

Fact 2. *Suppose that a class C is split into $C_1, \ldots, C_r$ such that $order(C_1) < \cdots < order(C_r)$. Then $rank(C_1) = rank(C)$ and $rank(C_k) = rank(C_{k-1}) + |C_{k-1}|$ for $2 \le k \le r$.*

Let $Iheight[k]$ be the longest common Iprefix of two adjacent Isuffixes $\alpha_{POS[k-1]}$ and $\alpha_{POS[k]}$ for $2 \le k \le n^2$. Fact 3 implies that we can get $Iheight[k]$ during the partitioning as a by-product.

Fact 3. *Suppose that a class C is split into $C_1, \ldots, C_r$ at stage l. Then $Iheight[rank(C_k)] = l$ for each C_k, $2 \le k \le r$.*

At stage l of the partitioning we will partition each class C of E_l with respect to a reference class X of C. When we say that we partition C *with respect to* X, we partition C into two subclasses C_1 and C_2 such that $C_1 = \{(i,j) \in C \mid r(i,j) \in X\}$ and $C_2 = \{(i,j) \in C \mid r(i,j) \notin X\}$.

Example 2. Fig. 4 shows positions in some equivalence classes. Consider an equivalence class $C = \{(2,1), (2,5), (3,9), (9,3), (9,9)\}$ of E_5. Because there exist

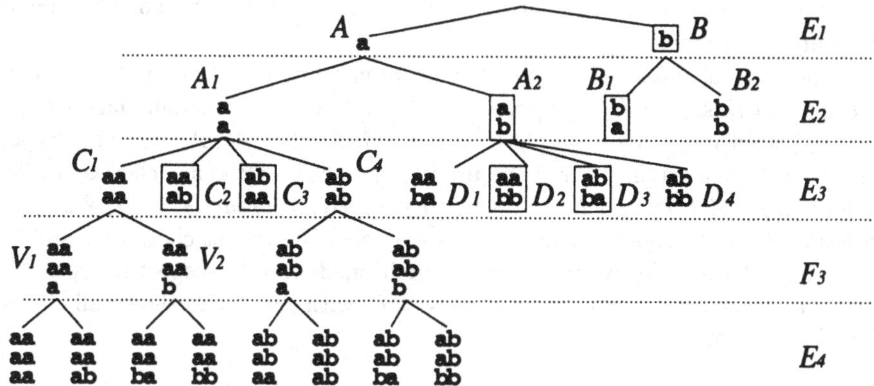

Fig. 5. Istrings of equivalence classes.

two distinct reference classes X and X' of C such that $\{(3,1),(3,5)\} \subset X$ and $\{(4,9),(10,3),(10,9)\} \subset X'$, we partition C with respect to X and X'. Thus, C can be split into two subclasses $C_1 = \{(2,1),(2,5)\}$ and $C_2 = \{(3,9),(9,3),(9,9)\}$ of E_6 at stage 5. Since $Istr(X) \prec Istr(X')$, $order(C_1) < order(C_2)$ by Fact 1. By Fact 2, $rank(C_1) = rank(C)$ and $rank(C_2) = rank(C_1) + |C_1|$. By Fact 3, $Iheight[rank(C_2)] = 5$.

Consider also class $C_2 = \{(3,9),(9,3),(9,9)\}$ of E_6. Since there are two reference classes Y and Y' of C_2 such that $\{(3,10),(9,4)\} \subset Y$ and $(9,10) \in Y'$, we can partition class C_2 into two subclasses $D_1 = \{(3,9),(9,3)\}$ and $D_2 = \{(9,9)\}$ of E_7. We also have $order(D_1) < order(D_2)$, $rank(D_1) = rank(C_2)$, $rank(D_2) = rank(D_1) + |D_1|$, and $Iheight[rank(D_2)] = 6$. □

3.2 Partitioning technique

We first describe main difficulties when we apply Hopcroft's partitioning technique to sorting the Isuffixes. Suppose that a class C of E_l is partitioned to subclasses $C_1, \ldots, C_r$ of E_{l+1}. We call C the *predecessor* of each C_k, $1 \le k \le r$, denoted by $pred(C_k)$. Among the subclasses $C_1, \ldots, C_r$, a largest one is called a *big* class of E_{l+1} (ties are broken arbitrarily); all the other classes are called *small* classes of E_{l+1}. The main idea of Hopcroft's partitioning is to partition with respect to small classes (i.e., with respect to small reference classes in our case). However, we cannot apply this idea to sorting the Isuffixes directly. In order to partition a class C by using the idea, the predecessors of all reference classes of C should be the same. (Otherwise, there may exist two or more reference classes that are not small classes.) Example 3 shows a counterexample.

Example 3. Fig. 5 shows Istrings of some classes of E_1 through E_4 where $\Sigma = \{a,b\}$. Let A and B be the classes of E_1 such that $Istr(A) = a$ and $Istr(B) = b$. Let A_1 and A_2 be the subclasses of A such that $Istr(A_1) = aa$ and $Istr(A_2) = ab$, and B_1 and B_2 be the subclasses of B such that $Istr(B_1) = ba$ and $Istr(B_2) = bb$. Suppose that A is big. At stage 1, A_2 is computed by partitioning A with respect to B, and B_2 is computed by partitioning B with respect to B. Then all classes of E_2 are determined.

However, all classes of E_3 cannot be computed by partitioning with respect to the small classes of E_2. Suppose that A_2 and B_1 are the small classes of E_2 and A_1 is split into classes C_1, C_2, C_3 and C_4. Note that A_1, A_2, B_1, and B_4 are the reference classes of A_1, and the predecessors of its reference classes not the same, i.e., $pred(A_1) = pred(A_2) = A$ and $pred(B_1) = pred(B_2) = B$. When we partition A_1 with respect to small classes A_2 and B_1, we get classes C_2 and C_3 only. To get C_1 and C_4, we should have partitioned A_1 with respect to A_1 or B_2. Hence, at stage $l > 1$ partitioning a class of E_l with respect to the small classes of E_l is not sufficient. □

To remedy this problem, we define *intermediate* equivalence relations F_l, $1 \leq l \leq 2n$. F_l is also defined on the set of start positions (i,j) of all Isuffixes α_{ij} of matrix T:

$$(i,j)\ F_l\ (u,v) \quad \text{if and only if} \quad \alpha_{ij}[1..l] = \alpha_{uv}[1..l] \ \text{ and } \ body(\alpha_{ij}[l+1]) = body(\alpha_{uv}[l+1]).$$

The $(l+1)$st Icharacter of an Isuffix α_{ij} is composed of $body(\alpha_{ij}[l+1])$ and $tail(\alpha_{ij}[l+1])$, which correspond to $\alpha_{r(i,j)}[l-1]$ and $tail(\alpha_{r(i,j)}[l])$, respectively. (See Fig. 1.) Hence, we will use the following relations.

$$(i,j)\ F_l\ (u,v) \quad \text{if and only if} \quad (i,j)\ E_l\ (u,v) \text{ and } r(i,j)\ E_{l-1}\ r(u,v).$$

$$(i,j)\ E_{l+1}\ (u,v) \quad \text{if and only if} \quad (i,j)\ F_l\ (u,v) \text{ and } r(i,j)\ E_l\ r(u,v).$$

In the sorting algorithm, we will divide each stage into two phases to perform the partitioning correctly. In the first phase we partition classes of E_l with respect to the small classes of E_{l-1} to get classes of intermediate relation F_l. In the second phase we partition classes of intermediate relation F_l with respect to the small classes of E_l, and we get all the classes of E_{l+1}. Suppose that a class D of E_l is split into classes $V_1, \ldots, V_r$ of intermediate relation F_l at the first phase of stage l. For all (i,j) in each intermediate class V_k, $1 \leq k \leq r$, there exist one class Y_k of E_{l-1} such that $r(i,j) \in Y_k$. We call Y_k the *1st-phase* reference class of D associate with V_k. At the second phase of stage l, suppose that an intermediate class V_k of F_l is again split into classes $C_1, \ldots, C_{r'}$ of E_{l+1}. For all (i,j) in each class $C_{k'}$, $1 \leq k' \leq r'$, there exist one class $X_{k'}$ of E_l such that $r(i,j) \in X_{k'}$. We call $X_{k'}$ the *2nd-phase* reference class of V_k associated with $C_{k'}$. Note that the reference class of D associated with $C_{k'}$, $1 \leq k' \leq r'$, is the 2nd-phase reference class of V_k associated with $C_{k'}$. Lemma 4 shows that we correctly partition the classes of E_l at stage l.

Lemma 4.

(1) Let $Y_1, \ldots, Y_r$ be the 1st-phase reference classes of a class D at stage l. Then $pred(Y_1) = \cdots = pred(Y_r)$.

(2) Let $X_1, \ldots, X_s$ be the 2nd-phase reference classes of an intermediate class V at stage l. Then, $pred(X_1) = \cdots = pred(X_s)$.

```
Procedure MAKE-POS
    compute and sort classes of E_1
    determine rank and Iheight for each class of E_1
    put all classes of E_1 except the big class into BUFFER
    SMALL ← φ, l ← 0
    while there is a non-singleton class of E_l do
        OLDSMALL ← SMALL, SMALL ← BUFFER, BUFFER ← φ
        l ← l + 1
        for each class Y of E_{l-1} in OLDSMALL do
            process Y
            for each split class D of E_l,
                set the relative order of new intermediate subclass of D od
        for each class X of E_l in SMALL do
            process X
            for each split intermediate class V of F_l,
                set the relative order of new subclass of V od
        for each split class D of E_l do
            assign order(C) in D for each subclass C of D od
        for each new class C of E_{l+1} do
            determine rank(C) and Iheight[rank(C)] = l
            if C is singleton class then assign the value of table POS fi
            if C is a small class then put C into BUFFER fi
        od
    od
end
```

Fig. 6. Procedure MAKE-POS that computes table *POS*.

Example 4. In Fig. 5, we show some classes of F_3 and E_4. The 1st-phase reference classes of C_1 are A_1 and A_2, and $pred(A_1) = pred(A_2) = A$. The 2nd-phase reference classes of V_1 are C_1 and C_2, and $pred(C_1) = pred(C_2) = C$. Similarly, the 2nd-phase reference classes of V_2 are D_1 and D_2, and $pred(D_1) = pred(D_2) = D$. □

4 The algorithm for Isuffix arrays

Fig. 6 shows MAKE-POS that sorts all Isuffixes of a square matrix. MAKE-POS maintains the following invariant:

At the beginning of stage l, OLDSMALL contains all small classes of E_{l-1} and SMALL contains all small classes of E_l. For all classes C of E_{l-1} and E_l, we know order(C) and rank(C).

Initially, the invariant is satisfied by the first few lines of MAKE-POS. Note that E_0 has only one class that contains all positions, and thus there are no small classes of E_0.

In each stage $l \geq 1$ we compute equivalence classes of E_{l+1}. In the first **for** loop, we perform the first phase of stage l in order to identify every intermediate class V of F_l and compute *order*(V). Recall that 1st-phase reference classes at stage l are classes of E_{l-1}. When we say that we *process* a small class Y of E_{l-1}, we partition with respect to Y every class D of E_l whose 1st-phase reference class is Y. We process all classes in OLDSMALL. Let $V_1, \ldots, V_r$ be the intermediate subclasses of a class D of E_l. Let Y_k, $1 \leq k \leq r$, be the 1st-phase reference classes of D associated with V_k. By definition of small classes and Lemma 4, $r-1$ or r 1st-phase reference classes of D are small classes depending on whether one or none of $Y_1, \ldots, Y_r$ is big. If none is big then we get every intermediate subclass V_k, $1 \leq k \leq r$, of D directly. If one (say, Y_s) is big, then we can compute the intermediate subclass V_s by $V_s = D - (V_1 + \cdots + V_{s-1} + V_{s+1} + \cdots + V_r)$. Therefore, we can identify every intermediate class of F_l in the first **for** loop. Since we know the relative order of all 1st-phase reference classes by the invariant, we can compute the relative order of the new intermediate classes $V_1, \ldots, V_r$. Similarly, in the second **for** loop we perform the second phase of stage l to compute all classes C of E_{l+1}.

Consider a class D of E_l that is split at stage l. Since we know the relative order of every intermediate subclass V of D and the relative order of every subclass C of V, we can assign *order*(C) in D for each subclass C of D, which is performed in the third **for** loop. For each new class C of E_{l+1}, we compute *rank*(C) and set $Iheight[rank(C)] = l$. If C is a singleton subclass that has one Isuffix α_{ij}, we determine the value of table POS: $POS[rank(C)] = (i, j)$. If C is a small subclass of D, put C to BUFFER. At the end of stage l, all small classes of E_{l+1} are in BUFFER, which makes the invariant hold in the next stage. The time complexity of procedure MAKE-POS is proportional to the number of positions that enter BUFFER. By definition of small classes, we have $|C| \leq |D|/2$ for each small subclass C of D. Hence one position cannot belong to BUFFER more than $\log n^2$ times. Since there are n^2 positions, procedure MAKE-POS takes $O(n^2 \log n)$ time. A simple one-dimensional version of this algorithm appeared in [13].

We now construct tables *Llcp* and *Rlcp*. Let *height*$[k]$, $2 \leq k \leq n^2$, be the longest common prefix of $\alpha_{POS[k-1]}$ and $\alpha_{POS[k]}$ as one-dimensional strings. Whenever we determine $Iheight[rank(C)]$ for each new class C of E_{l+1} in procedure MAKE-POS, $height[rank(C)]$ can be computed in $O(\log n)$ time using Manber and Myers's *interval trees* [14]. Since an internal node of the interval tree is associated with an interval (L, M, R) that can arise in a binary search, we can directly get tables *Llcp* and *Rlcp* from the interval tree after computing *height*$[k]$ for all $2 \leq k \leq n^2$.

Theorem 5. *The Isuffix array of an $n \times n$ square matrix can be constructed in $O(n^2 \log n)$ time.*

Size of text		200 x 200			400 x 400			600 x 600			800 x 800		
Size of alphabet		2	4	16	2	4	16	2	4	16	2	4	16
Time (sec)	Isuffix array	3.0	2.5	1.7	15.0	11.2	9.6	42.2	34.2	22.4	69.9	69.8	33.8
	Lsuffix tree	14.2	19.2	24.4	107.7	101.1	113.1						
Space (Mbytes)	Isuffix array	4.1	4.4	4.3	17.3	17.9	21.4	27.8	38.3	45.4	37.1	54.4	78.4
	Lsuffix tree	15.9	17.1	24.9	63.9	68.7	99.5						

Fig. 7. Experimental results for Isuffix arrays and Lsuffix trees.

5 Experimental Results

In this section we describe implementation details of Lsuffix trees and present experimental results.

Lsuffix trees: Giancarlo's construction algorithm [7] of Lsuffix trees inserts an Lsuffix L in each step which consists of two phases: the *rescanning* phase finds the extended locus of *head* of L; the *scanning* phase scans Lcharacter by Lcharacter and makes the leaf node corresponding to L. To perform each rescanning phase in $O(\log n)$ time, an Lsuffix tree is represented as a dynamic tree [19]. That is, an Lsuffix tree is partitioned into vertex-disjoint paths, which are represented as balanced binary search trees. We use splay trees [20] for the balanced binary search tree. There are two variations of dynamic trees by the way of partitioning the tree into vertex-disjoint paths: *naive partitioning* and *partitioning by size*. Since the naive partitioning method is simpler and faster, we use this method. To perform the scanning phase in $O(\log n)$ amortized time, we construct two suffix trees T_{rows} and T_{cols} [7] of two strings that are obtained by concatenating the rows and columns of the matrix, respectively. To query LCA nodes in $O(1)$ time, we augment both trees by using the algorithm of Schieber and Vishkin [18]. To encode outgoing edges of an internal node in suffix trees, three distinct methods are known: $|\Sigma|$-element vectors, linked lists, and binary search trees. We use $|\Sigma|$-element vectors for T_{rows} and T_{cols}. However, we cannot use $|\Sigma|$-element vectors for Lsuffix trees, since the size of $L\Sigma$ is not bounded. Thus we use top-down splay trees [20] in which search and insert operations take $O(\log n)$ time.

Experiments: The experimental results are summarized in Fig. 7. For experiments, we used random matrices. We varied $|\Sigma|$ and the size of matrices. We used a Sun Ultra-Sparc1 167MHz with 128 megabytes of main memory. The time in Fig. 7 did not include the time for refining Lcharacters [7] or Icharacters, which corresponds to computing *height* from *Iheight* in our case. The space is measured by the maximum amount of memory used during the construction of Lsuffix trees or Isuffix arrays. The blank in Fig. 7 means that we cannot construct the Lsuffix tree since it exceeds the size of main memory. In case of Isuffix arrays, if we increase $|\Sigma|$ then the number of partitioning stages is in general reduced, and thus the construction time decreased. Since we used $|\Sigma|$-element vectors for T_{rows} and T_{cols}, the space needed to construct Lsuffix trees increased

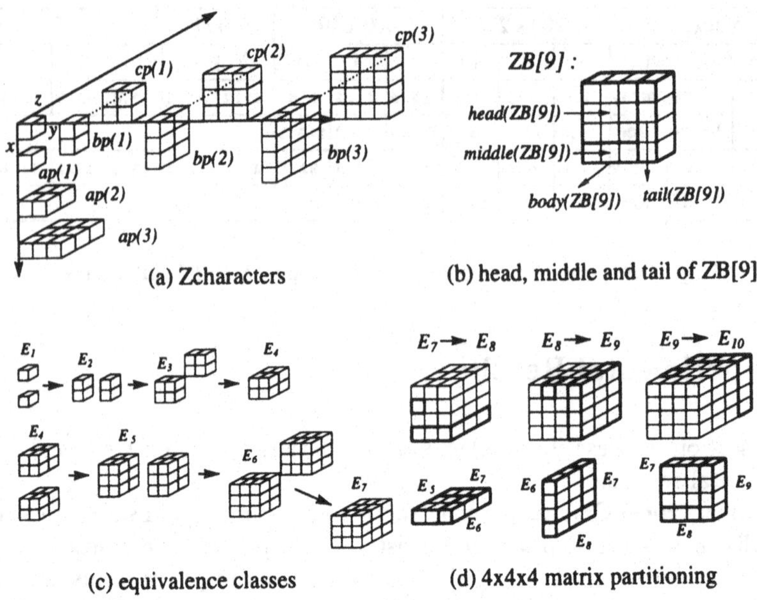

(a) Zcharacters

(b) head, middle and tail of ZB[9]

(c) equivalence classes

(d) 4x4x4 matrix partitioning

Fig. 8. Zstrings and Equivalence classes of a cubic matrix.

in proportion to $|\Sigma|$. The construction of Isuffix arrays is ten times faster and five times space-efficient than that of Lsuffix trees.

6 Three-Dimensional Suffix Arrays

We will construct a three-dimensional suffix array, the *Zsuffix array*, which is a generalization of the Isuffix array to three dimensions. In this section, we describe a linearization method of cubic matrices and briefly present how to sort the linearized strings.

Let $Z\Sigma = \bigcup_{i=1}^{\infty}\{\text{all Istrings of Ilength } i\}$. $Z\Sigma$ is called the *alphabet of Zcharacters*, and a linearized string of a cubic matrix $B[1:n, 1:n, 1:n]$ is called the *Zstring* of B. For $1 \leq i < n$, a Zstring is composed of three types of planes $ap(i)$, $bp(i)$, and $cp(i)$ that are represented as Istrings and perpendicular to x, y, and z-axis, respectively. The Zstring ZB of matrix B is the concatenation of Zcharacters $ZB[1], \ldots, ZB[3n-2]$ that are defined as follows: (See Fig. 8-(a).)

(i) $ZB[1] = B[1,1,1]$;
(ii) $ZB[3i-1] = ap(i)$, where $ap(i) = B[i+1, 1:i, 1:i]$ of Ilength $2i-1$;
(iii) $ZB[3i] = bp(i)$, where $bp(i) = B[1:i+1, i+1, 1:i]$ of Ilength $2i$;
(iv) $ZB[3i+1] = cp(i)$, where $cp(i) = B[1:i+1, 1:i+1, i+1]$ of Ilength $2i+1$.

For each Zcharacter $ZB[l]$, $3 < l \leq 3n-2$, $tail(ZB[l])$ is the last Icharacter of $ZB[l]$, $body(ZB[l])$ is the rest of $ZB[l]$, $middle(ZB[l])$ is the last Icharacter of $body(ZB[l])$, and $head(ZB[l])$ is the rest of $body(ZB[l])$. See Fig. 8-(b). The lth

Zprefix of an Zstring ZB, denoted by $ZB[1..l]$, is the concatenation of Zcharacters $ZB[1], \ldots, ZB[l]$.

Now we will define three-dimensional suffix arrays, *Zsuffix arrays*. For $1 \leq i, j, k \leq n$, the *suffix* SB_{ijk} *of* cubic matrix B is the largest cubic submatrix of B that starts at position (i, j, k) in B. For $1 \leq i, j, k \leq n$, the linearized Zstring of a suffix SB_{ijk} of B, denoted by β_{ijk}, is called a *Zsuffix* of B. Then, the *Zsuffix array* of B can be defined as the suffix array of all Zsuffixes of B.

We define equivalence relations E_l and two intermediate equivalence relations F_l and G_l: E_l, F_l and G_l are defined on the set of start positions (i, j, k) of all Zsuffixes β_{ijk} of a cubic matrix.

$$\begin{aligned}
(i,j,k)\ E_l\ (u,v,w) \text{ if and only if } & \beta_{ijk}[1..l] = \beta_{uvw}[1..l]. \\
(i,j,k)\ F_l\ (u,v,w) \text{ if and only if } & \beta_{ijk}[1..l] = \beta_{uvw}[1..l] \quad \text{and} \\
& head(\beta_{ijk}[l+1]) = head(\beta_{uvw}[l+1]). \\
(i,j,k)\ G_l\ (u,v,w) \text{ if and only if } & \beta_{ijk}[1..l] = \beta_{uvw}[1..l] \quad \text{and} \\
& body(\beta_{ijk}[l+1]) = body(\beta_{uvw}[l+1]).
\end{aligned}$$

We define the *reference* $r(i, j, k)$ of (i, j, k) in a class C of E_l as follows: if $l \bmod 3 = 1$, $r(i,j,k) = (i+1, j, k)$; if $l \bmod 3 = 2$, $r(i,j,k) = (i, j+1, k)$; if $l \bmod 3 = 0$, $r(i,j,k) = (i, j, k+1)$. The partitioning is based on:

$$(i,j,k)\ E_{l+1}\ (u,v,w) \text{ if and only if } (i,j,k)\ E_l\ (u,v,w) \text{ and } r(i,j,k)\ E_l\ r(u,v,w).$$

In Fig. 8-(c), the common Zprefixes of Zstrings in equivalence classes of E_1 through E_7 are shown. To apply the partitioning technique, we use the following relations.

$$\begin{aligned}
(i,j,k)\ F_l\ (u,v,w) \ \text{ iff } \ & (i,j,k)\ E_l(u,v,w) \text{ and } r(i,j,k)\ E_{l-2}\ r(u,v,w). \\
(i,j,k)\ G_l\ (u,v,w) \ \text{ iff } \ & (i,j,k)\ F_l\ (u,v,w) \text{ and } r(i,j,k)\ E_{l-1}\ r(u,v,w). \\
(i,j,k)\ E_{l+1}\ (u,v,w) \ \text{ iff } \ & (i,j,k)\ G_l\ (u,v,w) \text{ and } r(i,j,k)\ E_l\ r(u,v,w).
\end{aligned}$$

Fig. 8-(d) shows how to get equivalence classes of 4^3 cubic matrices using these relations.

Theorem 6. *The Zsuffix array of an $n \times n \times n$ cubic matrix can be constructed in $O(n^3 \log n)$ time.*

Corollary 7. *Three-dimensional suffix trees can be constructed in $O(n^3 \log n)$ time.*

Acknowledgement

We are grateful to Manber and Myers for providing us their implementation of suffix arrays.

References

1. A.V. Aho, J.E. Hopcroft and J.D. Ullman, The Design and Analysis of Computer Algorithms, *Addison-Wesley*, 1974.
2. A. Amir and M. Farach, Two-dimensional dictionary matching, *Inform. Processing Letters* 21 (1992), 233–239.
3. A. Amir, G. Benson and M. Farach, An alphabet independent approach to two-dimensional pattern matching, *SIAM J. Comput.* 23 (1994), 313–323.
4. A. Apostolico, The myriad virtues of subword trees, *Combinatorial Algorithms on Words*, Springer-Verlag, (1985), 85-95
5. M. Crochemore, An optimal algorithm for computing the repetitions in a word, *Inform. Processing Letters* 12 (1981), 244–250.
6. Z. Galil and K. Park, Alphabet-independent two-dimensional witness computation, *SIAM J. Comput.* 25, (1996), 907–935.
7. R. Giancarlo, A generalization of the suffix tree to square matrices with application, *SIAM J. Comput.* 24, (1995), 520–562.
8. D. Gusfield, Algorithms on Strings, Trees, and Sequences, *Cambridge Univ. Press*, 1997.
9. J.E. Hopcroft, An $n \log n$ algorithm for minimizing states in a finite automaton, in Kohavi and Paz, eds., *Theory of Machines and Computations,* Academic Press, New York, 1971.
10. C.S. Iliopoulos, D.W.G. Moore and K. Park, Covering a string, *Algorithmica* 16 (1996), 288–297.
11. J. Kärkkäinen, A cross between suffix tree and suffix array, *Symp. Combinatorial Pattern Matching* (1995), 191–204.
12. G.M. Landau and U. Vishkin, Fast parallel and serial approximate string matching, *J. Algorithms* 10 (1989), 157–169.
13. S.E. Lee and K. Park, A new algorithm for constructing suffix arrays, *J. Korea Information Science Society* 24 (1997), 697–703 (written in Korean).
14. U. Manber and G. Myers, Suffix arrays: A new method for on-line string searches, *SIAM J. Comput.* 22, (1993), 935–938.
15. E.M. McCreight, A space-economical suffix tree construction algorithms, *J. ACM* 23 (1976), 262–272.
16. R. Paige and R.E. Tarjan, Three partition refinement algorithms, *SIAM J. Comput.* 16, (1987), 973–989.
17. R. Paige, R.E. Tarjan and R. Bonic, A linear time solution to the single function coarsest partition problem, *Theoret. Comput. Sci.* 40, (1985), 67–84.
18. B. Schieber and U. Vishkin, On finding lowest common ancestors: simplification and parallelization, *SIAM J. Comput.* 17, (1988), 1253–1262.
19. D.D. Sleater and R.E. Tarjan, A data structure for dynamic trees, *J. Comput. System Sci.* 26, (1983), 362–391.
20. D.D. Sleater and R.E. Tarjan, Self-adjusting binary search trees, *J. ACM* 32, (1985), 652–686.
21. E. Ukkonen and D. Wood, Approximate string matching with suffix automata, *Algorithmica* 10 (1993), 353–364.
22. P. Weiner, Linear pattern matching algorithms, *Proc. 14th IEEE Symp. Switching and Automata Theory* (1973), 1–11.

Simple and Flexible Detection of Contiguous Repeats Using a Suffix Tree
(Preliminary Version)

Jens Stoye* and Dan Gusfield**

Department of Computer Science
University of California, Davis
Davis, CA 95616

Abstract. We study the problem of detecting all occurrences of (primitive) tandem repeats and tandem arrays in a string. We first give a simple time- and space- optimal algorithm to find all tandem repeats, and then modify it to become a time and space-optimal algorithm for finding only the primitive tandem repeats. Both of these algorithms are then extended to handle tandem arrays. The contribution of this paper is both pedagogical and practical, giving simple algorithms and implementations based on a suffix tree, using only standard tree traversal techniques.

1 Introduction

Suffix trees are a fundamental data structure supporting a wide variety of efficient string searching algorithms. Their "myriad virtues" are well known [1], and more than 30 non-trivial applications have been collected [5, 8]. Although alternative algorithms based on other data structures exist for many of these applications, it is remarkable that this single data structure allows so many efficient – and often surprisingly simple and elegant – solutions to so many string searching and matching problems. In particular, suffix trees are well known to allow efficient and simple solutions to many problems concerning the identification and location of repeated substrings, where the substrings are either *not* required to be contiguous, or where the substrings form the two halves of a palindrome (see [8] for a description of several of such problems). For example, the simple method described in [8] to enumerate occurrences of all maximal pairs of repeated substrings in time proportional to their number, has been independently found by several people [9, 11, 17].

---

* Research supported by the German Academic Exchange Service (DAAD). E-mail: `stoye@cs.ucdavis.edu`

** Research partially supported by grant DBI-9723346 from the National Science Foundation, and by grant DE-FG03-90ER60999 from the Department of Energy. E-mail: `gusfield@cs.ucdavis.edu`

Despite the enormous versatility of suffix trees and their natural application to problems concerning non-contiguous repeats and palindromes, problems concerning *contiguous* repeated substrings have not previously had simple, natural solutions based on suffix trees. This is both surprising and disappointing, making it more difficult to teach efficient algorithms for a wide range of string problems, and complicating the long-term project (at U.C. Davis) of building practical, easily understood software for many different string tasks, based around a single resident data structure, the suffix tree. Such tools are being developed for applications in bio-sequence analysis. The existing literature contains methods for locating certain contiguous repeats [3, 13, 14, 12] that are not based on suffix trees, although the method in [12] uses a suffix tree to solve certain subproblems. There are also two technically impressive papers, [10] and [2], which present time- and space-optimal methods using suffix trees for problems concerning contiguous repeated substrings. The methods in both of those papers are quite complex (in algorithmic detail, needed auxiliary data structures, embellishments required for optimal space use, or time and correctness proofs). The first of those papers concerns problems not addressed here, while the second paper does concern the same problems addressed here. The second paper processes a suffix tree from the bottom up and requires considerable auxiliary data structures.

In this paper we present simple, time- and space-optimal algorithms for problems of locating certain contiguous repeated substrings in a string S. Our methods only use standard tree traversal techniques, assuming the suffix tree for S is available. Our methods process a single suffix tree top down with only the addition of an array the size of the input string. These simple methods have both pedagogical and practical value. The algorithms are based on the fact that suffix trees allow the efficient location of what we call *branching* occurrences of tandem repeats in a string. Once these occurrences are found, almost all other repetitive structures of interest can be determined with little additional effort. Hence our various algorithms are not only simple, they are all derivatives of a single, basic algorithm.

In Sect. 2 we introduce our terminology and state basic facts about the repeated substrings we will search for. In Sect. 3 we present the basic algorithm and three extensions. In Sect. 4 we sketch a bound on the number of occurrences of primitive tandem arrays. Section 5 concludes with an open question.

2 Strings, Suffix Trees, and Tandem Arrays

2.1 Terminology and Basic Facts

We assume a finite alphabet Σ of a fixed size. Throughout this paper, a, b, c, x, and y denote single characters from Σ; S, w, α, β, γ, δ denote strings from Σ^*.

We fix attention to a string S of length $n = |S|$; for convenience, we assume S ends with a character '\$' not occurring elsewhere in S. For $1 \leq i \leq j \leq n$, $S[i..j]$ denotes the substring of S beginning with the ith and ending with the jth character of S; we say there is an *occurrence* of $S[i..j]$ at position i in S.

When the substring consists of only one letter we simply write $S[i]$ rather than $S[i..i]$.

A string w is a *tandem array* if it can be written as $w = \alpha^k$ for some $k \geq 2$; otherwise w is called *primitive*. An occurrence of a tandem array $w = \alpha^k = S[i..i+k|\alpha|-1]$ is represented by a triple (i, α, k). Such an occurrence is called *primitive* if α is primitive; it is called *right-maximal* if there is no additional occurrence of α immediately after w in S; it is called *left-maximal* if there is no additional occurrence of α immediately preceding w in S. A *tandem repeat* (in the literature also called a *square*) is a tandem array $w = \alpha^k$ with $k = 2$.

An occurrence $(i, \alpha, 2)$ of a tandem repeat is *branching* if and only if the character in S immediately to the right end of this occurrence, $S[i+2|\alpha|]$, differs from $S[i+|\alpha|]$ (which must equal $S[i]$, the first character of the repeat). Fig. 1 illustrates this definition.

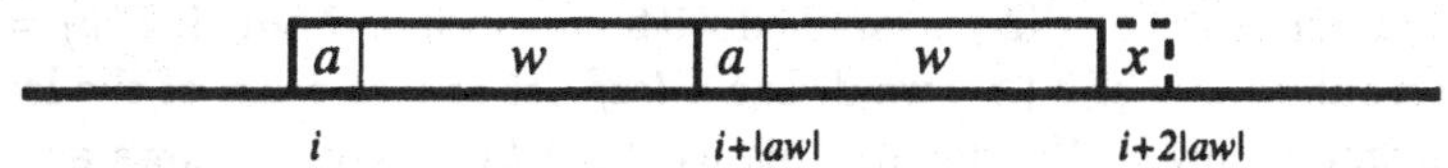

Fig. 1. Occurrences of branching and non-branching tandem repeats $(i, aw, 2)$; when $x = a$, the occurrence is non-branching, when $x \neq a$, the occurrence is branching

String aw is called the *left-rotation* of string wa.

Branching repeats and left-rotations are the keys to the algorithms presented in this paper. A first indication of their importance is contained in the following fact.

Lemma 1. *Any non-branching occurrence $(i, aw, 2)$ of a tandem repeat is the left-rotation of another tandem repeat, $(i+1, wa, 2)$, starting one place to its right. The tandem repeat $(i+1, wa, 2)$ may or may not be branching.*

By repeatedly applying Lemma 1, it follows that every tandem repeat is either branching, or is contained in a chain of tandem repeats created by successive left-rotations starting from a branching tandem repeat. (Recall that string S ends with a termination symbol \$). Furthermore, if $(i+1, wa, 2)$ is an occurrence of a tandem repeat (branching or not), then we can test in constant time if there is a tandem repeat of the same length starting at position i: simply test if $S[i] = a$. Hence, starting from a branching tandem repeat $(i+1, wa, 2)$, the chain of tandem repeats with $(i+1, wa, 2)$ at its right end can be determined in time proportional to the length of the chain (see Fig. 2).

The basic algorithm we will present in Sect. 3, first finds branching repeats, and then generates any desired non-branching repeats from the branching repeats. To prepare for that algorithm, we need to connect suffix trees with tandem repeats.

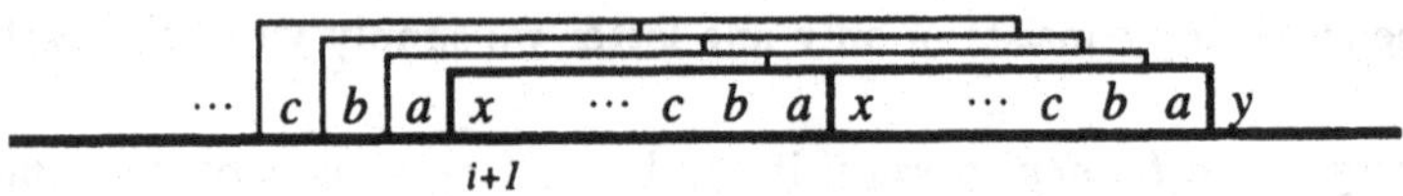

Fig. 2. Chain of non-branching tandem repeats

2.2 Suffix Trees and Tandem Repeats

We assume that the reader is familiar with the basic definitions of a suffix tree. Efficient, linear time methods are known to construct a suffix tree, e.g. [20, 16, 19, 7].

We denote by $T(S)$ the suffix tree of S, i.e., the compacted trie of all the suffixes of S; $L(v)$ denotes the *path-label* of node v in $T(S)$, i.e., the concatenation of the edge labels along the path from the root to v. $D(v) = |L(v)|$ is the *string-depth* of v. Each leaf v of $T(S)$ is labelled with index i if and only if $L(v) = S[i..n]$. At an internal node v of $T(S)$, we define a *leaf-list* of v as a list of the leaf-labels in the subtree below v. We denote this list by $LL(v)$. Fig. 3 shows an example of a suffix tree with its leaf-lists.

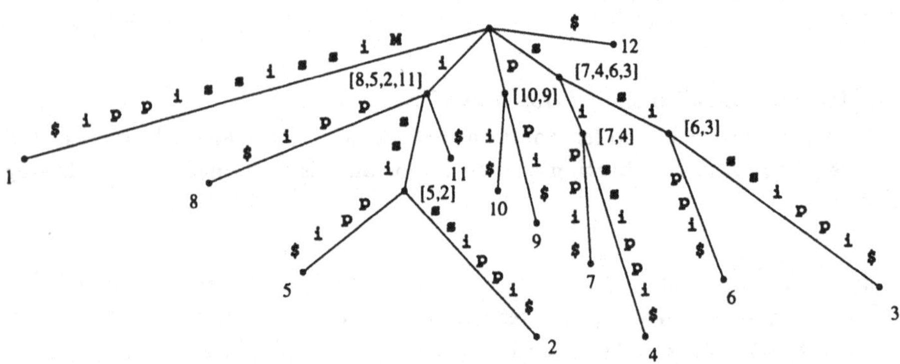

Fig. 3. Suffix tree of string `Mississippi` with leaf-list $LL(v)$ at each internal node

The following key fact about the relationship of tandem repeats and suffix trees follows easily from the definitions, and can be found (explicitly or implicitly) in [3, 2, 10, 8].

Lemma 2. *Consider two positions* i *and* j *of* S, $1 \le i < j \le n$, *let* $l = j - i$. *Then the following assertions are equivalent:*

(a) There is an occurrence of a tandem repeat of length $2l$ *starting at position* i *in* S;

(b) i *and* j *occur in the same leaf-list of some node* v *in* $T(S)$ *with depth* $D(v) \ge l$.

Lemma 2 is easily extended to characterize *branching* tandem repeats.

Lemma 3. *Consider two positions i and j of S, $1 \leq i < j \leq n$, let $l = j - i$. Then the following assertions are equivalent:*

(a) There is an occurrence of a branching tandem repeat of length $2l$ starting at position i in S;
(b) i and j occur in the same leaf-list of some node v in $T(S)$ with depth $D(v) = l$, but do not appear in the same leaf-list of any node with depth greater than l. Equivalently, they do not appear together in the leaf-list of any single child of v.

3 Algorithms

We will find all occurrences of branching tandem repeats in $O(n \log n)$ time, all occurrences of tandem repeats in $O(n \log n + z)$ time, where z is the number of occurrences, and all occurrences of primitive tandem repeats in $O(n \log n)$ time. All methods require just $O(n)$ space. With respect to worse case analysis, these bounds are time- and space optimal. All occurrences of tandem *arrays* of repeats (primitive or not) will be found in linear space, and in time equal or less than these bounds.

The basic algorithm and its variations are based on dividing the occurrences of tandem repeats in S into the two disjoint sets, the branching and non-branching occurrences. The branching occurrences of tandem repeats are found first, and then the non-branching occurrences are reported by successive left-rotations as suggested by Lemma 1.

3.1 The Basic Algorithm

Given Lemma 3, all occurrences of *branching* tandem repeats can be found in the following direct way:

Basic Algorithm. All nodes of $T(S)$ begin unmarked. Step 1 is repeated until all nodes are marked.

1. Select an unmarked internal node v. Mark v and execute steps 2a and 2b for node v.

2a. Collect the leaf-list, $LL(v)$, of v.

2b. For each leaf i in $LL(v)$, test whether leaf $j = i + D(v)$ is in $LL(v)$. If so, test whether $S[i] \neq S[i + 2D(v)]$. There is a branching tandem repeat of length $2D(v)$ starting at position i if and only if both tests return true. The first test determines if $L(v)^2$ is a tandem repeat and the second test determines if it is branching.

The leaf-list of v is collected via any linear time traversal of the subtree rooted at v. Assuming (as is standard) a representation of the suffix tree that allows the algorithm to move from a node to a child in constant time, that traversal takes time proportional to the size of $LL(v)$.

Given a leaf i in that leaf-list, we can test in constant time if $j = i + D(v)$ is also in $LL(v)$, provided we have preprocessed the suffix tree in the following standard way: During a depth-first traversal of the suffix tree (starting at the root), assign successive numbers (called *dfs numbers*) to the leaves in the order that they are encountered, and record these numbers in an array DFS, indexed by the original leaf numbers.[1] Additionally, when the depth-first traversal first visits an internal node v, record at v the next dfs number which will be given to a leaf, and when the depth-first traversal backs up from v, record at v the most recent dfs number assigned (see Fig. 4). It is well-known, and easy to establish, that all the leaves in $LL(v)$ are assigned dfs numbers (inclusively) between the two dfs numbers recorded at v. Hence to determine if a leaf $j = i + D(v)$ is in $LL(v)$ just check if $DFS[j]$ is between the two dfs numbers recorded at v.

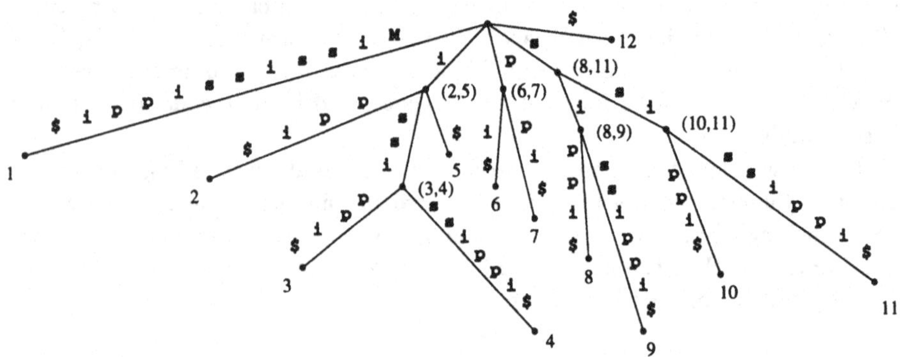

Fig. 4. Suffix tree of string `Mississippi` with dfs numbers at internal nodes

The above basic algorithm finds all occurrences of branching tandem repeats in time proportional to the total size of all the leaf-lists. That total size is $O(n^2)$. However, a simple modification leads to the desired time bound $O(n \log n)$.

3.2 Speeding Up the Basic Algorithm

For each node v, let v' denote the child of v whose leaf-list is largest over all the children of v. Let $LL'(v)$ denote the leaf-list of v minus the leaf-list of v', i.e., $LL'(v) = LL(v) - LL(v')$. By Lemma 3 (part b), if a branching tandem repeat starting at position i is detected by the basic algorithm during an examination of node v, then positions i and $j = i + D(v)$ must be in the leaf-lists of two

[1] As a side remark for those who know about suffix arrays [15], note that the array DFS is the inverse of the suffix array of S.

distinct children of v. Hence if one of those positions is in the leaf-list of v', the other position must be in $LL'(v)$. Therefore, we need execute step 2b of the basic algorithm only for each position in $LL'(v)$, provided we look both forward from that position (as in the above basic algorithm) and backward from it (as we will do below). These ideas are formalized in the following optimized basic algorithm.

Optimized Basic Algorithm. All nodes of $T(S)$ begin unmarked. Step 1 is repeated until all nodes are marked.

1. Select an unmarked internal node v. Mark v and execute steps 2a and 2b and 2c for node v.

2a. Collect the list $LL'(v)$ for v.

2b. For each leaf i in $LL'(v)$, test whether leaf $j = i + D(v)$ is in $LL(v)$, the leaf-list of v. If so, test whether $S[i] \neq S[i + 2D(v)]$. There is a branching tandem repeat of length $2D(v)$ starting at that position i if and only if both tests return true.

2c. For each leaf j in $LL'(v)$, test whether leaf $i = j - D(v)$ is in $LL(v)$. If so, test whether $S[i] \neq S[i + 2D(v)]$. There is a branching tandem repeat of length $2D(v)$ starting at that position i if and only if both tests return true.

Clearly, $LL'(v)$ can be found by a traversal from v that never visits v', and that traversal takes time proportional to the size of $LL'(v)$. Moreover, from the dfs numbers at each node, the size of that node's leaf-list can be obtained (it is simply the difference of the dfs numbers plus one), so that the child of any node v with the largest leaf-list can be easily identified when needed. Hence the time for the optimized algorithm is proportional to $\sum_v LL'(v)$. It is a well-known fact that this sum is at most $n \log_2 n$. To see this, note that if a leaf i is in $LL'(v)$ and is also in $LL'(u)$ for some ancestor u of v, then the size of $LL'(v)$ is at most half the size of $LL'(u)$. Hence, leaf i can be counted in $\sum_v LL'(v)$ at most $\log_2 n$ times. In summary,

Theorem 4. *All the branching tandem repeats are found in $O(n \log n)$ time and $O(n)$ space by the optimized basic algorithm.*

There are additional obvious ways to improve the running time of the algorithm in practice (such as combining traversals from the internal nodes). But for simplicity of exposition, and because these improvements don't reduce the worst case running time, we omit a discussion of them.

3.3 Finding All Occurrences of Tandem Repeats

From the set of branching occurrences of tandem repeats, the non-branching occurrences are obtained by a simple enumeration procedure, based on Lemma 1. In detail, the following is executed at each occurrence of a branching tandem repeat discovered by the optimized basic algorithm.

Starting with an occurrence $(i, wa, 2)$ of a branching tandem repeat, test if $S[i-1] = a$. If they are equal, $(i-1, aw, 2)$ is reported as a non-branching tandem repeat. This process, called the *rotation procedure*, is continued to the left until an inequality is observed, at which point the procedure stops. It is obvious that the additional time used by the rotation procedure is proportional to the total number, z, of occurrences of tandem repeats in S. Hence,

Theorem 5. *All occurrences of tandem repeats are found in $O(n \log n + z)$ time. No additional space is needed since all comparisons can be done directly on the string S.*

The same time and space bounds were also obtained for this problem, without the use of suffix trees, in [13, 14, 12].

3.4 Primitive Tandem Repeats

A tandem repeat $\alpha\alpha$ is called a *primitive* tandem repeat if string α is primitive, i.e., α cannot itself be expressed as the repeat of some substring. It is well known that there can be at most $O(n \log n)$ occurrences of primitive tandem repeats in a string of length n. We will sketch a proof of this fact in Sect. 4. Because the size of the output is smaller, and because any tandem repeat can be expressed as an array of primitive tandem repeats, it is often desirable to only report primitive tandem repeats. Prior algorithms which find all occurrences of primitive tandem repeats in $O(n \log n)$ time and linear space appear in [3] and [2].

We extend the basic algorithm of the previous section to report only the primitive tandem repeats. We begin by stating a general property of primitive strings.

Lemma 6. *A string wa is primitive if and only if its left-rotation aw is primitive. Hence, if $(i+1, wa, 2)$ is an occurrence of a* primitive *tandem repeat, and $(i, aw, 2)$ is also an occurrence of a tandem repeat, then $(i, aw, 2)$ is an occurrence of a* primitive *tandem repeat.*

Proof. If aw is non-primitive then $aw = \alpha^k$ for some α and $k > 1$. That means that each of the first $|\alpha|(k-1)$ characters in wa is equal to the character $|\alpha|$ places to its right. In particular, character $|\alpha| + 1$ in aw is a. Therefore, $wa = \beta^k$ where β consists of the last $k-1$ characters of α followed by character a. Hence wa is non-primitive.

The converse, that when wa is non-primitive, then aw is also primitive, is proved in essentially the same way. □

The algorithmic importance of Lemma 6 is that when the (optimized) basic algorithm identifies a branching tandem repeat associated with a node v, the tandem repeats generated by the rotation procedure at node v will either all be primitive, or will all be non-primitive. So to exclude all and only the non-primitive tandem repeats, it suffices to exclude every branching tandem repeat which is not primitive. Since branching tandem repeats are identified only at nodes, it suffices to identify every node u whose path-label $L(u) = \alpha^k$ for some $k \geq 2$, where α is primitive. Clearly, such a string α will be the path-label of some ancestor node v of u. Moreover, the basic algorithm will identify the primitive branching tandem repeat $L(v)^2 = \alpha^2$ at node v. We will show next that, at that point in its execution, the basic algorithm can be extended to efficiently locate and mark all nodes below node v whose path-labels are $L(v)^k = \alpha^k$ for $k \geq 2$. That extension will also identify some other nodes that may be marked for exclusion.

To exclude all non-primitive tandem repeats (but no primitive tandem repeats) we first modify the (optimized) basic algorithm to process the nodes in a top- down order, so that no node is selected in step 1 until all of its ancestors have been selected. This ensures that a node with path-label α will be selected before a node with path-label α^k for $k \geq 2$.

Second, we combine the rotation procedure with the (optimized) basic algorithm, so that when a branching primitive repeat $L(v)^2 = \alpha^2$ is found at a node v, the algorithm next executes a rotation procedure from each branching occurrence of α^2. Each such execution rotates left through each character in a chain of consecutive α's. As a side-effect of this computation, the algorithm can determine (in essentially no extra time) the largest value of k (call it k_v) such that α^k is a substring of S. Once k_v is determined, the algorithm walks from v to the end of the path labeled α^{k_v} in the suffix tree. That path exists (and will extend from v) since α^{k_v} is a substring in S. Moreover, since the path labeled α ends at a node (v), each string α^k, for $k < k_v$, will also end at a node. During the walk, the algorithm marks each node whose path-label is α^k, meaning that that node will not be selected in step 1 of the basic algorithm. (Recognizing that the node has that label is a trivial exercise.) This is a correct action because the path to any such marked node is either too long to be half of any tandem repeat, or it is the first half of a tandem repeat that is not primitive. Note that the number of steps in the walk from v is bounded by the number of left-rotations done in the rotation procedure that discovers k_v.

Clearly, any node corresponding to branching non-primitive tandem repeat will become marked in such a way, and hence never selected in step 1. Therefore the algorithm, as modified above, will enumerate all and only occurrences of primitive tandem repeats. The number of steps in all the extra walks is bounded by the number of left-rotations, and each left-rotation identifies a distinct occurrence of a primitive tandem repeat. Hence, the time for the algorithm is $O(n \log n + z)$, where z is the number of occurrences of primitive tandem repeats. However, it is known that z is $O(n \log n)$ in any string of length n. Hence,

Theorem 7. *The method described above finds all occurrences of primitive tandem repeats in $O(n \log n)$ time and $O(n)$ space.*

The time for the extra walks can be further reduced by using the skip/count trick that is well-known from suffix tree construction methods. That reduces the number of steps for a walk from the number of characters on the walk to the number of nodes on the walk, but, in this application, does not improve the worst case running time.

3.5 Primitive Tandem Arrays

Finally we extend the algorithm to locate all right-maximal occurrences of primitive tandem arrays. The idea is, for each branching primitive tandem repeat $(i, \alpha, 2)$ observed at a node v with $L(v) = \alpha$, successively test for $k = 1, 2, \ldots$ if leaf $i - k|\alpha|$ is also in the subtree below v. (Here it is not necessary to test explicitly if the tandem array is branching: From the fact that tandem repeat $(i, \alpha, 2)$ is branching, it follows immediately that all tandem arrays we find this way are also branching.) Each successful test corresponds to a branching tandem array $(i - k|\alpha|, \alpha, k + 2)$. Once the test fails, the procedure stops.

To also find the non-branching occurrences, the rotation procedure is applied to each of the branching occurrences $(i - k|\alpha|, \alpha, k + 2)$. If we stop the rotations after $|\alpha| - 1$ steps, all and only the right-maximal occurrences of primitive tandem arrays will be obtained; otherwise all occurrences of primitive tandem arrays are obtained, and there may be as many as $n(n-1)/2$ of these. Hence in the latter case the procedure runs in time $O(n \log n + z)$ where z is the output size.

The procedure can also easily be extended to find only those primitive tandem arrays which are simultaneously left- and right-maximal if for each of the chains of right-maximal primitive tandem repeats, only the last one (when the rotation procedure stops) is reported. This procedure takes time $O(n \log n)$ as well.

4 The Number of Occurrences of Primitive Tandem Repeats

In this section we sketch a proof that there can be at most $O(n \log n)$ occurrences of primitive tandem repeats in a string of length n. This fact is well established [3, 4, 6] (in fact, it is known [18] that the number of occurrences of primitive tandem repeats is bounded by $1.45(n+1)\log_2 n - 3.3n + 5.87$). We present here the $O(n \log n)$ bound to make the paper self-contained, and because the proof given here is simpler than previously published proofs.

We say two positions i and j in the leaf-list $LL(v)$ of some node v, are *adjacent in* $LL(v)$ if there is no position strictly between i and j that is also in $LL(v)$. The key fact we need is the following:

Lemma 8. *Assume $i < j = i + l$, and that there is an occurrence of a* primitive *tandem repeat of length $2l$ starting at position i in S. Then (a) i and j both occur in the leaf-list $LL(v)$ of some node v in $T(S)$ with depth $D(v) \geq l$, and (b) i and j are adjacent in $LL(v)$.*

Condition (a) simply repeats the necessary condition from Lemma 2 for an occurrence of a tandem repeat of length $2l$ starting at position i. Condition (b) distinguishes a primitive from a non-primitive tandem repeat. The key to proving this lemma is to show that if condition (a) is satisfied, and yet i and j are not adjacent in $LL(v)$, then the tandem repeat of length $2l$ starting at i is not primitive.

Proof (of Lemma 8). Let $\alpha\alpha$ be a tandem repeat of length $2l$ beginning at position i, and let $j = i + l$. Assume condition (a) is satisfied but (b) is not. That means there is another position k in $LL(v)$ strictly between i and j. So a copy of α occurs starting at position $k < i + l$. That copy of α can be expressed as a suffix, β, of α (from the copy starting at i) followed by a prefix, γ, of α (from the copy starting at j). It follows that $\alpha = \beta\gamma = \gamma\beta$, and by a well-known fact (Lemma 3.2.1 in [8]), α can be expressed as δ^q for some substring δ, and $q > 1$. Therefore, α is not primitive. □

A pair (i, j) is said to be an *adjacent pair* if there is some node v such that i and j are adjacent in $LL(v)$.

By Lemma 8, each occurrence of a primitive tandem repeat is associated with some adjacent pair. But each adjacent pair (i, j) is associated with at most one occurrence of a primitive tandem repeat, because that repeat is of length $2(j - i)$ and starts at i. Hence we can bound the number of occurrences of primitive tandem repeats in S by the total number of distinct adjacent pairs in all the leaf-lists of $T(S)$. For any node u, let $N(u)$ be the number of adjacent pairs that are in the leaf-list of u but not in the leaf-list of the parent of u. Define $N(r) = n - 1$, for the root r of $T(S)$. Any adjacent pair is adjacent in the leaf-lists of nodes that form a descending path in $T(S)$ (maybe only a single node in length), so the total number of distinct adjacent pairs is $\sum_u N(u)$.

Consider an internal node v' and its parent node v. Assume positions i and j are adjacent in $LL(v')$ but are not adjacent in $LL(v)$ (see Fig. 5). That means that in $LL(v)$ there is some position k strictly between i and j, and that k is not in $LL(v')$. So k must be contained in the leaf-list of some other child w of v. Since for each such pair (i, j) in $LL(v')$ there is a different such "witness" k, the value of $N(v')$ can not be larger than the number of entries in the lists $LL(w)$ summed over all children w of v other than v', so $N(v') \leq \sum_w |LL(w)| = |LL(v)| - |LL(v')|$.

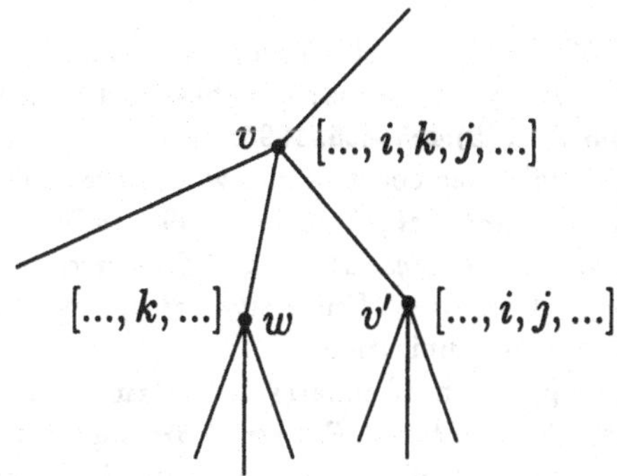

Fig. 5. Szenario where (i, j) is an adjacent pair in $LL(v')$ but not in $LL(v)$

Now for any internal node v, define (as in Sect. 3.2) v' to be the child of v with the largest leaf-list. It follows that $\sum_u N(u)$, and the total number of occurrences of tandem repeats, is bounded by $(n-1) + \sum_v |LL(v)| - |LL(v')|$. That sum is bounded by $O(n \log n)$ following the discussion in Sect. 3.2.

5 Summary and an Open Question

The time and space bounds for the methods presented here have been obtained earlier. Therefore, the contribution of this paper is the simplicity of the algorithms, which use only standard traversals of a suffix tree. The success of this effort must therefore be gauged by comparing the methods in this paper with earlier methods (particularly those in [2]) that use suffix trees to find contiguous repeated substrings.

We leave it as an open question whether the use of branching tandem repeats also allows linear-time solutions for related problems which are solvable within that time bound (e.g. the problem of finding the shortest tandem repeat beginning at each position of a string, cf. [10]). A positive indication is that the number of occurrences of branching tandem repeats in a string of length n seems to be bounded by n: we have experimentally verified this conjecture for all binary strings up to length 30 and for all ternary strings up to length 20.

References

1. A. Apostolico. The myriad virtues of subword trees. In A. Apostolico and Z. Galil, editors, *Combinatorial Algorithms on Words*, volume F12 of *NATO ASI Series*, pages 85–96. Springer Verlag, 1985.
2. A. Apostolico and F. P. Preparata. Optimal off-line detection of repetitions in a string. *Theor. Comput. Sci.*, 22:297–315, 1983.
3. M. Crochemore. An optimal algorithm for computing the repetitions in a word. *Inform. Process. Lett.*, 12(5):244–250, 1981.
4. M. Crochemore and W. Rytter. Periodic prefixes in texts. In R. Capodelli, A. De Santis, and U. Vaccaro, editors, *Sequences II*, pages 153–165. Springer Verlag, 1993.
5. M. Crochemore and W. Rytter. *Text Algorithms.* Oxford University Press, 1994.
6. M. Crochemore and W. Rytter. Squares, cubes, and time-space efficient string searching. *Algorithmica*, 13(5):405–425, 1995.
7. M. Farach. Optimal suffix tree construction with large alphabets. In *Proc. 38th Annu. Symp. Found. Comput. Sci., FOCS 97*, 1997. IEEE Press.
8. D. Gusfield. *Algorithms on Strings, Trees, and Sequences: Computer Science and Computational Biology.* Cambridge University Press, New York, NY, 1997.
9. R. W. Irving, Personal Communication.
10. S. R. Kosaraju. Computation of squares in a string. In M. Crochemore and D. Gusfield, editors, *Combinatorial Pattern Matching: 5th Annual Symposium, CPM 94. Proceedings*, number 807 in Lecture Notes in Computer Science, pages 146–150, 1994. Springer Verlag.
11. G. M. Landau, Personal Communication.

12. G. M. Landau and J. P. Schmidt. An algorithm for approximate tandem repeats. In A. Apostolico, M. Crochemore, Z. Galil, and U. Manber, editors, *Combinatorial Pattern Matching: 4th Annual Symposium, CPM 93. Proceedings*, number 684 in Lecture Notes in Computer Science, pages 120–133, 1993. Springer Verlag.
13. M. G. Main and R. J. Lorentz. An $O(n \log n)$ algorithm for finding all repetitions in a string. *J. Algor.*, 5:422–432, 1984.
14. M. G. Main and R. J. Lorentz. Linear time recognition of squarefree strings. In A. Apostolico and Z. Galil, editors, *Combinatorial Algorithms on Words*, volume F12 of *NATO ASI Series*, pages 271–278. Springer Verlag, Berlin, 1985.
15. U. Manber and E. W. Myers. Suffix arrays: A new method for on-line search. *SIAM J. Computing*, 22:935–948, 1993.
16. E. M. McCreight. A space-economical suffix tree construction algorithm. *Journal of the ACM*, 23(2):262–272, 1976.
17. J. P. Schmidt, Personal Communication.
18. P. F. Stelling. *Applications of Combinatorial Analysis to Repetitions in Strings, Phylogeny, and Parallel Multiplier Design*. Ph.d. dissertation, Department of Computer Science, University of California, Davis, 1995.
19. E. Ukkonen. On-line construction of suffix trees. *Algorithmica*, 14:249–260, 1995.
20. P. Weiner. Linear pattern matching algorithms. In *IEEE 14th Annual Symposium on Switching and Automata Theory*, pages 1–11. IEEE Press, 1973.

Comparison of Coding DNA

Christian N. S. Pedersen[1*], Rune Lyngsø[1], and Jotun Hein[2]

[1] BRICS**, Department of Computer Science, University of Aarhus, Ny Munkegade, DK-8000 Århus C, Denmark. E-mail: {cstorm,rlyngsoe}@brics.dk.

[2] Department of Ecology and Genetics, University of Aarhus, Ny Munkegade, DK-8000 Århus C, Denmark. E-mail: jotun@pop.bio.aau.dk

Abstract. We discuss a model for the evolutionary distance between two coding DNA sequences which specializes to the DNA/protein model proposed in Hein [4]. We discuss the DNA/protein model in details and present a quadratic time algorithm that computes an optimal alignment of two coding DNA sequences in the model under the assumption of affine gap cost. The algorithm solves a conjecture in [4] and we believe that the constant factor of the running time is sufficiently small to make the algorithm feasible in practice.

1 Introduction

A straightforward model of the evolutionary distance between two coding DNA sequences is to ignore the encoded proteins and compute the distance in some evolutionary model of DNA. We say that such a model is a DNA level model. The evolutionary distance between two sequences in a DNA level model can most often be formulated as a classical alignment problem and be efficiently computed using a dynamic programming approach [9, 11–13].

It is well known that proteins evolve slower than its coding DNA, so it is usually more reliable to describe the evolutionary distance based on a comparison of the encoded proteins rather than on a comparison of the coding DNA itself. Hence, most often the evolutionary distance between two coding DNA sequences is modeled in terms of amino acid events, such as substitution of a single amino acid and insertion-deletion of consecutive amino acids, necessary to transform the one encoded protein into the other encoded protein. We say that such a model is a protein level model. The evolutionary distance between two coding DNA sequences in a protein level model can most often be formulated as a classical alignment problem of the two encoded proteins. Even though a protein level model is usually more reliable than a DNA level model, it falls short because it postulates that all insertions and deletions on the underlying DNA occur at codon boundaries and because it ignores similarities on the DNA level.

* Supported by the ESPRIT Long Term Research Programme of the EU under project number 20244 (ALCOM-IT).

** Basic Research in Computer Science, Centre of the Danish National Research Foundation.

In this paper we present a model of the evolutionary distance between two coding DNA sequences in which a nucleotide event is penalized by the change it induces on the DNA as well as on the encoded protein. The model is a natural combination of a DNA level model and a protein level model. The DNA/protein model introduced in Hein [4, 6] is a biological reasonable instance of the general model in which the evolution of coding DNA is idealized to involve only substitution of a single nucleotide and insertion-deletion of a multiple of three nucleotides. Hein [4, 6] presents an $O(n^2m^2)$ time algorithm for computing the evolutionary distance in the DNA/protein model between two sequences of length n and m. This algorithm assumes certain properties of the cost function. We discuss these properties and present an $O(nm)$ time algorithm that solves the same problem under the assumption of affine gap cost.

The practicality of an algorithm not only depends on the asymptotic running time but also on the constant factor hidden by the use of O-notation. To determine the distance between two sequences of length n and m our algorithm computes $400nm$ table entries. Each computation involves a few additions, table lookups and comparisons. We believe the constant factor is sufficiently small to make the algorithm feasible in practice.

The problem of comparing coding DNA is also discussed by Arvestad [1] and Hua, Jiang and Wu [8]. The models discussed in these papers are inspired by the DNA/protein model in Hein [4, 6] but differ in the interpretation of gap cost. A heuristic algorithm for solving the alignment problem in the DNA/protein model is described In Hein [5]. A related problem of how to compare a coding DNA sequence with a protein has been discussed in [10, 14].

The rest of this paper is organized as follows: In Sect. 2 we introduce and discuss the DNA/protein model. In Sect. 3 we describe how to determine the cost of an alignment. In Sect. 4 we present the simple alignment algorithm of Hein [4]. In Sect. 5 we present a quadratic time alignment algorithm. Finally, in Sect. 6 we discuss future work.

2 The DNA/protein model

Let $a = a_1a_2a_3 \ldots a_{3n-2}a_{3n-1}a_{3n}$ be a coding sequence of DNA of length $3n$ with a reading frame starting at a_1. We introduce the notation $a_1^i a_2^i a_3^i$ to denote the ith codon $a_{3i-2}a_{3i-1}a_{3i}$ and the notation A_i to describe the amino acid coded by the ith codon. The amino acid sequence $A = A_1A_2 \ldots A_n$ describes the protein coded by a. Let $b = b_1b_2b_3 \ldots b_{3m-2}b_{3m-1}b_{3m}$, $b_1^i b_2^i b_3^i$ and $B = B_1B_2 \ldots B_m$ be defined similarly.

2.1 The general model

An evolutionary event e on the DNA that transforms a to a' will also change the encoded protein from A to A'. As some amino acids are coded by several codons, the proteins A and A' might be identical. The cost of e should reflect the changes on the DNA as well as the changes on the encoded protein.

$$cost(a \xrightarrow{e} a') = cost_d(a \xrightarrow{e} a') + cost_p(A \xrightarrow{e} A') \tag{1}$$

We say that $cost_d(a \stackrel{e}{\to} a')$ is the DNA level cost of e and that $cost_p(A \stackrel{e}{\to} A')$ is the protein level cost of e. In this paper we assume that the DNA level cost and the protein level cost are combined by addition but other combination functions $f : \mathbb{R} \times \mathbb{R} \to \mathbb{R}$ could of course also be considered.

The cost of a sequence E of evolutionary events $e_1, e_2, \ldots, e_k$ transforming $a^{(0)}$ to $a^{(k)}$ as $a^{(0)} \stackrel{e_1}{\to} a^{(1)} \stackrel{e_2}{\to} a^{(2)} \stackrel{e_3}{\to} \cdots \stackrel{e_k}{\to} a^{(k)}$ is defined as some function of the costs of each event. In the rest of this paper we will assume that this function is the sum of the costs of each event.

$$cost(a^{(0)} \stackrel{E}{\to} a^{(k)}) = \sum_{i=1}^{k} cost(a^{(i-1)} \stackrel{e_i}{\to} a^{(i)}) \tag{2}$$

We define the distance between two coding sequences of DNA a and b according to the parsimony principle as the minimum cost of a sequence of evolutionary events which transforms a to b.

$$dist(a, b) = \min\{cost(a \stackrel{E}{\to} b) \mid E \text{ is a sequence of events}\} \tag{3}$$

In order to compute $dist(a, b)$ we have to specify the set of allowed evolutionary events and define the cost of each event on the DNA level as well as on the protein level. The choice of evolutionary events and cost function influences both the biological relevance of the distance measure and the computational complexity of computing the distance.

2.2 The specific model

The DNA/protein model introduced in [4] can be described as an instance of the general model where the evolution of coding DNA is idealized to involve only substitutions of single nucleotide and insertion-deletions of a multiple of three consecutive nucleotides. The DNA level cost of an event is defined in the classical way by specifying a substitution cost and a gap cost. The protein level cost of an event that changes the encoded protein from A to A' is defined to reflect the difference between protein A and protein A'.

- The DNA level cost $cost_d(a \stackrel{e}{\to} a')$ depends on e. The cost of substituting a nucleotide σ with σ' is $c_d(\sigma, \sigma')$ for some metric c_d on nucleotides. The cost of inserting or deleting $3k$ consecutive nucleotides is $g_d(3k)$ for some subadditive[1] function $g_d : \mathbb{N} \to \mathbb{R}^+$.
- The protein level cost $cost_p(A \stackrel{e}{\to} A')$ is defined as the distance $dist_p(A, A')$ between A and A'. The distance $dist_p(A, A')$ is the minimum cost of a distance alignment of A and A' where we allow substitution of a single amino acid and insertion-deletion of consecutive amino acids. The substitution cost is given by a metric c_p on amino acids and the gap cost is given by a subadditive function $g_p : \mathbb{N} \to \mathbb{R}^+$. Additional restrictions will be given in Sect. 2.3.

[1] A function is subadditive if $f(i+j) \leq f(i) + f(j)$. A subadditive gap cost function implies that an insertion-deletion of a consecutive block of nucleotides is best explained as a single event.

The reason why gap lengths are restricted to a multiple of three is that an insertion or deletion of length not divisible by three changes the reading frame. This is called a frame shift and it may change the entire remaining amino acid sequence as illustrated in Fig. 1. Frame shifts are believed to be rare biological events, so it is not unreasonable to leave them out of the model.

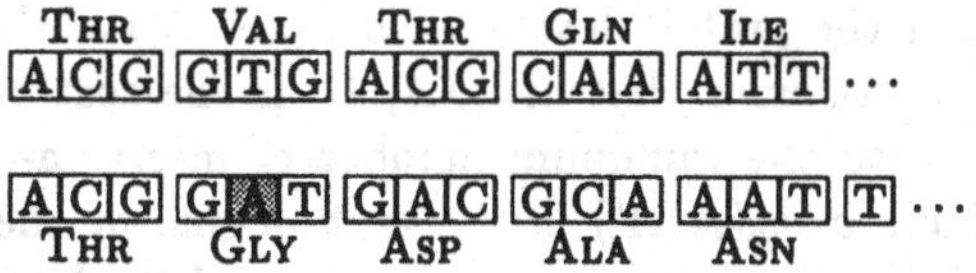

THR VAL THR GLN ILE
ACG GTG ACG CAA ATT ...

ACG GAT GAC GCA AAT T ...
THR GLY ASP ALA ASN

Fig. 1. An insertion-deletion of length not divisible by three changes the reading frame.

Except for the restriction on insertion-deletion length the DNA/protein model allows the traditional set of symbol based nucleotide events. This allows us to use the notion of an alignment. An alignment of two sequences describes a set of substitution or insertion-deletion events necessary to transform one of the sequences into the other sequence. The set of events is usually described by a matrix or a path in a graph as illustrated in Fig. 2. The cost of an alignment is the optimal cost of any sequence of the events described by the alignment. Hence, the evolutionary distance $dist(a, b)$ in the DNA/protein model between two coding DNA sequences a and b is the cost of an optimal alignment of a and b in the DNA/protein model. In the rest of this paper we will address the problem of computing the cost of an optimal alignment in the DNA/protein model.

$$\begin{bmatrix} T\,T\,G\,C\,T\,-\,-\,-\,C \\ T\,-\,-\,-\,C\,A\,T\,G\,C \end{bmatrix}$$

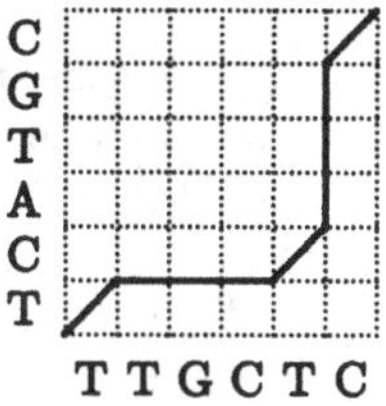

C
G
T
A
C
T
T T G C T C

Fig. 2. An alignment can be described by a matrix or a path in the alignment graph. The above alignment describes three matches and two gaps of combined length six.

If the cost of any sequence of events is independent of the order but only depends on the set of events, then an optimal alignment can be computed efficiently using dynamic programming [9, 11–13]. In the DNA/protein model the cost of an event is the sum of the DNA level cost and the protein level cost. We observe that the DNA level cost of a sequence of events is independent of the order but that the protein level cost is not [4, Fig. 2]. This implies that we cannot use a classical alignment algorithm to compute an optimal alignment in the DNA/protein model. In order to formulate an efficient alignment algorithm in the DNA/protein model we must examine the protein level cost further.

2.3 Restrictions on the cost function

A single nucleotide event affects nucleotides in one or more consecutive codons. Since a nucleotide event in the DNA/protein model cannot change the reading frame then only the amino acids encoded by the affected codons are affected by the nucleotide event. A nucleotide event thus changes protein $A = UXV$ to protein $A' = UX'V$ where X and X' are the amino acids affected by nucleotide event. Let us consider the protein level cost $dist_p(A, A')$ of the event.

Hein [4] implicitly assumes that $dist_p(A, A')$ is the cost of a distance alignment of X and X' with the minimum number of insertion-deletions. This assumption implies that $dist_p(A, A')$ is the cost of one of the alignments of A and A' shown in Fig. 3. This assumption is essential to the formulation of our alignment algorithms in Sect. 4 and 5.

If the cost of alignment of A and A' implied by Fig. 3 is not minimal then the assumption conflicts with our previous definition of $dist_p(A, A')$ as being the minimum cost of an alignment of A and A'. Lemma 1 states restrictions on c_p and g_p that prevent this conflict. If we assume an affine gap cost function $g_p(k) = \alpha_p + \beta_p k$ for some $\alpha_p, \beta_p \geq 0$ (and define $g_p(0)$ to be zero), then the restrictions on c_p and g_p becomes $c_p(\sigma, \tau) + \alpha_p + \beta_p k \leq 2\alpha_p + \beta_p(k + 2l)$ for any amino acids σ, τ and all lengths $0 < l \leq n - k$. This simplifies to $c_p(\sigma, \tau) \leq \alpha_p + 2\beta_p$ for all amino acids σ, τ which is a biological reasonable restriction as insertions and deletions are rare events compared to substitutions.

$$\begin{bmatrix} A_1 \ A_2 \cdots A_{i-1} & A_i & A_{i+1} \cdots A_n \\ A_1 \ A_2 \cdots A_{i-1} & A'_i & A_{i+1} \cdots A_n \end{bmatrix}$$

(a) A substitution in the ith codon. The cost is $c_p(A_i, A'_i)$.

$$\begin{bmatrix} A_1 \ A_2 \cdots A_{i-1} \ A_i & A_{i+1} \cdots A_{i+k} & A_{i+k+1} \cdots A_n \\ A_1 \ A_2 \cdots A_{i-1} \ A_i & - \cdots - & A_{i+k+1} \cdots A_n \end{bmatrix}$$

(b) An insertion-deletion of $3k$ nucleotides affecting exactly k codons. The cost is $g_p(k)$.

$$\begin{bmatrix} A_1 \ A_2 \cdots A_{i-1} & A_i \cdots A_{j-1} \ A_j \ A_{j+1} \cdots A_{i+k} & A_{i+k+1} \cdots A_n \\ A_1 \ A_2 \cdots A_{i-1} & - \cdots - \ \ v \ \ - \cdots - & A_{i+k+1} \cdots A_n \end{bmatrix}$$

(c) An insertion-deletion of $3k$ nucleotides affecting $k+1$ codons. The remaining amino acid v is matched with one of the amino acids affected by the deletion. The cost is $\min_{j=0,1,\ldots,k} \{g_p(j) + c_p(A_{i+j}, v) + g_p(k-j)\}$.

Fig. 3. The protein level cost of a nucleotide event can be determined by considering only the amino acids affected by the event.

Lemma 1. *Assume a nucleotide event changes $A = UXV$ to $A' = UX'V$. Let $n = |A|$ and $k = ||A| - |A'||$. If there for any amino acids σ, τ and for all $0 < l \leq n-k$ exists $0 \leq j \leq k$ such that $c_p(\sigma,\tau)+g_p(j)+g_p(k-j) \leq g_p(l)+g_p(l+k)$, then $dist_p(A, A')$ is the cost of an alignment describing exactly k insertions or deletions. Furthermore $dist_p(A, A')$ only depends on X and X'.*

Proof. We note that the alignments in Fig. 3 all describe the minimum number of insertion-deletions and that only the sub-alignment of X and X', as illustrated by the shaded parts, contributes to the cost. We will argue that the assumption on c_p and g_p stated in the lemma ensures that $dist_p(A, A')$ is the cost of one of the alignments in Fig. 3. We split the argumentation depending on the event. Since $dist_p(A, A')$ is equal to $dist_p(A', A)$ the cost of an insertion transforming A to A' is equal to the cost of a deletion transforming A' to A. We thus only consider substitutions and deletions.

A substitution of a nucleotide in the ith codon of A transforms A_i to A'_i. The alignment in Fig. 3(a) describes no insertion-deletions and has cost $c_p(A_i, A'_i)$. Any other alignment of A and A' must describe an equal number of insertions and deletions, so by subadditivity of g_p the cost is at least $2g_p(l)$ for some $0 < l \leq n$. The assumption in the lemma implies that $c_p(A_i, A'_i) \leq 2g_p(l)$ for any $0 < l \leq n$. The alignment in Fig. 3(a) is thus optimal and the protein level cost of the substitution is $c_p(A_i, A'_i)$.

A deletion of $3k$ nucleotides affecting k codons transforms $A = A_1A_2 \cdots A_n$ to $A' = A_1A_2 \cdots A_iA_{i+k+1}A_{i+k+2} \cdots A_n$. Any alignment of A and A' must describe l insertions and $l+k$ deletions for some $0 \leq l \leq n-k$, so the cost is at least $g_p(l) + g_p(l+k)$. The alignment in Fig. 3(b) describes k deletions and has cost $g_p(k)$. The assumption in the lemma and the sub-additivity of g_p implies that $g_p(k) \leq g_p(j) + g_p(k-j) \leq g_p(l) + g_p(l+k)$ for all $l > 0$. The alignment in Fig. 3(b) is thus optimal and the protein level cost of the deletion is $g_p(k)$.

A deletion of $3k$ nucleotides affecting $k+1$ codons, say a deletion of the $3k$ nucleotides $a_3^i a_1^{i+1} a_2^{i+1} a_3^{i+1} \cdots a_1^{i+k} a_2^{i+k}$, transforms $A = A_1A_2 \cdots A_n$ to $A' = A_1A_2 \cdots A_{i-1}vA_{i+k+1} \cdots A_n$ where v is the amino acid coded by $a_1^i a_2^i a_3^{i+k}$. We say that v is the remaining amino acid and $a_1^i a_2^i a_3^{i+k}$ is the remaining codon. Any alignment of A and A' describing exactly k deletions must align v with A_{i+j} for some $0 \leq j \leq k$, so by subadditivity of g_p the cost is at least $g_p(j)+c_p(A_{i+j}, v)+g_p(k-j)$. The alignment in Fig. 3(c) illustrates one of the $k+1$ alignments of A and A' where v is aligned with an affected amino acid and all non-affected amino acids are aligned. Such an alignment describes exactly k deletions and the cost of the optimal alignment among them has cost

$$\min_{j=0,1,\ldots,k} \{g_p(j) + c_p(A_{i+j}, v) + g_p(k-j)\}, \tag{4}$$

and is thus optimal for any alignment describing exactly k deletions. Any other alignment of A and A' must describe l insertions and $l+k$ deletions for some $0 < l \leq n-k$, so the cost is at least $g_p(l) + g_p(l+k)$. The assumption in the lemma implies that the cost given by (4) is less than or equal to $g_p(l)+g_p(l+k)$. The protein level cost of the deletion is thus given by (4). □

The assumption in Lemma 1 is sufficient to ensure that we can compute the protein level cost of a nucleotide event efficiently, but the formulation of the lemma is to general to make the assumption necessary. The following example however suggests when the assumption is necessary. Consider a deletion of three nucleotides that transforms the six amino acids ABEFCD to ABGCD, i.e. $X =$ EF and $X' =$ G. Consider the two alignments shown in Fig. 4.

$$\begin{bmatrix} \mathrm{A\,B\,E\,F\,C\,D} \\ \mathrm{A\,B\,G} - \mathrm{C\,D} \end{bmatrix} \qquad \begin{bmatrix} \mathrm{A\,B\,E\,F} - \mathrm{C\,D} \\ \mathrm{A\,B} - - \mathrm{G\,C\,D} \end{bmatrix}$$

Fig. 4. Two alignments of the amino acids ABEFCD and ABGCD.

If we assume that $c_p(\mathrm{E}, \mathrm{G}) \leq c_p(\mathrm{F}, \mathrm{G})$ then the cost of the alignment in Fig. 4 (left) is $c_p(\mathrm{E}, \mathrm{G}) + g_p(1)$ while the cost of the alignment in Fig. 4 (right) is $g_p(2)+g_p(1)$. If the assumption in lemma 1 does not hold then $g_p(2)+g_p(1)$ might be less than $c_p(\mathrm{E}, \mathrm{G}) + g_p(1)$ because $c_p(\mathrm{E}, \mathrm{G})$ can be arbitrary large. Hence, the protein level cost of the deletion would not be the cost of an alignment describing the minimum number of insertion-deletions.

3 The cost of an alignment

Before we can describe how to compute the cost of an optimal alignment of two sequence in the DNA/protein model, we need to know how to compute the cost of a given alignment in the model.

An alignment of two sequences describes a set of events necessary to transform one of the sequences into the other sequence but it does not describe the order of the events. As observed in Sect. 2.2 the DNA level cost of an alignment is independent of the order of the events, while the protein level cost of an alignment depends on the order of the events. This implies that the DNA level cost of an alignment is just the sum of the DNA level cost of the events described by the alignment, while the protein level cost of the same alignment is somewhat harder to determine. An obvious way to determine the protein level cost of an alignment is to minimize over all possible sequences of the events described by the alignment. This method is however not feasible in practice due to the factorial number of possible sequences one has to consider.

If Lemma 1 is fulfilled then we know that the protein level cost of a nucleotide event only depends on the affected codons. We can use this property to decompose the computation of the protein level cost of an alignment into smaller subproblems. The idea is to decompose the alignment into *codon alignments*. A codon alignment is a minimal part of the alignment that corresponds to a path connecting two nodes $(3i', 3j')$ and $(3i, 3j)$ in the alignment graph. We can decompose an alignment uniquely into codon alignments as illustrated in Fig. 5.

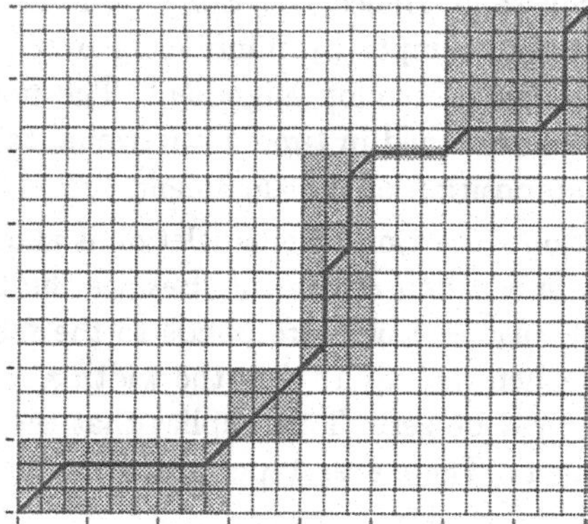

Fig. 5. An alignment of two sequences decomposed into codon alignments.

The assumption that the length of an insertion or deletion is a multiple of three implies that a codon alignment either describes no substitutions (see Type 2 and 3 in Fig. 7) or exactly three substitutions. If a codon alignment describes exactly three substitutions then it can also describe one or more insertion-deletions. More precisely, between any two consecutive substitutions in the codon alignment there can be an alternating sequence of insertions and deletions each with length a multiple of three. Such a sequence of insertion-deletions corresponds to a "staircase" in the alignment graph. The set of codon alignments that describe three substitutions and at most one insertion and deletion between two consecutive substitutions, i.e. the "staircase" is limited to at most one "step", is illustrated in Fig. 6. A particular codon alignment in this set corresponds to a choice of which sides of the two rectangles to traverse.

Now consider the decomposition of an alignment into codon alignments (this could be as illustrated in Fig. 5). We observe that nucleotide events described by two different codon alignments in the decomposition do not affect the same codons. Hence, the protein level cost of a codon alignment can be computed independently of the other codon alignments in the decomposition. We can thus compute the protein level cost of an alignment as the sum of the protein level cost of each of the codon alignments in the decomposition.

Since a codon alignment can describe an alternating sequence of insertion-deletions between two consecutive substitutions it is possible that a decomposition of an alignment of two sequences of length n and m contains codon alignments describing $\Theta(n + m)$ events. This implies that the problem of computing the cost of the codon alignments in a decompsition of an alignment is, in the worst case, not any easier than computing the cost of the alignment itself. One way to circumvent this problem is to only consider alignments that can be decomposed into (or built of) codon alignments with at most some maximum number of insertion-deletions between two consecutives substitutions. This upper bounds the number of events described by any codon alignment by some constant. Hence, we can determine the cost of a codon alignment in constant time simply by minimizing over all possible sequences of the events described by the codon alignment. The protein level cost of an alignment can thus be determined in time proportional to the number of codon alignments in the decomposition of the alignment.

In Hein [4] at most one insertion *or* one deletion is allowed between two consecutive substitutions in a codon alignment. Besides the two codon alignments that describe no substitutions, this corresponds to the set of codon alignments we obtain from Fig. 6 when we require that the width and/or the height of each of the two rectangles must be zero. This implies that there are eleven different types of codon alignments. The eleven types are shown in Fig. 7.

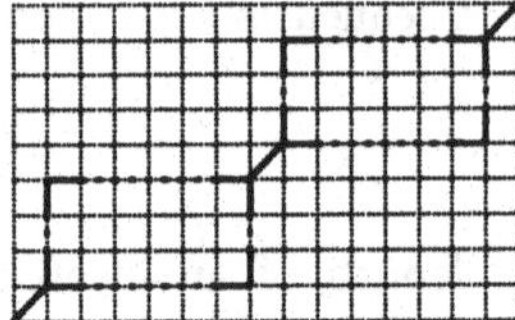

Fig. 6. A summary of all possible codon alignments with at most one insertion and one deletion between two consecutive substitutions.

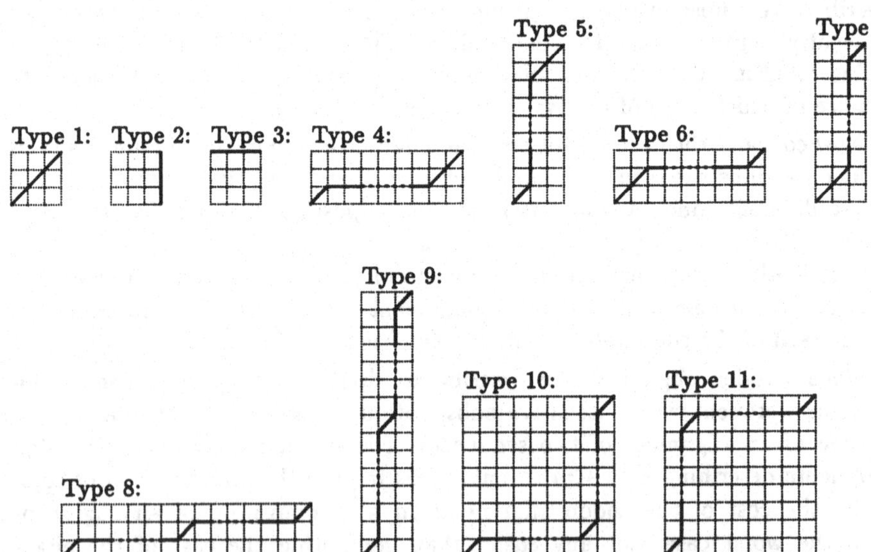

Fig. 7. The eleven types of codon alignments with at most one inseriton or one deletion between two consecutive substitutions.

4 A simple alignment algorithm

Let $a_1a_2\cdots a_{3n}$ and $b_1b_2\cdots b_{3m}$ be two coding sequences of DNA. Hein [4] describes how the decomposition into codon alignments makes it possible to compute the cost of an optimal alignment of a and b in the DNA/protein model in time $O(n^2m^2)$. The algorithm assumes that Lemma 1 is fulfilled and that we only allow codon alignments with some maximum number of insertion-deletions between two consecutive substitutions, e.g. the eleven types of codon alignments in Fig. 7. The algorithm can be summarized as follows.

Let $D(i,j)$ denote the cost of an optimal alignment of $a_1a_2\cdots a_{3i}$ and $b_1b_2\cdots b_{3j}$. We define $D(0,0)$ to be zero and $D(i,j)$ to be infinity for $i<0$ or $j<0$. An optimal alignment of $a_1a_2\cdots a_{3i}$ and $b_1b_2\cdots b_{3j}$ can be decomposed into codon alignments $ca_1, ca_2, \ldots, ca_k$. We say that ca_k is the last codon alignment and that $ca_1, ca_2, \ldots, ca_{k-1}$ is the remaining alignment.

If the last codon alignment ca_k is an alignment of $a_{3i'+1}a_{3i'+2}\cdots a_{3i}$ and $b_{3j'+1}b_{3j'+2}\cdots b_{3j}$ for some $(i',j')<(i,j)^2$, then $D(i,j)$ is equal to $D(i',j')+cost(ca_k)$. This is the cost of the last codon alignment plus the cost of the remaining alignment. We can compute $D(i,j)$ by minimizing the expression $D(i',j')+cost(ca)$ over all $(i',j')<(i,j)$ and all possible codon alignments ca of $a_{3i'+1}a_{3i'+2}\cdots a_{3i}$ and $b_{3j'+1}b_{3j'+2}\cdots b_{3j}$.

The upper bound on the number of insertion-deletions in a codon alignment implies that the number of possible codon alignments of $a_{3i'+1}a_{3i'+2}\cdots a_{3i}$ and $b_{3j'+1}b_{3j'+2}\cdots b_{3j}$, for all $(i',j')<(i,j)$, is upper bounded by some constant. For example, if we only consider the eleven types of codon alignments in Fig. 7 then there are at most three possible codon alignments of $a_{3i'+1}a_{3i'+2}\cdots a_{3i}$ and $b_{3j'+1}b_{3j'+2}\cdots b_{3j}$. Hence, if we assume that $D(i',j')$ is known for all $(i',j')<(i,j)$ then we can compute $D(i,j)$ in time $O(ij)$. By dynamic programming this implies that we can compute $D(n,m)$ in time $O(n^2m^2)$ and space $O(nm)$. By back-tracking we can also get the optimal alignment (and not only the cost) within the same time and space bound.

5 An improved alignment algorithm

Let a and b be coding sequences of DNA as introduced in the previous section. We will describe how to compute the cost of an optimal alignment of a and b in the DNA/protein model in time $O(nm)$ and space $O(n)$. Besides the assumptions of the simple algorithm described in the previous section we also assume that the function $g(k)=g_d(3k)+g_p(k)$ is affine $\alpha+\beta k$ for some $\alpha,\beta\geq 0$. We say that g is the *combined gap cost function*.

The idea behind the improved algorithm is similar to the idea behind the simple algorithm in the sense that we compute the cost of an optimal alignment by minimizing the cost over all possible last codon alignments. We define $D^t(i,j)$ to be the cost of an optimal alignment of $a_1a_2\cdots a_{3i}$ and $b_1b_2\cdots b_{3j}$ under the

[2] We say that $(i',j')<(i,j)$ iff $i'\leq i \wedge j'\leq j \wedge (i'\neq i \vee j'\neq j)$.

assumption that the last codon alignment is of type t. Remember that we only allow codon alignments with some maximum number of insertion-deletions between two consecutive substitutions. This implies that the number of possible codon alignment, i.e. types of codon alignments, is upper bounded by some constant. We define $D^t(0,0)$ to be zero and $D^t(i,j)$ to be infinity for $i < 0$ or $j < 0$. We can compute $D(i,j)$ as

$$D(i,j) = \min_t D^t(i,j). \tag{5}$$

Lemma 1 ensures that $D^t(i,j)$ is the cost of some last codon alignment (of type t) plus the cost of the corresponding remaining alignment. This allows us to compute $D^t(i,j)$ by minimizing the cost over all possible last codon alignments of type t. The assumption that g is affine makes it possible to compute $D^t(i,j)$ in constant time if $D^t(k,l)$, for all t, is known for some $(k,l) < (i,j)$. Since we only consider a constant number of possible codon alignments (e.g. the types in Fig. 7) this implies that we can compute $D(n,m)$ in time $O(nm)$ and space $O(n)$. By adapting the technique in Hirschberg [7] or the variant described in Durbin et al. [2] we can also get the optimal alignment (and not only the cost) within the same time and space bound.

The method we use to compute $D^t(i,j)$ can be used for any type of last codon alignment but the bookkeeping and thereby the constant overhead increases with the number of gaps described by the last codon alignment. By restricting ourselves to the eleven types of codon alignments shown in Fig. 7 we can compute the cost of an optimal alignment with a reasonable constant overhead. In the rest of this paper we therefore focus on these eleven types of codon alignments. We divide the explanation of how to compute $D^t(i,j)$ for $t = 1, 2, \ldots, 11$ according to the number of gaps within a codon (denoted *internal gaps*) described by a codon alignment of type t. Codon alignments of type 1–3 describe no internal gaps, codon alignments of type 4–7 describe one internal gap and codon alignments of type 8–11 describe two internal gaps.

For use in the explanation we introduce $c_p^* : \{A, C, G, T\}^3 \times \{A, C, G, T\}^3 \to \mathbb{R}$. We define $c_p^*(\sigma_1\sigma_2\sigma_3, \tau_1\tau_2\tau_3)$ as the distance between $\sigma_1\sigma_2\sigma_3$ and $\tau_1\tau_2\tau_3$ in the DNA/protein model. This is the minimum over the cost[3] of the six possible sequences of the three substitutions $\sigma_1 \to \tau_1$, $\sigma_2 \to \tau_2$ and $\sigma_3 \to \tau_3$.

5.1 Codon alignments with no internal gaps

The cost $D^1(i,j)$ is the cost of the last codon alignment of type 1 plus the cost of the remaining alignment. The last codon alignment is an alignment of $a_1^i a_2^i a_3^i$ and $b_1^j b_2^j b_3^j$. By definition of c_p^* the cost is $c_p^*(a_1^i a_2^i a_3^i, b_1^j b_2^j b_3^j)$. The cost of the remaining alignment is $D(i-1, j-1)$.

$$D^1(i,j) = D(i-1, j-1) + c_p^*(a_1^i a_2^i a_3^i, b_1^j b_2^j b_3^j) \tag{6}$$

[3] We use the term *cost* to denote the DNA level cost plus the protein level cost.

A codon alignment of type 2 or type 3 describes a gap between codons. Since the combined gap cost function is affine we can use the technique introduced in [3] saying that a gap ending in (i, j) is either a continuation of an existing gap ending in $(i-1, j)$ or $(i, j-1)$, or the start of a new gap.

$$D^2(i,j) = \min\{D(i,j-1) + \alpha + \beta, D^2(i,j-1) + \beta\} \tag{7}$$

$$D^3(i,j) = \min\{D(i-1,j) + \alpha + \beta, D^3(i-1,j) + \beta\} \tag{8}$$

5.2 Codon alignments with one internal gap

We describe how to compute $D^6(i,j)$. The other cases where the last codon alignment describes one internal gap are handled similarly. The cost $D^6(i,j)$ is the cost of the last codon alignment of type 6 plus the cost of the remaining alignment. The last codon alignment of type 6 describes three substitutions and one deletion. If the deletion has length k (a deletion of $3k$ nucleotides) then the cost of the remaining alignment is $D(i-k-1, j-1)$ and the last codon alignment is an alignment of $a_1^{i'} a_2^{i'} a_3^{i'} \cdots a_1^i a_2^i a_3^i$ and $b_1^j b_2^j b_3^j$ where $i' = i - k$. This is illustrated in Fig. 8.

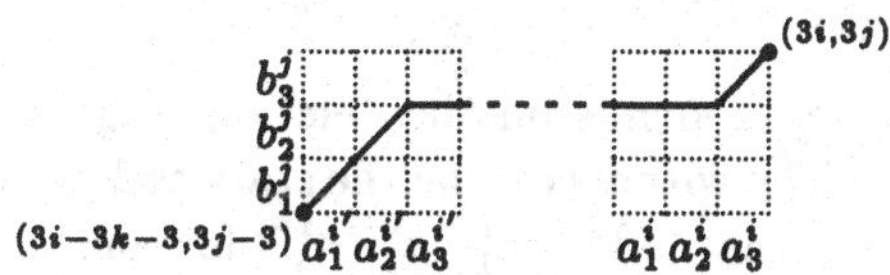

Fig. 8. The last codon alignment of type 6.

The cost of the last codon alignment is the minimum cost of a sequence of the three substitutions and the deletion. Any sequence of these four events can be divided into three steps: The substitutions occurring before the deletion, the deletion and the substitutions occurring after the deletion. Figure 9 illustrates the three steps of the evolution of $a_1^{i'} a_2^{i'} a_3^{i'} \cdots a_1^i a_2^i a_3^i$ to $b_1^j b_2^j b_3^j$. The nucleotides x_1, x_2 and x_3 are the result of the up to three substitutions before the deletion. For example, if the substitution $a_1^{i'} \to b_1^j$ occurs before the deletion, then x_1 is b_1^j, otherwise it is $a_1^{i'}$. We say that $x_1 \in \{a_1^{i'}, b_1^j\}$, $x_2 \in \{a_2^{i'}, b_2^j\}$ and $x_3 \in \{a_3^i, b_3^j\}$ are the status of the three substitutions before the deletion. We observe that $x_1 x_2 x_3$ is the remaining codon of the deletion.

The substitutions occurring before the deletion change codon $a_1^{i'} a_2^{i'} a_3^{i'}$ to $x_1 x_2 a_3^{i'}$ and codon $a_1^i a_2^i a_3^i$ to $a_1^i a_2^i x_3$. The substitutions occurring after the deletion change codon $x_1 x_2 x_3$ to $b_1^j b_2^j b_3^j$. We recall that the cost of changing codon $\sigma_1 \sigma_2 \sigma_3$ to codon $\tau_1 \tau_2 \tau_3$ by a sequence of the substitutions $\sigma_1 \to \tau_1$, $\sigma_2 \to \tau_2$ and $\sigma_3 \to \tau_3$ is $c_p^*(\sigma_1 \sigma_2 \sigma_3, \tau_1 \tau_2 \tau_3)$. Since an identical substitution has cost zero then the cost of the three substitutions in the last codon alignment is equal to the

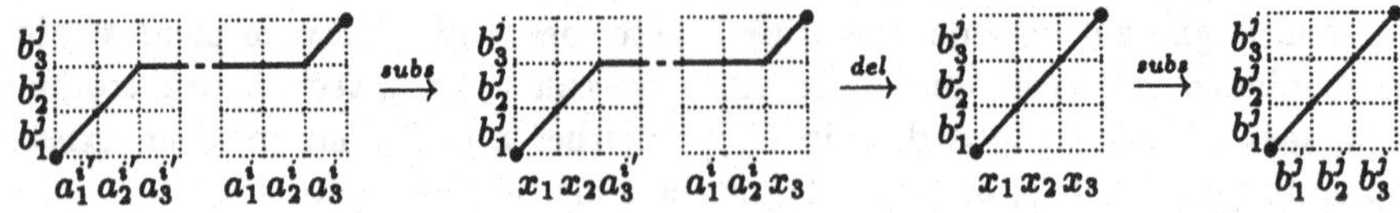

Fig. 9. The evolution of $a_1^{i'} a_2^{i'} a_3^{i'} \cdots a_1^i a_2^i a_3^i$ to $b_1^j b_2^j b_3^j$ described by the last codon alignment.

cost of the induced codon changes. The cost is

$$cost(subs) = c_p^*(a_1^{i'} a_2^{i'} a_3^{i'}, x_1 x_2 a_3^{i'}) + c_p^*(a_1^i a_2^i a_3^i, a_1^i a_2^i x_3) + c_p^*(x_1 x_2 x_3, b_1^j b_2^j b_3^j). \quad (9)$$

The cost of the deletion of $3k$ nucleotides in the last codon alignment is the sum of the DNA level cost $g_d(3k)$ and the protein level cost as given by (4). By using the combined gap cost function $g(k) = g_d(3k) + g_p(k) = \alpha + \beta k$ and our knowledge of the remaining codon $x_1 x_2 x_3$ of the deletion, we can formulate this sum as

$$cost(del) = \min \begin{cases} \alpha + \beta k + c_p(a_1^i a_2^i x_3, x_1 x_2 x_3)^4 \\ \alpha_p + \alpha + \beta k + \min_{0<l<k} c_p(a_1^{i-l} a_2^{i-l} a_3^{i-l}, x_1 x_2 x_3) \\ \alpha + \beta k + c_p(x_1 x_2 a_3^{i'}, x_1 x_2 x_3) \end{cases} . \quad (10)$$

The cost of the deletion depends on the deletion length, the remaining codon $x_1 x_2 x_3$ and a witness. The witness can be the start-codon $x_1 x_2 a_3^{i'}$, the end-codon $a_1^i a_2^i x_3$ or one of the internal codons $a_1^{i-l} a_2^{i-l} a_3^{i-l}$ for some $0 < l < k$. The witness encodes the amino acid aligned with the remaining amino acid.

We can now compute $D^6(i,j)$ under the assumption of a certain deletion length k and remaining codon $x_1 x_2 x_3$ of the deletion in the last codon alignment as the sum $cost(subs) + cost(del) + D(i-k-1, j-1)$. We can thus compute $D^6(i,j)$ by minimizing this sum over all possible combinations of deletion length k and remaining codon $x_1 x_2 x_3$. A combination of deletion length k and remaining codon $x_1 x_2 x_3$ is possible if $x_1 \in \{a_1^{i'}, b_1^j\}$, $x_2 \in \{a_2^{i'}, b_2^j\}$ and $x_3 \in \{a_3^i, b_3^j\}$ where $i' = i - k$. The terms $c_p^*(a_1^i a_2^i a_3^i, a_1^i a_2^i x_3)$ and $c_p^*(x_1 x_2 x_3, b_1^j b_2^j b_3^j)$ of $cost(subs)$ do not depend on the deletion length, so we can split the minimization as

$$D^6(i,j) = \min_{x_1 x_2 x_3} \{c_p^*(a_1^i a_2^i a_3^i, a_1^i a_2^i x_3) + c_p^*(x_1 x_2 x_3, b_1^j b_2^j b_3^j) + D^6_{x_1 x_2 x_3}(i,j)\} \quad (11)$$

where

$$D^6_{x_1 x_2 x_3}(i,j) = \min_{0<k<i} \{D(i-k-1, j-1) + c_p^*(a_1^{i-k} a_2^{i-k} a_3^{i-k}, x_1 x_2 a_3^{i-k}) + cost(del)\} \quad (12)$$

[4] We use $c_p(\sigma_1 \sigma_2 \sigma_3, \tau_1 \tau_2 \tau_3)$ as a convenient notation for $c_p(\sigma, \tau)$ where σ and τ are the amino acids coded by the codons $\sigma_1 \sigma_2 \sigma_3$ and $\tau_1 \tau_2 \tau_3$ respectively.

is the minimum cost of the terms that depend on both the deletion length and the remaining codon under the assumption that the remaining codon is $x_1x_2x_3$. If we expand the term *cost*(*del*) we get

$$D^6_{x_1x_2x_3}(i,j) = \min_{0<k<i}\{\text{len}^6_{x_1x_2}(i,j,k) + \min\left\{\begin{array}{l} c_p(a_1^i a_2^i x_3, x_1x_2x_3) \\ \alpha_p + \min_{0<l<k} c_p(a_1^{i-l} a_2^{i-l} a_3^{i-l}, x_1x_2x_3) \\ c_p(x_1x_2a_3^{i-k}, x_1x_2x_3) \end{array}\right\}\} \quad (13)$$

where

$$\text{len}^6_{x_1x_2}(i,j,k) = D(i-k-1, j-1) + c_p^*(a_1^{i-k} a_2^{i-k} a_3^{i-k}, x_1x_2a_3^{i-k}) + \alpha + \beta k \quad (14)$$

is the cost of the remaining alignment plus the part of the cost of the last codon alignment that does not depend on the codon $a_1^i a_2^i a_3^i$ and the witness. The cost $\text{len}^6_{x_1x_2}(i,j,k)$ is defined if $x_1 \in \{a_1^{i-k}, b_1^j\}$ and $x_2 \in \{a_2^{i-k}, b_2^j\}$. The cost $D^6_{x_1x_2x_3}(i,j)$ is defined if there exists a deletion length k such that k and $x_1x_2x_3$ is a possible combination of deletion length and remaining codon.

We observe that there are at most 32 possible remaining codons $x_1x_2x_3$. The observation follows because we known that x_3 must be one of the two known nucleotides a_3^i or b_3^j. If we can compute $D^6_{x_1x_2x_3}(i,j)$ in constant time for each of the possible remaining codons then we can also compute $D^6(i,j)$ in constant time. To compute $D^6_{x_1x_2x_3}(i,j)$ we must determine a combination of witness and deletion length that minimizes the cost. We say that we must determine the witness and deletion length of $D^6_{x_1x_2x_3}(i,j)$.

It is easy to see that witness and deletion lenghth of $D^6_{x_1x_2x_3}(i,j)$ must be one of the four combinations illustrated in Fig. 10. We can thus compute $D^6_{x_1x_2x_3}(i,j)$ as the minimum over the cost of the four cases illustrated in Fig. 10. The cost of case 1–3 is computed by simplifying (13) for a particular witness and deletion length. The cost of case 4 cannot be computed this way because both the witness and the deletion length are unknown.

Case 1. The end-codon is the witness and the deletion length is at least one. The cost is $\min_{0<k<i} \text{len}^6_{x_1x_2}(i,j,k) + c_p(a_1^i a_2^i x_3, x_1x_2x_3)$.

Case 2. The last internal codon is the witness and the deletion length is at least two. The cost is $\min_{1<k<i} \text{len}^6_{x_1x_2}(i,j,k) + \alpha_p + c_p(a_1^{i-1}a_2^{i-1}a_3^{i-1}, x_1x_2x_3)$.

Case 3. The start-codon is the witness and the deletion length is one. The cost is $\text{len}^6_{x_1x_2}(i,j,1) + c_p(x_1x_2a_3^{i-1}, x_1x_2x_3)$.

Case 4. The witness is neither the end-codon nor the last internal codon and the deletion length is at least two. We observe that if the witness of $D^6_{x_1x_2x_3}(i-1,j)$ is not the end-codon $a_1^{i-1}a_2^{i-1}x_3$ then by optimality of $D^6_{x_1x_2x_3}(i-1,j)$ this witness must also be the witness of case 4. If this is the case then the cost of case 4 is $D^6_{x_1x_2x_3}(i-1,j) + \beta$.

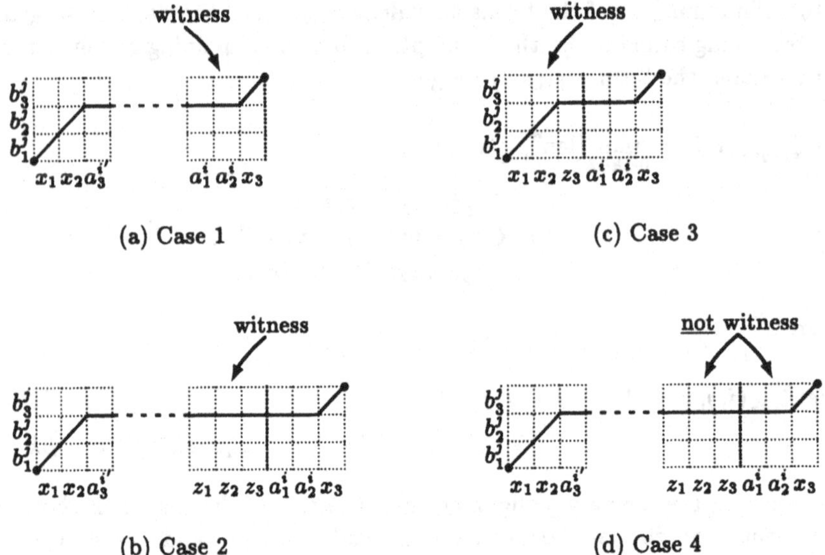

(a) Case 1

(c) Case 3

(b) Case 2

(d) Case 4

Fig. 10. The four cases in the computation of $D^6_{x_1x_2x_3}(i,j)$. We use $z_1z_2z_3$ as notation for $a_1^{i-1}a_2^{i-1}a_3^{i-1}$.

The observation in case 4 suggests that we can use dynamic programming to keep track of $D^6_{x_1x_2x_3}(i,j)$ under the assumption that the end-codon is not the witness, i.e. use dynamic programming to keep track of the minimum cost of case 2–4. We introduce tables $F^6_{x_1x_2x_3}$ corresponding to the 64 combinations of $x_1x_2x_3$. We maintain that if $x_1x_2x_3$ is a possible remaining codon and the end-codon $a_1^i a_2^i x_3$ is not the witness of $D^6_{x_1x_2x_3}(i,j)$, then $F^6_{x_1x_2x_3}(i,j)$ is equal to $D^6_{x_1x_2x_3}(i,j)$. If we define $F^6_{x_1x_2x_3}(0,j)$ to infinity, then

$$F^6_{x_1x_2x_3}(i,j) = \min \begin{cases} \text{cost of } \textit{Case 2} \\ \text{cost of } \textit{Case 3} \\ F^6_{x_1x_2x_3}(i-1,j) + \beta \end{cases} \tag{15}$$

In order to compute the cost of case 1 and 2 in constant time we maintain the minimum of $\text{len}^6_{x_1x_2}(i,j,k)$ over k by dynamic programming. We introduce tables $L^6_{x_1x_2}$ corresponding to the 16 combinations of x_1x_2 such that $L^6_{x_1x_2}(i,j)$ is equal to $\min_{0<k<i} \text{len}^6_{x_1x_2}(i,j,k)$. If we define $L^6_{x_1x_2}(0,j)$ to infinity, then

$$L^6_{x_1x_2}(i,j) = \min \begin{cases} \text{len}^6_{x_1x_2}(i,j,1) \\ L^6_{x_1x_2}(i-1,j) + \beta \end{cases} \tag{16}$$

We can now compute $D^6_{x_1x_2x_3}(i,j)$ in constant time as the minimum cost of case 1–4. The cost of case 1 is $L^6_{x_1x_2}(i,j) + c_p(a_1^i a_2^i x_3, x_1x_2x_3)$ and the minimum cost of case 2–4 is $F^6_{x_1x_2x_3}(i,j)$, so

$$D^6_{x_1x_2x_3}(i,j) = \min \begin{cases} L^6_{x_1x_2}(i,j) + c_p(a^i_1 a^i_2 x_3, x_1x_2x_3) \\ F^6_{x_1x_2x_3}(i,j) \end{cases} \tag{17}$$

The computation of $D^6(i,j)$ by (11) requires us to compute $D^6_{x_1x_2x_3}(i,j)$ for each of the 32 possible remaining codons. To do this we must compute entry (i,j) in the 16 tables $L^6_{x_1x_2}$ and entry (i,j) in the 64 tables $F^6_{x_1x_2x_3}$. As explained in this section all this can be done in constant time.

The other three cases where the last codon alignment describes one internal gap (type 4, 5 and 7) are handled similarly. However, if the last codon alignment is of type 4 or 5, then only the first nucleotide x_1 in the remaining codon depends on the deletion (or insertion) length. This limits the number of possible remaining codons to 16 and implies that only four tables are needed to keep track of $\min_{0<k<i} \text{len}^t_{x_1}(i,j,k)$ for $t = 4,5$. Hence, to compute $D^t(i,j)$ for $t = 4,5,6,7$, we compute $2 \cdot 4 + 2 \cdot 16 + 4 \cdot 64 = 296$ table entries in total.

5.3 Codon alignments with two internal gaps

We describe how to compute $D^8(i,j)$. The other cases where the last codon alignment describes two internal gaps are handled similarly. The cost $D^8(i,j)$ is the cost of the last codon alignment of type 8 plus the cost of the remaining alignment. The last codon alignment of type 8 describes three substitutions and two deletions. If the first deletion has length k' and the second deletion has length k then the cost of the remaining alignment is $D(i-k-k'-1, j-1)$ and the last codon alignment is an alignment of $a^{i''}_1 a^{i''}_2 a^{i''}_3 \cdots a^{i'}_1 a^{i'}_2 a^{i'}_3 \cdots a^i_1 a^i_2 a^i_3$ and $b^j_1 b^j_2 b^j_3$ where $i' = i-k$ and $i'' = i'-k'$. This is illustrated in Fig. 11.

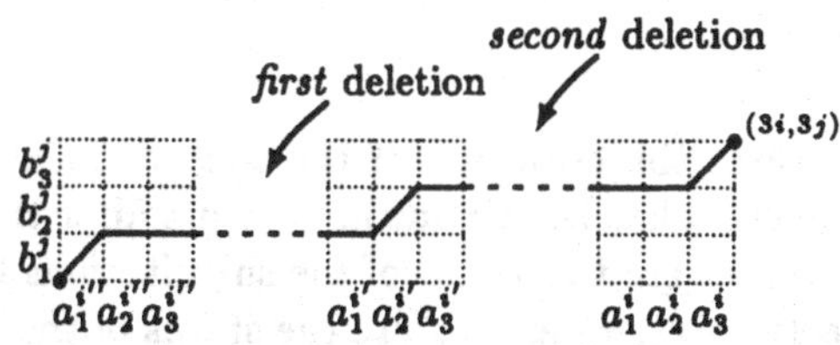

Fig. 11. The last codon alignment of type 8

We will compute $D^8(i,j)$ as we computed $D^6(i,j)$ by minimizing the cost over all possible combinations of deletion length k and remaining codon $x_1x_2x_3$ of the second deletion. This reduces the problem of computing $D^8(i,j)$ to computing $D^8_{x_1x_2x_3}(i,j)$, the cost under the assumption of a certain remaining codon of the second deletion, for each of the 32 possible remaining codons of the second deletion. We can compute $D^8_{x_1x_2x_3}(i,j)$ similar to the way we computed $D^6_{x_1x_2x_3}(i,j)$. An inspection of (15), (16) and (17) reveals that all we essentially have to do is to replace $\text{len}^6_{x_1x_2}(i,j,1)$ with the corresponding part of $D^8(i,j)$. We denote this part of the cost $\text{len}^8_{x_1x_2x_3}(i,j,1)$.

More precisely. If we assume that the second deletion has length k and remaining codon $x_1x_2x_3$ then $\text{len}^8_{x_1x_2x_3}(i,j,k)$ is the part of $D^8(i,j)$ that does not depend on the codon $a_1^i a_2^i a_3^i$ and the witness of the second deletion. This cost depends on the order of the two deletions in the last codon alignment. Hence, we introduce $\text{len}^{8'}_{x_1x_2x_3}(i,j,k)$ and $\text{len}^{8''}_{x_1x_2x_3}(i,j,k)$ to denote the cost when the first deletion occurs before the second deletion and vice versa. We define $\text{len}^8_{x_1x_2x_3}(i,j,k)$ as $\min\{\text{len}^{8'}_{x_1x_2x_3}(i,j,k), \text{len}^{8''}_{x_1x_2x_3}(i,j,k)\}$. Since we only have to compute $\text{len}^8_{x_1x_2x_3}(i,j,1)$ we can restrict ourselves to the case where the second deletion has length one and the first deletion has length k'. In the rest of this section we use the notation that $i' = i-1$ and $i'' = i' - k'$. We split into two cases depending on the order of the first and second deletion.

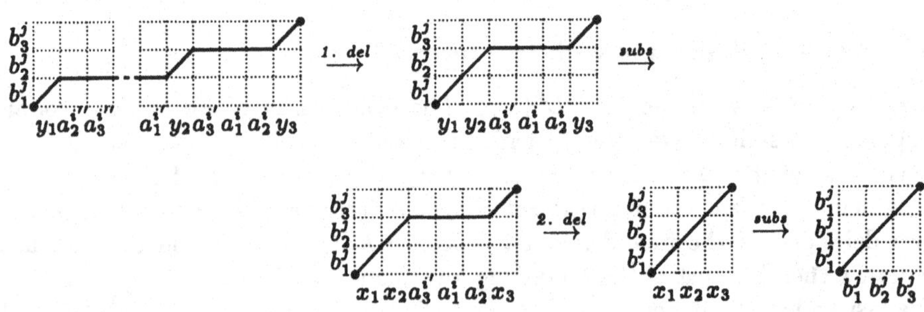

Fig. 12. The first deletion occurs before the second deletion and the second deletion has length one.

Figure 12 illustrates the evolution of the last codon alignment (of type 8) when the second deletion has length one and occurs after the first deletion. The nucleotides y_1, y_2 and y_3 are the status of the substitutions before the first deletion and the nucleotides x_1, x_2 and x_3 are the status of the substitutions before the second deletion. Similar to (9) we compute the cost of the three substitutions in the last codon alignment as the cost of the induced codon changes. An inspection of Fig. 12 reveals that the cost of the three substitutions is

$$\begin{aligned} cost(subs) = {} & c_p^*(a_1^{i''} a_2^{i''} a_3^{i''}, y_1 a_2^{i''} a_3^{i''}) + c_p^*(a_1^{i'} a_2^{i'} a_3^{i'}, a_1^{i'} y_2 a_3^{i'}) + \\ & c_p^*(a_1^i a_2^i a_3^i, a_1^i a_2^i y_3) + c_p^*(y_1 y_2 a_3^{i'}, x_1 x_2 a_3^{i'}) + \\ & c_p^*(a_1^i a_2^i y_3, a_1^i a_2^i x_3) + c_p^*(x_1 x_2 x_3, b_1^j b_2^j b_3^j). \quad (18) \end{aligned}$$

We can compute the cost of the two deletions similar to (10). Remember that the first deletion has length k' and the second deletion has length one and that we use the notation $i' = i - 1$ and $i'' = i' - k'$.

$$cost(del_1) = \min \begin{cases} \alpha + \beta k' + c_p(a_1^{i'} y_2 a_3^{i'}, y_1 y_2 a_3^{i'}) \\ \alpha_p + \alpha + \beta k' + \min_{0<l<k'} c_p(a_1^{i'-l} a_2^{i'-l} a_3^{i'-l}, y_1 y_2 a_3^{i'}) \\ \alpha + \beta k' + c_p(y_1 a_2^{i''} a_3^{i''}, y_1 y_2 a_3^{i'}) \end{cases} \tag{19}$$

$$cost(del_2) = \alpha + \beta + \min \begin{cases} c_p(x_1 x_2 a_3^{i'}, x_1 x_2 x_3) \\ c_p(a_1^i a_2^i x_3, x_1 x_2 x_3) \end{cases} \tag{20}$$

If we assume that the first deletion occurs before the second deletion and that the second deletion has length one and remaining codon $x_1x_2x_3$ then $D^8(i,j)$ is given by the sum $cost(subs) + cost(del_1) + cost(del_2) + D(i'-k'-1, j-1)$ minimized over all possible combinations of $y_1y_2y_3$ and k'. Remember that $\text{len}^{8'}_{x_1x_2x_3}(i,j,1)$ is the part of this minimum that does not depend on $a_1^i a_2^i a_3^i$ or the witness of the second deletion. An inspection of the above expressions reveals that $\text{len}^{8'}_{x_1x_2x_3}(i,j,1)$ includes everything but $c_p^*(a_1^i a_2^i a_3^i, a_1^i a_2^i y_3)$ and $c^(a_1^i a_2^i y_3, a_1^i a_2^i x_3)$ of $cost(subs)$ and $\min\{c_p(x_1x_2a_3^{i'}, x_1x_2x_3), c_p(a_1^i a_2^i x_3, x_1x_2x_3)\}$ of $cost(del_2)$. It is easy to verify that $D^4_{y_1y_2a_3^{i'}}(i',j)$ is equal to the sum $D(i'-k'-1, j-1) + c_p^*(a_1^{i''} a_2^{i''} a_3^{i''}, y_1 a_2^{i''} a_3^{i''}) +$ $cost(del_1)$ minimized over the deletion length k' of the first deletion. This observation makes it possible to compute $\text{len}^{8'}_{x_1x_2x_3}(i,j,1)$ as

$$\begin{aligned} \text{len}^{8'}_{x_1x_2x_3}(i,j,1) &= \alpha + \beta + c_p^*(x_1x_2x_3, b_1^j b_2^j b_3^j) + \\ &\min_{y_1y_2}\{c_p^*(a_1^{i'} a_2^{i'} a_3^{i'}, a_1^{i'} y_2 a_3^{i'}) + D^4_{y_1y_2a_3^{i'}}(i',j) + c_p^*(y_1y_2a_3^{i'}, x_1x_2a_3^{i'})\} \end{aligned} \tag{21}$$

where we minimize over $y_1 \in \{a_1^{i''}, x_1\}$ and $y_2 \in \{a_2^{i'}, x_2\}$. The cost $\text{len}^{8'}_{x_1x_2x_3}(i,j,1)$ is defined if $x_1x_2x_3$ allows the second deletion to have length one, i.e. if $x_1 \in \{a_1^{i''}, b_1^j\}$, $x_2 \in \{a_2^{i'}, b_2^j\}$ and $x_3 \in \{a_3^i, b_3^j\}$. The nucleotide $a_1^{i''}$ depends on the unknown length of the first deletion, so we must assume that it can be any of the four nucleotides.

Figure 13 illustrates the evolution of the last codon alignment when the second deletion has length one and occurs before the first deletion. The nucleotides z_1, x_2 and x_3 are the status of the substitutions before the second deletion and the nucleotides y_1, y_2 and y_3 are the status of the substitutions before the first deletion. Observe that x_1 is just $a_1^{i'}$. The cost of this case can be described as above. This would reveal that $\text{len}^{8''}_{x_1x_2x_3}(i,j,1)$ includes everything but $c_p^*(a_1^i a_2^i a_3^i, a_1^i a_2^i x_3) + c_p^*(a_1^i a_2^i x_3, a_1^i a_2^i y_3) + c_p(x_1x_2x_3, w_1w_2w_3)$ where $w_1w_2w_3$ is the witness of the second deletion. Furthermore it would reveal that $D^4_{y_1y_2y_3}(i',j)$ is equal to the sum of the cost of the remaining alignment, the cost of the first deletion and the cost of changing codon $a_1^{i''} a_2^{i''} a_3^{i''}$ to $z_1 a_2^{i''} a_3^{i''}$ to $y_1 a_2^{i''} a_3^{i''}$ minimized over the deletion length k' of the first deletion. This makes it possible to compute $\text{len}^{8''}_{x_1x_2x_3}(i,j,1)$ as

$$\begin{aligned} \text{len}^{8''}_{x_1x_2x_3}(i,j,1) &= \alpha + \beta + c_p^*(a_1^{i'} a_2^{i'} a_3^{i'}, x_1x_2a_3^{i'}) + \\ &\min_{y_1y_2y_3}\{c_p^*(a_1^{i'} x_2x_3, a_1^{i'} y_2y_3) + D^4_{y_1y_2y_3}(i',j) + c_p^*(y_1y_2y_3, b_1^j b_2^j b_3^j)\} \end{aligned} \tag{22}$$

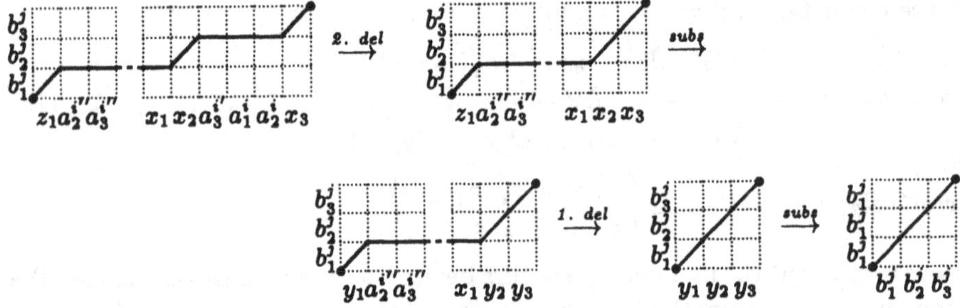

Fig. 13. The second deletion occurs before the first deletion and the second deletion has length one.

where we minimize over $y_1 \in \{z_1, b_1^j\}$, $y_2 \in \{x_2, b_2^j\}$ and $y_3 \in \{x_3, b_3^j\}$. The nucleotide z_1 depends on the unknown length of the first deletion, so we must assume that z_1 can be any of the four nucleotides. The cost $\text{len}^{8''}_{x_1x_2x_3}(i,j,1)$ is defined if $x_1x_2x_3$ allows the second deletion to have length one, i.e. if $x_1 = a_1^{i'}$, $x_2 \in \{a_2^{i'}, b_2^j\}$ and $x_3 \in \{a_3^i, b_3^j\}$.

We are finally in a position where we can describe how to use the method from the previous section to compute $D^8(i,j)$. The cost $\text{len}^8_{x_1x_2x_3}(i,j,k)$ depends on x_1, x_2 and x_3, so instead of 16 tables we need 64 tables $L^8_{x_1x_2x_3}$ to keep track of $\min_{0<k<i} \text{len}^8_{x_1x_2x_3}(i,j,k)$. We still need 64 tables $F^8_{x_1x_2x_3}$ to keep track of the cost under the assumption that the end-codon $a_1^i a_2^i x_3$ is not the witness (of the second deletion). We compute table entry (i,j) in these tables as

$$L^8_{x_1x_2x_3}(i,j) = \min \begin{cases} \text{len}^8_{x_1x_2x_3}(i,j,1) \\ L^8_{x_1x_2x_3}(i-1,j) + \beta \end{cases} \tag{23}$$

$$F^8_{x_1x_2x_3}(i,j) = \min \begin{cases} L^8_{x_1x_2x_3}(i-1,j) + \beta + \alpha_p + c_p(a_1^{i-1}a_2^{i-1}a_3^{i-1}, x_1x_2x_3) \\ \text{len}^8_{x_1x_2x_3}(i,j,1) + c_p(x_1x_2a_3^{i-1}, x_1x_2x_3) \\ F^8_{x_1x_2x_3}(i-1,j) + \beta \end{cases} \tag{24}$$

We compute $D^8_{x_1x_2x_3}(i,j)$ using the above tables and we compute $D^8(i,j)$ by minimizing over the 32 possible remaining codons of the second deletion.

$$D^8_{x_1x_2x_3}(i,j) = \min \begin{cases} L^8_{x_1x_2x_3}(i,j) + c_p(a_1^i a_2^i x_3, x_1x_2x_3) \\ F^8_{x_1x_2x_3}(i,j) \end{cases} \tag{25}$$

$$D^8(i,j) = \min_{x_1x_2x_3} \{c_p^*(a_1^i a_2^i a_3^i, a_1^i a_2^i x_3) + D^8_{x_1x_2x_3}(i,j)\} \tag{26}$$

This constant time computation of $D^8(i,j)$ requires us to compute entry (i,j) in 128 tables. The other three cases where the last codon alignment describes two internal gaps are handled similarly. Hence, to compute $D^t(i,j)$ for $t = 8, 9, 10, 11$ we compute $4 \cdot 128 = 512$ table entries in total.

5.4 Combining the computation

We observe that the only real difference between the computation of $D^6(i,j)$ and $D^8(i,j)$ is between $\text{len}^6_{x_1x_2}(i,j,1)$ and $\text{len}^8_{x_1x_2x_3}(i,j,1)$. The similarity stems from the fact that a codon alignment of type 6 and type 8 ends in the same way. By "end in the same way" we mean that the events described on the codon $a^i_1a^i_2a^i_3$ are the same. Figure 7 reveals that a codon alignment of type 11 also ends in the same way as codon alignments of type 6 and 8.

The similarity between the computation of $D^t(i,j)$ for $t = 6, 8, 11$ makes it possible to combine the computation of the three costs and thereby reduce the number of tables. We can replace the three tables $L^6_{x_1x_2}$, $L^8_{x_1x_2x_3}$ and $L^{11}_{x_1x_2x_3}$ with one table $L^{6,8,11}_{x_1x_2x_3}$ where $L^{6,8,11}_{x_1x_2x_3}(i,j)$ is the minimum of entry (i,j) in the three tables it replaces. Similarly we can replace $F^6_{x_1x_2x_3}$, $F^8_{x_1x_2x_3}$ and $F^{11}_{x_1x_2x_3}$ with $F^{6,8,11}_{x_1x_2x_3}$. We can compute $L^{6,8,11}_{x_1x_2x_3}(i,j)$ and $F^{6,8,11}_{x_1x_2x_3}(i,j)$ by expressions similar to (23) and (24). All we essentially have to do is to replace $\text{len}^8_{x_1x_2x_3}(i,j,1)$ by

$$\text{len}^{6,8,11}_{x_1x_2x_3}(i,j,1) = \min_{t=6,8,11} \text{len}^t_{x_1x_2x_3}(i,j,1) \tag{27}$$

where we in order to ensure that $\text{len}^t_{x_1x_2x_3}(i,j,1)$ for $t = 6, 8, 11$ describes the same part of the total cost must redefine $\text{len}^6_{x_1x_2x_3}(i,j,1)$ as $\text{len}^6_{x_1x_2}(i,j,1) + c^*_p(x_1x_2x_3, b^j_1b^j_2b^j_3)$.

We introduce $D^{6,8,11}(i,j)$ as the minimum of $D^t(i,j)$ over $t = 6, 8, 11$. We can compute $D^{6,8,11}(i,j)$ by using $L^{6,8,11}_{x_1x_2x_3}$ and $F^{6,8,11}_{x_1x_2x_3}$ in expressions similar to (25) and (26). The computation of $D^{6,8,11}(i,j)$ requires us to compute only $64 + 64 = 128$ table entries while the individual computation of $D^t(i,j)$ for $t = 6, 8, 10$ requires us to compute $80 + 128 + 128 = 336$ table entries. Figure 7 also reveals that codon alignments of type 7, 9 and 10 end in the same way. Hence, we can also combine the computation of $D^t(i,j)$ for $t = 7, 9, 10$.

Finally, to compute $D(i,j)$ by (5) we must minimize over $D^1(i,j)$, $D^2(i,j)$, $D^3(i,j)$, $D^4(i,j)$, $D^5(i,j)$, $D^{6,8,11}(i,j)$ and $D^{7,9,10}(i,j)$. In total this computation requires us to compute $1 + 7 + 68 + 68 + 128 + 128 = 400$ table entries.

6 Future work

We are working on implementing the alignment algorithm described in the previous section in order to compare it to the heuristic alignment algorithm described in [5]. The heuristic algorithm allows frame shifts, so an obvious extension of our exact algorithm would be to allow frame shifts, e.g. to allow insertion-deletions of arbitrary length. This however makes it difficult to split the evaluation of the alignment cost into small independent subproblems (codon alignments) of known size.

Another interesting extension would be to annotate the DNA sequence with more information. For example, if the DNA sequence codes in more than one reading frame (overlapping reading frames) then the DNA sequence should be annotated with all the amino acid sequences encoded and the combined cost of a nucleotide event should summarize the cost of changes induced on all the amino acid sequences encoded by the DNA sequence. This extension also makes it difficult to split the evaluation of the alignment cost into small independent subproblems. To implement these extensions efficiently it might be fruitful to investigate reasonable restrictions of the cost functions.

References

1. L. Arvestad. Aligning coding DNA in the presence of frame-shift errors. In *Proceedings of the 8th Annual Symposium of Combinatorial Pattern Matching (CPM 97)*, volume 1264 of *Lecture Notes in Computer Science*, pages 180–190, 1997.
2. R. Durbin, R. Eddy, A. Krogh, and G. Mitchison. *Biological Sequence Analysis: Probalistic Models of Proteins and Nucleic Acids.* Cambrigde University Press, 1998.
3. O. Gotoh. An improved algorithm for matching biological sequences. *Journal of Molecular Biology*, 162:705–708, 1981.
4. J. Hein. An algorithm combining DNA and protein alignment. *Journal of Theoretical Biology*, 167:169–174, 1994.
5. J. Hein and J. Støvlbæk. Genomic alignment. *Journal of Molecular Evolution*, 38:310–316, 1994.
6. J. Hein and J. Støvlbæk. Combined DNA and protein alignment. *Methods in Enzymology*, 266:402–418, 1996.
7. D. S. Hirschberg. A linear space algortihm for computing maximal common subseqeunce. *Communication of the ACM*, 18(6):341–343, 1975.
8. Y. Hua, T. Jiang, and B. Wu. Aligning DNA sequences to minimize the change in protein. Accepted for CPM 98.
9. S. B. Needleman and C. D. Wunsch. A general method applicable to the search for similarities in the amino acid seqeunce of two proteins. *Journal of Molecular Biology*, 48:433–443, 1970.
10. H. Peltola, H. Söderlund, and E. Ukkonen. Algorithms for the search of amino acid patterns in nucleic acid sequences. *Nuclear Acids Research*, 14(1):99–107, 1986.
11. D. Sankoff. Matching sequences under deletion/insertion constraints. In *Proceedings of the National Acadamy of Science USA*, volume 69, pages 4–6, 1972.
12. P. H. Sellers. On the theory and computation of evolutionary distance. *SIAM Journal of Applied Mathematics*, 26:787–793, 1974.
13. R. A. Wagner and M. J. Fisher. The string to string correction problem. *Journal of the ACM*, 21:168–173, 1974.
14. Z. Zhang, W. R. Pearson, and W. Miller. Aligning a DNA sequence with a protein sequence. *Journal of Computational Biology*, 4(3):339–349, Fall 1997.

Fixed Topology Alignment with Recombination

Bin Ma[1], Lusheng Wang[2] and Ming Li[3]

[1] Department of Mathematics, Peking University, Beijing 100871, P.R. China
[2] Department of Computer Science, City University of Hong Kong, 83 Tat Chee Avenue, Kowloon, Hong Kong
[3] Department of Computer Science, University of Waterloo, Waterloo, Ont. N2L 3G1, Canada

Abstract. In this paper, we study a new version of multiple sequence alignment, *fixed topology alignment with recombination*. We show that it can not be approximated within any constant ratio unless $P = NP$. For a more restricted version, we show that the problem is MAX-SNP-hard. This implies that there is no PTAS for this version unless $P = NP$. We also propose approximation algorithms for a special case, where each internal node has at most one recombination child and any two *merge paths* for different recombination nodes do not share any common node.

1 Introduction

Multiple sequence alignment is the most critical cutting-edge tool for extracting and representing biologically important commonalities from a set of sequences. It plays an essential role in the solution of many problems such as searching for highly conserved subregions among a set of biological sequences and inferring the evolutionary history of a family of sequences [5, 24]. Many versions have been proposed [1, 5, 24]. *Tree alignment* is one of the most famous versions. It was first proposed by Sankoff in [14].

For tree alignment, we are given k sequences and a tree $Tree$ of k leaves, each of which is labeled with a unique given sequences. The goal is to construct a sequence for each internal node in $Tree$ such that the cost of the tree is minimized. The cost of an edge in a tree is defined as the edit distance between the two sequences assigned to the two ends of the edge. The cost of a tree is the total cost of edges in the tree. Once each internal node is assigned a sequence, one can produce a multiple sequence alignment by optimally aligning two sequences on every edge. An alternative definition of tree alignment is to find an optimal multiple sequence alignment with *tree score*. Given a multiple sequence alignment $\mathcal{A}$, the score $\mu(s_1(i), s_2(i), \ldots, s_k(i))$ on the i-th column of $\mathcal{A}$, where $s_j(i)$ is the letter (possibly a space) from sequence s_j in the i-th column of $\mathcal{A}$, is defined as follows:

Let tree $Tree = (V, E)$, where V and E are the sets of nodes and edges in $Tree$, have k leaves. Let $k+1$, $k+2$, ..., $k+m$ be the internal nodes on T. For each internal node l, reconstruct a letter (possibly a space) $s_l(i)$ such that $\sum_{(p,q)\in E} \mu(s_p(i), s_q(i))$, where $\mu((s_p(i), s_q(i))$ is the score for the pair of letters $s_p(i)$ and $s_q(i)$, is minimized. The score $\mu(s_1(i), s_2(i), \ldots, s_k(i))$ of the i-th column is thus defined as:

$$\mu(s_1(i), s_2(i), \ldots, s_k(i) = \sum_{(p,q)\in E} \mu(s_p(i), s_q(i)). \qquad (1)$$

The tree score for $\mathcal{A}$ is the total scores of all columns in $\mathcal{A}$. It is known that the two definitions of tree alignment are equivalent.

Tree alignment was proved to be NP-hard [20]. Many heuristic algorithms have been proposed in the literature [1, 6, 15, 16]. Some approximation algorithms with guaranteed relative error bounds have been reported recently. In particular, a polynomial time approximation scheme (PTAS) is presented in [21] and improved versions are given in [22, 23].

Multigene families, viruses, and alleles from within populations experience recombinations. When recombinations occur, the evolutionary history can not be represented as a tree. Hein first studied the method to reconstruct the history of sequences subject to recombination [7, 8]. Hein observed that the evolution of a sequence with k recombinations could be described by k recombination points and $k+1$ trees describing the evolution of the $k+1$ intervals, where two neighboring trees were either identical or differed by one subtree transfer operation [7, 8, 9, 3, 2]. A heuristic method was proposed to find the most parsimonious history of the sequences in terms of mutation and recombination operations. Another strike was given by Kececioglu and Gusfield [12]. They introduced two new problems, *recombination distance*, and *bottleneck recombination history*. They tried to include higher-order evolutionary events such as block insertions and deletions [4], and tandem repeats [11, 13].

In this paper, we propose a model called *fixed topology alignment with recombination* (FTAR for short) which is analogous to tree alignment. The difference is that in the given topology, some nodes (called *recombination nodes*) have two parents instead of one. A recombination node obtains its ancestral material from both parents. Moreover, the given topology may have more than one root. As for tree alignment, there are two ways to define the problem. One is to reconstruct a sequence for each internal node such that the total cost of the topology is minimized. Another is to construct a multiple sequence alignment with minimum score. The first version is called fixed topology history with recombination (FTHR) and the second version is called fixed topology alignment with recombination (FTAR). When recombination is considered, the two versions are no longer equivalent. We show that, in general, both versions do not have a constant ratio approximation algorithm. Even for a special case, where each internal node has at most one recombination child and there are at most 6 parents of recombination nodes in any path from a root to a leaf in the given topology, they do not has a PTAS. The above results show that the new problems are much harder than the normal tree alignment problem, in terms of approximation. However, since recombination rarely occur in practice, it is interesting to study special cases. We design a ratio-3 algorithm for both FTAR and FTHR in a more restricted case, where each internal node has at most one recombination child and any two merge paths for different recombination nodes do not share any common node. We then extend the algorithm into a PTAS for FTAR in the case, where each internal node has at most one recombination child, any two merge paths for different recombination nodes do not share any common node, and there is a constant number of crossovers for each recombination node.

2 Fixed topology alignment with recombination – the model

When recombination happens, the ancestral material on the present sequence s_1 is located on two sequences s_2 and s_3. s_2 and s_3 can be cut at k locations (break points) into $k+1$ pieces, where $s_2 = s_{2,1}s_{2,2}\ldots s_{2,l+1}$ and $s_3 = s_{3,1}s_{3,2}\ldots s_{3,l+1}$. s_1 can be represented as $s_{\hat{2},1}s_{\hat{3},2}s_{\hat{2},3}\ldots s_{\hat{2},i}s_{3,\hat{i}+1}\ldots$, where subsequences $s_{\hat{2},i}$ and $s_{3,\hat{i}+1}$ differ from the corresponding $s_{2,i}$ and $s_{3,i+1}$ by insertion, deletion, and substitution of letters. k, the number of times s_1 switches between s_2 and s_3, is called the number of *crossovers*. The cost of the recombination is

$$dist(s_{1,1}, s_{\hat{1},1}) + dist(s_{2,2}, s_{\hat{2},2}), \ldots dist(s_{1,i}, s_{\hat{1},i}) + dist(s_{2,i+1}, s_{2,\hat{i}+1}) + \ldots + k\chi,$$

where $dist(s_{2,i+1}, s_{2,\hat{i}+1})$ is the edit distance between the two sequences $s_{2,i+1}$ and $s_{2,\hat{i}+1}$, k is the number of crossovers and χ is the crossover penalty. The *recombination* distance to produce s_1 from s_2 and s_3 is the cost of a recombination that has the smallest cost among all possible recombinations. We use $r\_dist(s_1, s_2, s_3)$ to denote the recombination distance. For more details, see [12, 19].

When recombination occurs, the given topology is no longer a binary tree. Instead, some nodes, called *recombination nodes*, in the given topology may have two parents[7, 8]. In a more general case as described in [12], the topology may have more than one root. The set of roots is called a *protoset*. The edges incident to recombination nodes are called *recombination* edges. See Figure 1 (b). A node/edge is *normal* if it is not a recombination node/edge.

The cost on a pair of recombination edges is the recombination distance to produce the sequence on the recombination node from the two sequences on its parents. The cost on other normal edges is the edit distance between two sequences. A topology is *fully labeled* , if every node on the topology is labeled. For a fully labeled topology, the cost of the topology is the total cost of edges in the topology. Each node in the topology with degree greater than 1 is an internal node. Each leaf/terminal (degree 1 node) in the topology is labeled with a given sequence. The goal here is to construct a sequence for each internal node such that the cost of the topology is minimized. We call this problem *fixed topology history with recombination* (FTHR).

Obviously, this problem is a generalization of tree alignment. The difference is that the given topology is no longer a binary tree. Instead, there are some recombination nodes which have two parents instead of one. Moreover, there may be more than one root in the topology.

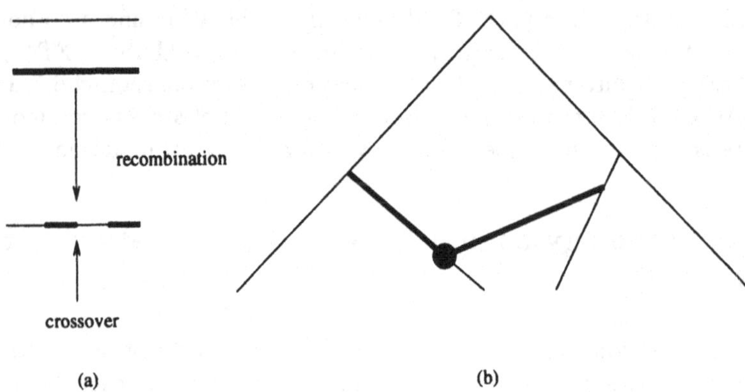

Fig. 1. (a) Recombination operation. (b) The topology. The dark edges are recombination edges. The circled node is a recombination node.

As for tree alignment, there is an alternative definition. Let the topology $T = (V, E)$, where V and E are the sets of nodes and edges in T, have k leaves (nodes without any child). Let $k+1$, $k+2$, ..., $k+m$ be the internal nodes on T. Let $E(q_1, \ldots, q_m)$ be the subset of edges in E such that each pair of recombination edges has exactly one edge in $E(q_1, \ldots, q_m)$ and the choices are indicated by q_i's (q_i indicates that either the left or right edge is selected). Given a multiple sequence alignment $\mathcal{A}$ of N columns, the score $\mu(s_1(i), s_2(i), \ldots, s_k(i), q_1, \ldots q_m)$ on the i-th column, where $s_j(i)$ is the letter (possibly a space) from sequence s_j in the i-th column, is defined as follows:
For each internal node l, reconstruct a letter (possibly a space) $s_l(i)$ such that

$$\sum_{(p,q)\in E(q_1,\ldots,q_m)} \mu(s_p(i), s_q(i)),$$

where $\mu((s_p(i), s_q(i))$ is the score for the pair of letters $s_p(i)$ and $s_q(i)$, is minimized. The score $\mu(s_1(i), s_2(i), \ldots, s_k(i), q_1, \ldots q_m)$ of the i-th column with the choice $(q_1, q_2, \ldots, q_m)$ is thus defined as:

$$\mu(s_1(i), s_2(i), \ldots, s_k(i), q_1, \ldots q_m) = \sum_{(p,q)\in E(q_1,\ldots,q_m)} \mu(s_p(i), s_q(i)). \quad (2)$$

The cost of $\mathcal{A}$ is defined as:

$$\min_{q_1(1),\ldots q_m(1), q_1(2),\ldots q_n(2),\ldots,q_1(N),\ldots,q_m(N)} \{\sum_{i+1}^{N} \{\mu(s_1(i), s_2(i), \ldots, s_k(i), q_1(i), \ldots q_m(i)) + f(i)\chi\}\}, \quad (3)$$

where $f(1) = 0$ and $f(i) = c$ if there are c $q_l(i)$'s that are different from $q_l(i-1)$'s. This version is called *fixed topology alignment with recombination* (FTAR).

Unlike tree alignment, the above two definitions are not equivalent. That is, the cost of an optimal solution for FTHR might be smaller than that for FTAR. For example, the topology and the sequences are given as in Figure 2 (a). If a mismatch costs 1 and the crossover penalty is $\chi = 0.5$, the sequences assigned to internal nodes should be as shown in Figure 2 (a). Thus, the cost of an optimal solution for FTHR is $3 + \chi$. However, an optimal solution for FTAR costs $4 + \chi$ as shown in Figure 2 (a).

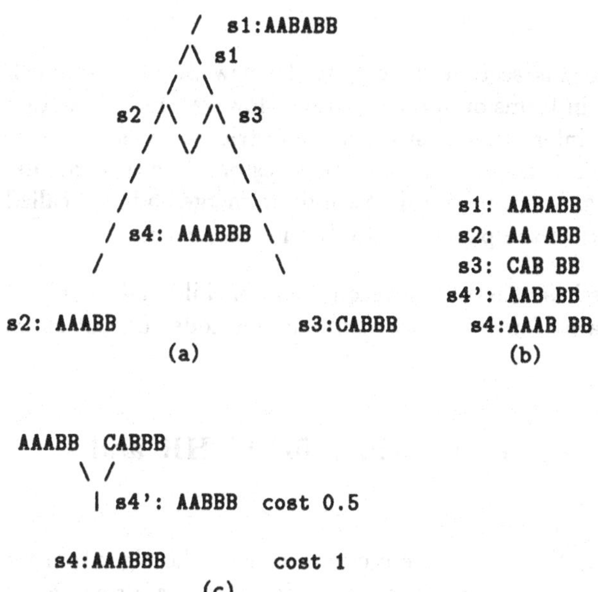

Fig. 2. An optimal alignment.

The reason is that the alignment in Figure 2 (b) imposes that a crossover produces an intermediate sequence s4':AABBB. (See Figure 2 (c).) Moreover, the example demonstrates that a solution for FTHR might not lead to an alignment.

3 Hardness results

In this section, we show that FTHR and FTAR cannot be approximated within any constant performance ratio in general. We then consider a restricted version, where each internal node has at most one recombination child and there is at most 6 parents of recombination nodes in any path from the root to a leaf in the given topology. We show that the restricted version for both FTHR and FTAR do not have a PTAS.

Theorem 1. *FTHR and FTAR cannot be approximated within any constant performance ratio unless $P = NP$.*

Now, we consider a more restricted case, where each internal node has at most one recombination child and there are at most 6 parents of recombination nodes in any path from the root to a leaf in the given topology.

Theorem 2. *The restricted version for both FTHR and FTAR is MAX-SNP-hard. That is, there is no polynomial time approximation scheme unless $P = NP$.*

The hardness results in this section show that the new problem is much harder than tree alignment in terms of approximation. However, recombination occur infrequently. So, it is interesting to study some restricted cases. A *merge node* of recombination node v is the lowest common ancestor of v's two parents. The two different paths from a recombination node to its merge node are called *merge paths*. In the rest part of the paper, we study the case, where

- (C1) each internal node has at most one recombination child and
- (C2) any two merge paths for different recombination nodes do not share any common node.

4 Ratio-4 approximation algorithms for FTHR and FTAR

We first consider FTHR. At the end of the section, we give the algorithm for FTAR with fixed number of crossovers. The basic idea of our approximation algorithms is to combine a method to deform the layout of the given topology with the uniform lifting method for tree alignment in [22].

Deforming the layout Given a topology T satisfies conditions (C1) and (C2), we extend some of the edges in T such that the two parents of a recombination node are at the same level. Figure 3 gives an example.

Uniform Lifting For a deformed layout of the topology, ignoring the recombination edges, treating the topology as a set of binary trees (denoted as *forest* F), and keeping the level positions of all binary trees in the layout, we can uniformly lift the sequences to internal nodes. The lifting choice for all the nodes (even in different binary trees) at the same level is uniform. Since we do not delete any node, each internal node gets a sequence. We thus get a fully labeled topology. Figure 3 (b) gives an example. A topology obtained in this way is called a *uniform lifted* topology.

We can show that there is a uniform lifted topology with cost at most 4 times of the optimum. First, let us consider some definitions.

Given a binary tree *Tree*, we arbitrarily extend the edges in *Tree* to form an *extended layout* such that the two parents of every recombination node are at the same level. We then uniformly lift the sequences from leaves to the internal

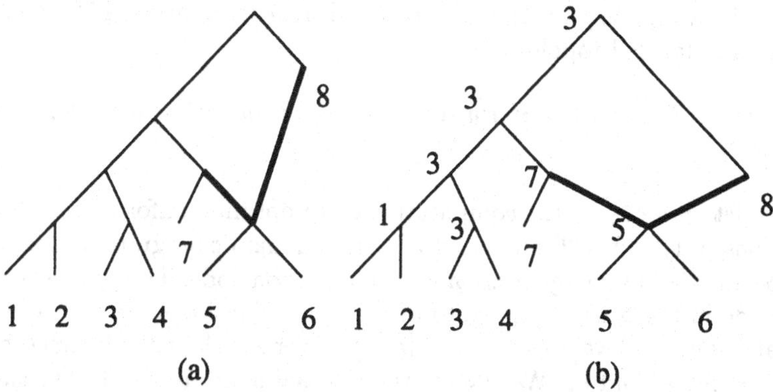

Fig. 3. (a) The original topology. (b) The deformed layout. The numbers stand for given sequences and they are uniformly lifted to their ancestors. Since we ignore recombination edges, 7 is lifted to its parent.

nodes in the layout. The fully labeled tree obtained in this way is called an *extended uniform lifted* tree.

Let $Tree^{min}$ be an optimal fully labeled tree for a binary tree $Tree$. Let d be the depth of the extended layout of $Tree$, there are 2^d extended uniform lifted trees $Tree(1)$, $Tree(2)$, ..., $Tree(2^d)$. Let $v(i)$ be the sequence lifted to v in the extended uniform lifted tree $Tree(i)$ for $i = 1, 2, \ldots, 2^d$. $\hat{v(i)}$ is the sequence lifted to v's child which is different from $v(i)$ in $Tree(i)$. For each node v, $S(v)$ denotes the set of given sequences assigned to the descendent leaves of v. Let μ_v^s be the cost of the path in $Tree^{min}$ from the internal node v to its descendent leaf which is labeled with the given sequence $s \in S(v)$. Let $V(Tree)$ be the set of internal nodes (degree-3) in $Tree$.

Similar to the uniform lifted trees in [22], the total cost of the 2^d extended uniform lifted trees is bounded as follows:

Lemma 3.

$$\sum_{i=1}^{2^d} C(Tree(i)) \leq \sum_{i=1}^{2^d} \{2 \times \{ \sum_{v \in V(Tree)} \mu_v^{v(i)}\} + \mu_r^{r(i)} + \mu_r^{\hat{r(i)}}\}$$

$$= \sum_{i=1}^{2^d} \{2 \times \{ \sum_{v \in V(Tree)} \mu_v^{\hat{v(i)}}\} + \mu_r^{r(i)} + \mu_r^{\hat{r(i)}}\}$$

$$= 2 \times 2^d C(Tree^{min}) - \sum_{i=1}^{2^d} \{\mu_r^{r(i)} + \mu_r^{\hat{r(i)}}\} \quad (4)$$

$$= 2 \times 2^d C(Tree^{min}) - 2 \times \sum_{i=1}^{2^d} \mu_r^{r(i)}. \quad (5)$$

Let T be a topology for tree alignment with recombination. T^{min} denotes an optimal fully labeled topology.

Theorem 4. *There is a uniform lifted topology with cost at most* 4 *times of the optimum.*

Now, let us focus on the computation of an optimal uniform lifted topology. Conditions (C1) and (C2) allow us to design a dynamic programming algorithm. Let v be a node in a merge path of a recombination node. The *parallel* node u of v is a node in the other merge path of the recombination node such that u and v are at the same level. We use $dist[s, s']$ to denote the edit distance between two sequences s_1 and s_2. We classify the internal nodes in T into five cases:

Case 1: Let v be a node such that (1) there is no descendent recombination node or (2) it reachs a merge node first before it reachs the corresponding recombination node if there is any descendent recombination node of v. Note that, v could be a normal node or a recombination node. Define $d[v, s]$ as the smallest cost of the sub-topology rooted at v such that sequence s is uniformly lifted to v.

$$d[v, s] = \min_{all\ possible\ s_2} d[v_1, s_1] + d[v_2, s] + dist(s_1, s), \tag{6}$$

where v_1 and v_2 are the two children of v, v_1 is assigned sequence s_1 and v_2 is assigned the sequence s. The choice of s_1 may not be unique. (See Figure 4 (a).)

Case2: Let v be the parent of a recombination node. We define $d[v, s]$ as the smallest cost of the sub-topology such that the two parents are the only nodes at the top level of the sub-topology and s is uniformly lifted to v. Let v' be the other parent of the recombination node and sequence s' be the lifted sequence on v'. $d[v, s]$ can be computed as follows:

$$d[v, s] = \min_{all\ possible\ s'\ and\ s_3} d[v_1, s] + d[v_2, s'] + d[v_3, s_3] + r\_dist(s, s', s_3), \tag{7}$$

where v_3 is the recombination node, v_1 is the normal child of v, v_2 is the normal child of v', and s_3 is uniformly lifted to v_3. Note that, the sequences s and s' are also lifted to v_1 and v_2, respectively. (See Figure 4 (b).)

Case 3: Let v be a normal node between the merge node and the parent of the recombination node such that v has a parallel node v'. Define $d[v, s]$ as the smallest cost of the sub-topology with nodes v and v' at the top and containing all the descendent nodes of v and v'. (See Figure 4 (c).) $d[v, s]$ can be computed as follows:

$$\begin{aligned} d[v, s] = \min_{all\ possible\ s,\ s_1, s_2, s_3\ and\ s_4} & dist(s, s_1) + dist(s, s_2) + dist(s', s_3) \\ & + dist(s', s_4) + d[v_1, s_1] + d[v_2, s_2] + d[v_3, s_3], \end{aligned} \tag{8}$$

where v_1 and v_2 are the two children of v that are assigned s_1 and s_2, respectively; v_3 and v_4 are the two children of v' that are assigned s_3 and s_4, respectively;

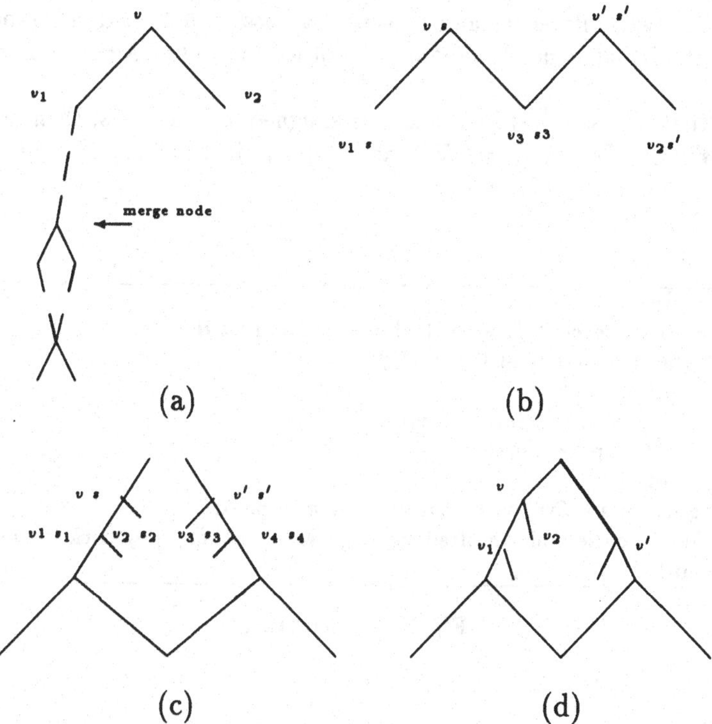

Fig. 4. (a) Case 1. (b) Case 2. (c) Case 3. (d) Case 4. The heavy edge is an extended edge.

and v' is assigned the sequence s'. Note that s is either s_1 or s_2 and s' is either s_3 or s_4.

Case 4: Let v be the node in a merge path of a recombination node such that v does not have a parallel node. Corresponding to v, there is an extended edge in the other merge path of the recombination node. Let v' be the node at the lower end of the extended edge. (See Figure 4 (d).) Define $d[v, s]$ as the smallest cost of the sub-topology rooted at nodes v and v' and containing all their descendent nodes.

$$d[v,s] = \min_{\textit{all possible } s_1 \textit{ and} s_2} d[v_1, s_1] + d[v_2, s_2] + \textit{dist}(s_1, s_2), \tag{9}$$

where v_1 and v_2 are the two children of v that are assigned sequences s_1 and s_2, respectively. s is either s_1 or s_2.

Case 5: Let v be a merge node. $d[v, s]$ be the smallest cost of sub-topology rooted at v such that v is assigned the sequence s. $d[v, s]$ can be computed as:

$$d[v,s] = \min_{\textit{all possible } s_1 \textit{ and } s_2} d[v_1, s_1] + \textit{dist}(s, s_1) + \textit{dist}(s, s_2). \tag{10}$$

where v has two children v_1 and v_2 which are assigned s_1 and s_2, respectively. v_1 is the child that is not lower than the other child of v. Again, s is either s_1 or s_2.

Note that, if v is a leaf in T and v is assigned a sequence s, then $d[v, s] = 0$ and $d[v, s'] = \infty$ for any $s' \neq s$. We can compute all the $d[v, s]$'s bottom up. The algorithm is shown in Figure 5.

```
begin
for each level of T, with the bottom level first do
  for each node v at the level do
  begin
    for each sequence s ∈ S(v),
      Compute d[v, s].
  end
Select an s ∈ S(r) such that d[r, s] is minimized.
Compute the uniform lifted topology with cost d[r, s] by back-tracing.
end.
```

Fig. 5. Algorithm 1

The running time depends on the time to compute each $d[v, s]$. From equations (6)–(10), the runing time could be as high as $O(k^6)$ if the involved $dist(s, s')$s and $r\_dist(s, s_1, s_2)$'s are previously computed. In fact, using the data structure called *extended tree* in [22], the running time of the algorithm purely depends on the number of $dist(s_1, s_2)$'s and $r\_dist(s_1, s_2, s_3)$'s used in (6)-(10). Similar to [22], we can reduce the total number of $dist(s_1, s_2)$'s and $r\_dist(s_1, s_2, s_3)$'s used in the algorithm to $O(kd)$, where k is the number of given sequences and d is the depth of the topology. Thus, we can get an $O(kd)$ algorithm, if all involved $dist(s_1, s_2)$'s and $r\_dist(s, s_1, s_2)$'s have been previously computed. The computation of each $dist(s_1, s_2)$ needs $O(n^2)$ time if the length of sequences is n. The computation of $r\_dist(s, s_1, s_2)$ needs $O(n^3)$ time and $O(n^2)$ space using the method in [12].

Now, we can conclude that

Theorem 5. *When the given topology satisfies (C1) and (C2), there is a ratio-3 approximation algorithm that takes $O(kd + kdn^3)$ time.*

FTAR with fixed number of crossovers

The above algorithm does not always give an alignment. Thus it only works for FTHR. However, the bound developed in Theorem 4 also holds for FTAR. Now, we modify Algorithm 1 to work for FTAR when each recombination contains at most a constant number, say, c, of crossovers. The key observation is that when the top node v has a parallel node v', i.e., Cases 2 and 3, the recombination imposes c fixed pairs of indexes for the two sequences s and s' that are assigned to v and v', respectively, in the resulting alignment. Thus, the recurrence equations are similar to those in (6)-(10) except that we have to keep c pairs of fixed indexes.

For simplity, we give the recurrence equations for the case, where each recombination contains one crossover. For Case 1, the equation is the same as (6).

Case 2: Let $d[v, s, s', i, j]$ be the smallest cost of sub-topology rooted at v such that v is assigned the sequence s and v' is assigned the sequence s' and the crossover breaks s and s' into two parts, $s[1...i]$ and $s[i+1...n]$, and $s'[1...j]$ and $s'[j+1...n]$. (See Figure 4 (b).)

$$\mathrm{d[v,s,s',i,j]} = \min_{\mathrm{allpossibles_3}} \mathrm{d[v_1,s] + d[v_2,s'] + d[v_3,s_3] + r\_dist(s_3,s,s',i,j)}, \quad (11)$$

where $r\_dist(s_3, s, s', i, j)$ is the recombination distance such that there is one crossover and it breaks s and s' into two parts, $s[1...i]$ and $s[i+1...n]$, and $s'[1...j]$ and $s'[j+1...n]$.

Case 3: Let $d[v, s, s', i, j]$ be the smallest cost of sub-topology rooted at v such that v is assigned the sequence s and v' is assigned the sequence s' and the crossover divides s_1 and s_2 into two parts $s_1[1...i]$ and $s[i+1...n]$, and $s_2[1...j]$ and $s_2[j+1...n]$. (See Figure 4 (c).)

$$\begin{aligned} d[v,s,s',i,j] = & \min_{all\ possible\ s,\ s_1,s_2,s_3\ ,\ s_4,\ i'\ and\ j'} dist(s,s_1,i,i') + dist(s,s_2) \\ & + dist(s',s_4) + dist(s',s_3,j,j') + d[v_1,s_1,s_4,i',j'] + d[v_2,s_2] \\ & + d[v_3,s_3], \end{aligned} \quad (12)$$

where $dist(s_1s_2, i, j) = dist(s_1[1..i], s_2[1...j]) + dist(s_1[i+1...n], s_2[j+1...n])$.

Case 4: Let $d[v, s, s', i, j]$ be the smallest cost of sub-topology rooted at v such that v is assigned the sequence s and v' is assigned the sequence s' and the crossover breaks s and s' into two parts, $s[1...i]$ and $s[i+1...n]$, and $s'[1...j]$ and $s'[j+1...n]$. (See Figure 4 (d).)

$$\mathrm{d[v,s,,s',i,j]} = \min_{\mathrm{all\ possible\ s_1\ s_2\ ,and\ i'}} \mathrm{d[v_1,s_1,s',i',j] + d[v_2,s_2] + dist(s_1,s_2)} \quad (13)$$

where v_1 and v_2 are the two children of v that are assigned sequences s_1 and s_2, respectively; v' is assigned the sequence s'; and s is either s_1 or s_2.

Case 5: Let v be a merge node. $d[v, s]$ be the smallest cost of sub-topology rooted at v such that v is assigned the sequence s. $d[v, s]$ can be computed as:

$$d[v,s] = \min_{all\ possible\ s_1\ ,\ s_2,\ i\ and\ j} \{d[v_1, s_1, s_2, i, j] + dist(s_1, s_2, i, j)\}. \quad (14)$$

where v has two children v_1 and v_2 which are assigned s_1 and s_2, respectively. v_1 is the child that is not lower than the other child of v. Again, s is either s_1 or s_2.

Again, using the techniques in [22], computing the value in (12) requires the highiest time complexity, i.e., $O(n^2)$. Therefore, we can get an algorithm with time complexity $O(kdn^3n^{4c})$ for the case where each recombination contains at most c crossovers.

5 A PTAS

We can extend the ratio-3 algorithm for FTAR with bounded number of crossovers to a PTAS. The basic idea is to partition the extended topology into a set of slides, each of which contains t levels of nodes. We number the levels from top to bottom. The root is at level 1. The top slide can contain 1, 2, ..., t levels of nodes. Thus, there are t different partitions $P_1, P_2, \ldots, P_t$, where the top slide of P_p contains p levels. (See Figure 6.) Each P_p partitions the topology into *small topologies*, each of which contains at most t levels of nodes. A *boundary node* for P_p is a node at the j-th level where j mod t = p. Since in the extended topology some edges cross more than one level, they might be cut by all the t different partitions. (This case happens when an edge crosses more than t level.) If so, it will result in a constant ratio.

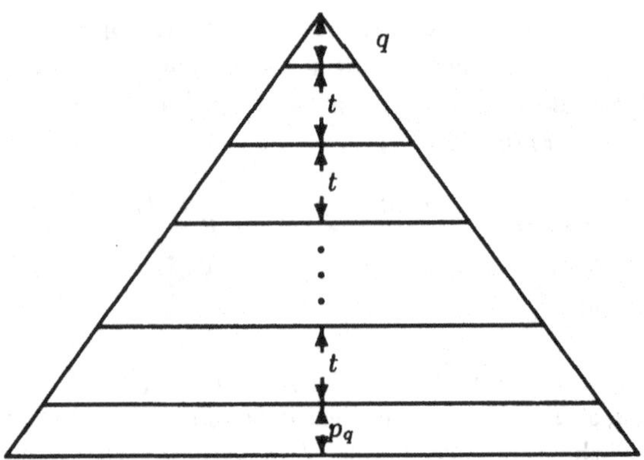

Fig. 6. The partition P_q.

An edge that crosses more than one level is called a *long* edge, whereas other edges are called *short* edges. For a fixed partition P_p, if two small topologies obtained from P_p share a common long edge, we construct an edge (u, v), where u stands for the upper small topology and v stands for the lower small topology. In this way, we can get a graph $G(p)$, called a *component* graph for partition P_p, to describe the relationships among small topologies. From (C2), it is easy to see that $G(p)$ is a forest (a set of trees). Thus, we can organize the nodes in $G(p)$ level by level. For each $G(p)$, we further cut the nodes (small topologies in T) for every t levels and obtain t different partitions $P_{p,q}$ for $q = 1, 2, \ldots, t$. Here we number the levels of $G(p)$ from top to bottom and the root is at level 1. Therefore, we have t^2 different partitions. A *boundary node* v for $P_{p,q}$ is a boundary node for P_p that is not incident to any long edge, or if v is incident to a long edge then the small topology that v is in is at level j in $G(p)$ such that j mod t = q. A *boundary edge* for $P_{p,q}$ is an edge such that at least one of its ends is a boundary node for $P_{p,q}$.

We then (1) uniformly lift the sequences to the internal nodes in T, (2) keep the lifted sequences on boundary nodes for $P_{p,q}$, and (3) reconstruct the sequences for the rest of the nodes in the topology such that the total cost of the topology is minimized.

A fully labeled topology thus obtained is called a $(t-t)$-uniform lifted topology. Since there are 2^d uniform lifted choices and t^2 different partitions , there are $2^d \times t^2$ different $(t-t)$-uniform lifted topologies. We use $T_{p,q}(i)$ to denote the $(t-t)$-topology obtained from partition $P_{p,q}$ and uniform lifting choice i. The total cost of those $2^d \times t^2$ $(t-t)$-uniform lifted topologies is bounded as follows:

Theorem 6.

$$\sum_{p=1}^{T}\sum_{q=1}^{t}\sum_{i=1}^{2^d} C(T_{p,q}(i)) \leq 2^d \times t^2 C(T^{min}) + 4 \times 2^d \times tC(T^{min}). \tag{15}$$

Corollary 7. *There exists a $(t-t)$-uniform lifted topology with cost at most $(1+4/t)C(T^{min})$.*

Note that, Theorem 6 and Corollary 7 hold for both FTHR and FHAR. However, it is hard to find a polynomial time algorithm to compute an optimal $(t-t)$-uniform lifted topology. Fortunately, we can give a dynamic programming algorithm to compute an optimal $(t-t)$-uniform lifted topology for FTAR when each recombination contains a bounded number of crossovers.

Theorem 8. *There is a polynomial time approximation scheme (PTAS) for FTAR when every recombination contains a bounded number of crossovers and the given topology satisfies (C1) and (C2).*

6 Remarks

The time complexity for the PTAS proposed here is too high so that we can not try the PTAS even for short sequences of 100 letters. It might be possible to improve the ratio of PTAS to $1+\frac{5}{t}-\frac{4}{t2^t}$ or $1+\frac{4}{t+1}+\frac{1}{t}$ with the same time complexity using the methods in [22, 23]. However, even if we can do so, the time complexity is still too high and it is impossible to try the PTAS for sequences of 100 letters.

7 Acknowledgement

We thank Tao Jiang for discussions on the topics studied in this paper.

References

1. S. Altschul and D. Lipman, Trees, stars, and multiple sequence alignment, *SIAM Journal on Applied Math.*, 49 (1989), pp. 197-209.
2. B. DasGupta, X. He, T. Jiang, M. Li, J. Tromp and L. Zhang, On distances between phylogenetic trees, *Proc. 8th Annual ACM-SIAM Symposium on Discrete Algorithms*, Jan. 1997, New Orleans.
3. B. DasGupta, X. He, T. Jiang, M. Li, and J. Tromp, On the linear-cost subtree-transfer distance, submitted to *Algorithmica*, 1997.
4. Z. Galil and R. Ciancarlo, 11Speeding up dynamic programming with applications to molecular biology", Theoretical Computer Science **64**, pp. 107-118, 1989.
5. D. Gusfield, *Algorithms on Strings, Trees, and Sequences: Computer Science and Computational Biology*, Cambridge University Press, 1997.
6. J. Hein, A new method that simultaneously aligns and reconstructs ancestral sequences for any number of homologous sequences, when the phylogeny is given, *Mol. Biol. Evol.* 6 (1989), 649-668.
7. J. Hein, Reconstructing evolution of sequences subject to recombination using parsimony, *Math. Biosci.* 98(1990), 185-200.
8. J. Hein, A heuristic method to reconstruct the history of sequences subject to recombination, *J. Mol. Evo.* 36(1993) 396-405.
9. J. Hein, T. Jiang, L. Wang, and K. Zhang, On the complexity of comparing evolutionary trees, *Discrete Applied Mathematics*, 71(1996), 153-169.
10. D. Hochbaum, *Approximation Algorithms for NP-hard Problems*, PWS, to appear.
11. S. K. Kannan and E. W. Myers, "An algorithm for locating non-overlapping regions of maximum alignment score", CPM93, pp. 74-86, 1993.
12. J. Kececioglu and D. Gusfield, " Reconstructing a history of recombinations from a set of sequences", 5th Annual ACM-SIAM Symposium on Discrete Algorithms, Arlington, Virginia, pp. 471-480, January 1994
13. G. M. Landau and J. P. Schmidt, "An algorithm for approximate tandem repeats, CPM'93, pp. 120-133, 1993.
14. D. Sankoff, Minimal mutation trees of sequences, *SIAM J. Applied Math.* 28(1975), 35-42.
15. D. Sankoff, R. J. Cedergren and G. Lapalme, Frequency of insertion-deletion, transversion, and transition in the evolution of 5S ribosomal RNA, *J. Mol. Evol.* 7(1976), 133-149.
16. D. Sankoff and R. Cedergren, Simultaneous comparisons of three or more sequences related by a tree, In D. Sankoff and J. Kruskal, editors, *Time warps, string edits, and macromolecules: the theory and practice of sequence comparison*, pp. 253-264, Addison Wesley, 1983.
17. F. W. Stahl, "Genetic recombination", Scientific American, 90-101, February 1987.
18. D. Swofford and G. Olson, Phylogenetic reconstruction, in *Molecular Systemtics*, D. Hillis and C. Moritz (eds), Sinauer Associates, Sunderland, MA, 1990.
19. J. D. Watson, N. H. Hopkins, J. W. Roberts, J. A. Steitz, A. M. Weiner, "Molecular Biology of the gene, 4th edition, Benjamin-Cummings, Menlo Park, California, 1987.
20. L. Wang and T. Jiang, On the complexity of multiple sequence alignment, *Journal of Computational Biology*, 1(1994), 337-348.
21. L. Wang, T. Jiang and E.L. Lawler, Approximation algorithms for tree alignment with a given phylogeny, *Algorithmica*, 16(1996), 302-315.

22. L. Wang and D. Gusfield, Improved approximation algorithms for tree alignment, *Journal of Algorithms*, vol. 25, pp. 255-173, 1997.
23. L. Wang, T. Jiang, and Dan Gusfield, A more efficient approximation scheme for tree alignment, *Proceedings of the First Annual International Conference on Computational Molecular Biology*, pp. 310-319, 1997.
24. M.S. Waterman, *Introduction to Computational Biology: Maps, sequences, and genomes*, Chapman and Hall, 1995.

Aligning Alignments

John D. Kececioglu* Weiqing Zhang†

Abstract While the area of sequence comparison has a rich collection of results on the alignment of two sequences, and even the alignment of multiple sequences, there is little known about the alignment of two *alignments*. The problem becomes interesting when the alignment objective function counts gaps, as is common when aligning biological sequences, and has the form of the sum-of-pairs objective. We begin a thorough investigation of aligning two alignments under the sum-of-pairs objective with general linear gap costs when either of the two alignments are given in the form of a *sequence* (a degenerate alignment containing a single sequence), a *multiple alignment* (containing two or more sequences), or a *profile* (a representation of a multiple alignment often used in computational biology). This leads to five problem variations, some of which arise in widely-used heuristics for multiple sequence alignment, and in assessing the relatedness of a sequence to a sequence family. For variations in which exact gap counts are computationally difficult to determine, we offer a framework in terms of *optimistic* and *pessimistic* gap counts. For optimistic and pessimistic gap counts we give efficient algorithms for the sequence vs. alignment, sequence vs. profile, alignment vs. alignment, and profile vs. profile variations, all of which run in essentially $O(mn)$ time for two input alignments of lengths m and n. For exact gap counts, we give the first provably efficient algorithm for the sequence vs. alignment variation, which runs in essentially $O(mn \log n)$ time using the candidate-list technique developed for convex gap-costs, and we conjecture that the alignment vs. alignment variation is NP-complete.

Keywords Sequence comparison, sum-of-pairs alignment, affine gap costs, quasi-natural gap costs, profiles

1 Introduction

While the comparison of two or more *sequences* is a richly developed area that has now made its way into several texts [19, 24, 20, 12], it is interesting that the seemingly similar problem of comparing two *alignments* has received little attention, even though it arises naturally when constructing approximate multiple sequence alignments [2, 18], and evaluating the relatedness of a sequence

*Department of Computer Science, The University of Georgia, Athens, GA 30602-7404. Electronic mail: kece@cs.uga.edu. Research supported in part by a National Science Foundation CAREER Award, Grant DBI-9722339.

†Department of Medicinal Chemistry, The University of Georgia, Athens, GA 30602. Electronic mail: wzhang@rx.uga.edu

to a sequence family [25]. Given that an alignment can be viewed as a kind of sequence where each element is a column of characters, it may seem at first glance that the problem of aligning two alignments is in principle no harder than aligning two sequences, yet the problem becomes challenging when the objective function scores the resulting multiple alignment under the commonly-used sum-of-pairs measure with gaps having a cost that is a general linear function of their length.[1] In this context, the well-known dynamic programming solution for two sequences [7], which corresponds to a system of three mutual recurrences, breaks down.

In aligning two alignments under the sum-of-pairs objective with linear gap costs, we consider three forms in which an alignment may be given: (1) as an ordinary *sequence*, which is really a degenerate multiple alignment, (2) as a nondegenerate *alignment* of two or more sequences, or (3) as a *profile* [10], a representation of a multiple alignment often used in computational biology in which each column of the alignment is reduced to a distribution giving the frequency of each character of the alphabet at the particular column. Combining all pairs of forms gives rise to five new problem variations, and we consider each variation in turn, except for the alignment vs. profile variation which does not appear relevant. For many variations, accurately counting gaps is either difficult or impossible, and in this context we revisit the "quasi-natural gap costs" of Altschul [1] within a uniform framework of *optimistic* and *pessimistic* gap counts, which generalize to variations involving profiles. For optimistic and pessimistic gap counts, we give efficient algorithms for all problem variations that run in essentially $O(mn)$ time for two input alignments of lengths m and n. For exact gap counts, we give the first provably efficient algorithm for the sequence vs. alignment variation, which runs in essentially $O(mn \log n)$ time and corresponds to solving a system of 13 recurrence equations.

Our initial motivation for this study came from the problem in computational biology of assessing the relatedness, or homology, of a given sequence to a known sequence family. The second author and E. Will Taylor approached the first author with the problem of objectively assessing the significance of homology between an aligned protein family and a proposed member where the pairwise similiarity between the sequence and any isolated member of the family was quite weak. (Specifically, the problem was to determine whether a newly-sequenced HIV gene was homologous to the selenium dependent glutathione peroxidase protein family, which if true would implicate selenium deficiency in the onset of AIDS [21].) The traditional solution to this problem for two sequences is to quantify the significance of their homology by measuring a shuffling statistic, which tests how many standard deviations their optimal alignment score is above a sample of optimal alignment scores for random sequences having the

[1] The *sum-of-pairs* objective function scores a multiple alignment by the sum of the scores of its induced pairwise alignments. A *gap* in a pairwise alignment is a maximal run of insertions or deletions. We say the alignment scoring function uses *linear gap costs* when a gap of length $x > 0$ costs $a + bx$ for constants a and b, which is sometimes referred to as *affine gap costs* in the literature.

same composition of letters. Since pairwise similarity between the sequence and the family members was low, standard programs that evaluate the shuffling statistic between two sequences gave low measures of significance of homology. A much more sensitive approach is to use an algorithm for optimal alignment of the sequence to the family (which we call here the sequence vs. alignment variation) when evaluating the shuffling statistic. This in fact gave a significance of homology that was sufficiently high to support membership of the HIV gene in the peroxidase family. (A full study measuring sequence-family homology using the algorithm in this paper for the sequence vs. alignment variation with pessimistic gap costs, with a comparison among several definitions of the shuffling statistic, is reported in [25].)

The plan of the paper is as follows. In the first section we review the solution to the sequence vs. sequence problem with linear gap costs. The following sections generalize this to sequence vs. alignment, alignment vs. alignment, sequence vs. profile, and profile vs. profile variations, discussing optimistic, pessimistic, and exact gap counts. For each problem, we present the solution in terms of dynamic programming recurrences with complete boundary conditions in a form that can be readily implemented. (In this proceedings we omit proofs of correctness for the recurrences and time bounds for reasons of limited space.) We then conclude with some open problems.

2 Sequence vs. sequence

We first review how to efficiently compute an optimal alignment of a sequence A with a sequence B with linear gap costs, which was solved by Gotoh [7]. To specify the input, we denote A by the sequence $a_1 a_2 \cdots a_m$ and B by the sequence $b_1 b_2 \cdots b_n$. We refer to a *substring* of a sequence by subscripting with the range of positions: $A_{i,j} := a_i a_{i+1} \cdots a_j$.

To describe the alignment scoring function, we use δ to denote the *substitution costs* (such as the PAM [4] or BLOSUM [13] matrices in the form of dissimilarity scores), and we write $\delta(a, b)$ for the cost of substituting character a by character b. For the gap costs we use γ to denote the *gap start-up cost* and λ to denote the *gap extension cost.* Thus a gap of length $x > 0$ has cost $\gamma + \lambda x$. We denote the *null character* (usually represented in an alignment by a dash) by ϵ, and assume that for any non-null character a, $\delta(a, \epsilon) = \delta(\epsilon, a) = \lambda$, while $\delta(\epsilon, \epsilon) = 0$. Together, δ, γ, and λ specify the alignment scoring function.

Recall that the key to solving the alignment problem with linear gap costs is to record how the optimal alignment over a prefix of A and a prefix of B *ends*, since the gap cost associated with a given alignment column is completely determined by the preceding column. An optimal alignment can end in one of three ways. Let

- $D(i, j)$ be the cost of an optimal aligment of $A_{1,i}$ versus $B_{1,j}$ that ends by substituting a_i with b_j (which corresponds to a *diagonal* edge from node $(i-1, j-1)$ to (i, j) in the standard dynamic programming graph),

- $V(i,j)$ be the cost of an optimal alignment of $A_{1,i}$ versus $B_{1,j}$ that ends by deleting a_i (which corresponds to a *vertical* edge into node (i,j)), and
- $H(i,j)$ be the cost of an optimal alignment of $A_{1,i}$ versus $B_{1,j}$ that ends by inserting b_j (which corresponds to a *horizontal* edge into node (i,j)).

Then the cost of an optimal alignment of A and B is

$$\min\{D(m,n), V(m,n), H(m,n)\}.$$

We now give the recurrences for D, V, and H, with complete boundary conditions. In the recurrences, cost γ is incurred once at the start of a gap.

For $i > 0$ and $j > 0$,

$$H(i,j) = \lambda + \min\begin{cases} H(i,j-1), \\ V(i,j-1) + \gamma, \\ D(i,j-1) + \gamma \end{cases} \tag{1}$$

$$V(i,j) = \lambda + \min\begin{cases} H(i-1,j) + \gamma, \\ V(i-1,j), \\ D(i-1,j) + \gamma \end{cases} \tag{2}$$

$$D(i,j) = \delta(a_i,b_j) + \min\begin{cases} H(i-1,j-1), \\ V(i-1,j-1), \\ D(i-1,j-1) \end{cases} \tag{3}$$

For $i = 0$ and $j > 0$,

$$H(0,j) = H(0,j-1) + \lambda, \tag{4}$$
$$V(0,j) = H(0,j) + \gamma, \tag{5}$$
$$D(0,j) = H(0,j). \tag{6}$$

For $i > 0$ and $j = 0$,

$$V(i,0) = V(i-1,0) + \lambda, \tag{7}$$
$$H(i,0) = V(i,0) + \gamma, \tag{8}$$
$$D(i,0) = V(i,0). \tag{9}$$

For $i = 0$ and $j = 0$,

$$H(0,0) = \gamma, \tag{10}$$
$$V(0,0) = \gamma, \tag{11}$$
$$D(0,0) = 0 \tag{12}$$

Recurrences (1) through (3) are often presented in the literature in an optimized form. The recurrence for $D(i,j)$ is replaced by a recurrence for a quantity $S(i,j)$ which represents $\min\{D(i,j), V(i,j), H(i,j)\}$. This yields a system that references only seven terms at an (i,j)-position, instead of nine as above.

Interestingly, this optimization does not generalize to the alignment problems considered in this paper. Hence we present the basic sequence-sequence recurrences in the above unoptimized form as this makes comparison with the more general recurrences more straightforward.

Evaluating the above recurrences in three $(m+1) \times (n+1)$ tables that correspond to H, V, and D above, and tracing back through the tables to recover the optimal alignment once the solution value is known, gives the following result.

Theorem 1 (Gotoh [7]) *An optimal alignment of two sequences of lengths m and n with linear gap costs can be computed in $O(mn)$ time.*

Myers and Miller [17] show that an optimal alignment with linear gap costs can be found in the above time using only $O(\min\{m, n\})$ working space, by applying a technique of Hirschberg [14].

3 Sequence vs. alignment

As soon as one generalizes from comparing two sequences to comparing several sequences or, as considered in this paper, to comparing two objects one or both of which may be alignments, the problem of optimal alignment with linear gap costs becomes more difficult. When aligning several sequences or two alignments, the output is a multiple sequence alignment, and a common objective function on multiple alignments is the *sum-of-pairs* objective [3], which takes as the cost of a multiple alignment the sum of the costs of all induced pairwise alignments. If in the sum-of-pairs objective pairwise alignments are scored with linear gap costs, the problem of counting gaps in the dynamic programming recurrences arises. Unfortunately in this more general setting it is no longer possible to determine when a gap is started by keeping track of the preceding column of the alignment, or for that matter any fixed number of preceding columns. To deal with this difficulty, Altschul [1] introduced what were termed "quasi-natural gap costs." This is a practical suggestion, in which an induced pairwise alignment that ends with the pattern $\begin{smallmatrix} -\mathtt{X} \\ -- \end{smallmatrix}$ or $\begin{smallmatrix} -- \\ -\mathtt{X} \end{smallmatrix}$, where `X` represents a non-null character, is considered to start a gap, even though it may not. (For all other patterns there is no difficulty in determining the start of a gap.) While this will not give exact gap counts, it allows gap start-ups to be counted on the basis of the preceding column alone as in Gotoh's solution of Section 2, and hopefully overestimates the true count only rarely.

In this paper we refer to this type of compromise in terms of *optimistic* and *pessimistic* gap counts. Pessimistic gap counts count one gap start-up in the situation $\begin{smallmatrix} -\mathtt{X} \\ -- \end{smallmatrix}$ and $\begin{smallmatrix} -- \\ -\mathtt{X} \end{smallmatrix}$, while optimistic gap counts do not. Pessimistic counts are used in the multiple alignment program `MSA` [15, 11] which scores alignments under the sum-of-pairs objective; pessimistic counts are needed in this context instead of optimistic counts so that the sum of all optimal pairwise alignments yields a lower bound on the objective function. However, pessimistic gap counts can overestimate the exact count by an arbitrarily large value, and it can be argued that optimistic counts are likely to yield a more accurate estimate.

The next section develops recurrences for the sequence vs. alignment variation with optimistic or pessimistic gap counts. Following this we give a more involved system of recurrences that solves the problem with *exact* gap counts.

3.1 Optimistic and pessimistic gap counts

In the sequence vs. alignment variation, the input is a sequence A and an alignment B. We again denote A by the sequence $a_1 \cdots a_m$, but now B is represented by a $k \times n$ matrix (b_{ij}), where each character b_{ij} is either a symbol from the alphabet Σ or the null character ϵ. (Thus B is a multiple alignment of length n of k sequences.) The problem is to align the characters of A to the columns of B so as to minimize the sum over all rows of B of the cost of the induced pairwise alignment between A and the given row of B, where pairwise alignments are scored with linear gap costs. This differs from general sum-of-pairs multiple alignment [3] (which is intractable [22]) in that the alignment on the rows of B is fixed. The alignment cost function is again given by substitution costs δ, gap start-up cost γ, and gap extension cost λ.

To count whether or not an element b_{ij} of alignment B is the null character, we define

$$\beta_{ij} := \begin{cases} 1, & b_{ij} \neq \epsilon \text{ and } j > 0; \\ 0, & b_{ij} = \epsilon \text{ and } j > 0; \\ 1, & j = 0. \end{cases}$$

(This treats alignment B as starting with an artificial 0th column of non-null characters for the boundary conditions of the recurrences.) The number of non-null characters in column j of B is then $f_1(j) := \sum_{1 \le i \le k} \beta_{ij}$.

To count gap start-ups we use a function $\Gamma(a_1 a_2, b_1 b_2)$ given in tabular form below. Essentially, $a_1 a_2$ is a binary string that represents whether the last two entries of the row for A are non-null characters in the alignment, while $b_1 b_2$ is a binary string that represents whether the last two entries of a given row for B are non-null characters in the alignment. In the table for Γ, rows are indexed by $a_1 a_2$, and columns by $b_1 b_2$.

	00	01	10	11
00	0	$\{0,1\}$	0	0
01	$\{0,1\}$	0	1	0
10	0	1	0	1
11	0	0	1	0

The two entries of Γ that differ for optimistic and pessimistic gap counts contain the set $\{0,1\}$; with optimistic counts both entries are 0, while for pessimistic counts both entries are 1.

We can now write down recurrences for the value of an optimal alignment with optimistic or pessimistic gap costs. As in Section 2, let $H(i,j)$, $V(i,j)$, and $D(i,j)$ be the cost of an optimal alignment of the first i characters of A against the first j columns of B that ends with an insertion, deletion, or substitution, respectively. For $i > 0$ and $j > 0$,

$$H(i,j) = \lambda f_1(j) + \min \begin{cases} H(i,j-1) + \gamma \sum_{1 \le r \le k} \Gamma(00, \beta_{rj-1}\beta_{rj}), \\ V(i,j-1) + \gamma \sum_{1 \le r \le k} \Gamma(10, 0\beta_{ij}), \\ D(i,j-1) + \gamma \sum_{1 \le r \le k} \Gamma(10, \beta_{rj-1}\beta_{rj}) \end{cases} \tag{13}$$

$$V(i,j) = \lambda k + \min \begin{cases} H(i-1,j) + \gamma \sum_{1 \le r \le k} \Gamma(01, \beta_{rj}0), \\ V(i-1,j) + \gamma \sum_{1 \le r \le k} \Gamma(11, 00), \\ D(i-1,j) + \gamma \sum_{1 \le r \le k} \Gamma(11, \beta_{rj}0) \end{cases} \tag{14}$$

$$D(i,j) = \sum_{1 \le r \le k} \delta(a_i, b_{rj}) + \min \begin{cases} H(i-1,j-1) + \gamma \sum_{1 \le r \le k} \Gamma(01, \beta_{rj-1}\beta_{rj}), \\ V(i-1,j-1) + \gamma \sum_{1 \le r \le k} \Gamma(11, 0\beta_{rj}), \\ D(i-1,j-1) + \gamma \sum_{1 \le r \le k} \Gamma(11, \beta_{rj-1}\beta_{rj}) \end{cases} \tag{15}$$

The recurrences can be simplified using the following counts. Let $f_b(j)$ count the number of occurrences of character $b \in \Sigma$ in column j of B, and for $x, y \in \{0,1\}$, let us count patterns of null- and non-null characters by

$$\begin{aligned} f_x(j) &:= \sum_{1 \le r \le k} \xi(x, \beta_{rj}), \\ f_{xy}(j) &:= \sum_{1 \le r \le k} \xi(x, \beta_{rj-1})\, \xi(y, \beta_{rj}), \end{aligned}$$

where $\xi(x,y)$ is 1 if $x = y$ and 0 otherwise. Finally, let $f_{\perp 1}(j)$ denote the number of non-null characters in column j of B that are the *first* non-null character in their row. The recurrences in terms of these functions, together with complete boundary conditions, are given below. The boundary conditions count gaps exactly.

For $i > 0$ and $j > 0$,

$$H(i,j) = \lambda f_1(j) + \min \begin{cases} H(i,j-1) + \gamma \begin{Bmatrix} 0, & \text{opt;} \\ f_{01}(j), & \text{pess.} \end{Bmatrix}, \\ V(i,j-1) + \gamma f_1(j), \\ D(i,j-1) + \gamma f_1(j) \end{cases} \tag{16}$$

$$V(i,j) \quad = \quad \lambda k \; + \; \min \begin{cases} H(i-1,j) \; + \; \gamma \begin{Bmatrix} f_1(j), & \text{opt;} \\ k, & \text{pess.} \end{Bmatrix}, \\ V(i-1,j), \\ D(i-1,j) \; + \; \gamma f_1(j) \end{cases} \tag{17}$$

$$D(i,j) \quad = \quad \sum_{b \,\in\, \Sigma \cup \{\epsilon\}} f_b(j)\, \delta(a_i, b) \; + \; \min \begin{cases} H(i-1,j-1) \; + \; \gamma \begin{Bmatrix} f_{10}(j), & \text{opt;} \\ f_0(j), & \text{pess.} \end{Bmatrix}, \\ V(i-1,j-1), \\ D(i-1,j-1) \; + \; \gamma f_{10}(j) \end{cases} \tag{18}$$

For $i = 0$ and $j > 0$,

$$H(0,j) \quad = \quad H(0,j-1) \; + \; \lambda f_1(j) \; + \; \gamma f_{\perp 1}(j), \tag{19}$$
$$V(0,j) \quad = \quad H(0,j) \; + \; \gamma k, \tag{20}$$
$$D(0,j) \quad = \quad V(0,j). \tag{21}$$

For $i > 0$ and $j = 0$,

$$V(i,0) \quad = \quad V(i-1,0) \; + \; \lambda k, \tag{22}$$
$$H(i,0) \quad = \quad V(i,0) \; + \; \gamma f_1(1), \tag{23}$$
$$D(i,0) \quad = \quad H(i,0). \tag{24}$$

For $i = 0$ and $j = 0$,

$$V(0,0) \quad = \quad \gamma k, \tag{25}$$
$$H(0,0) \quad = \quad 0, \tag{26}$$
$$D(0,0) \quad = \quad 0. \tag{27}$$

It is straightforward to verify that recurrences (16) through (18) are faithful to (13) through (15), and to check boundary conditions (19) through (27). Evaluating these recurrences in three tables, and tracing back to recover the optimal alignment, gives the following result.

Theorem 2 *An optimal alignment of a sequence versus an alignment under the sum-of-pairs objective with optimistic or pessimistic gap counts can be computed in time*

$$O(kn \, + \, \min\{s,k\} mn),$$

where m is the length of the input sequence, n is the length of the input alignment, k is the number of sequences in the input alignment, and s is the size of the alphabet.

The factor of $\min\{s,k\}$ in the running time comes from evaluating the substitution cost for $D(i,j)$ by summing over the s symbols in the alphabet, as shown in (18), or by summing over the k entries in the column of B.

3.2 Exact gap counts

Examining the optimistic-pessimistic recurrences, the only source of inaccuracy is in counting gap start-ups in the HH, HV, and HD cases of equations (16) through (18), in which the second-to-last operation is an insertion of a column of B, corresponding to the H-term in the minimum. These are the only cases in which the pattern $\genfrac{}{}{0pt}{}{\texttt{--}}{\texttt{-x}}$ or $\genfrac{}{}{0pt}{}{\texttt{-x}}{\texttt{--}}$ can arise, which are the only situations in which the optimistic-pessimistic gap counts are inaccurate. We now develop recurrences that handle these three cases exactly.

Developing an exact system of recurrences

The key observation is the following. In the HH, HV, and HD cases, the insertion that is the second-to-last operation may, in an optimal alignment, be itself preceded by an insertion, giving rise to cases HHH, HHV, and HHD. This third-to-last insertion may also contain the pattern $\genfrac{}{}{0pt}{}{\texttt{-}}{\texttt{-}}$, and may, in an optimal alignment, be preceded by yet another insertion. Notice, however, that in these cases with an HH $\cdots$ H run, eventually the run of insertions must be preceded by (1) a substitution, corresponding to $\cdots$ DHH $\cdots$ H, (2) a deletion, corresponding to $\cdots$ VHH $\cdots$ H, or (3) go all the way back to the beginning of B, corresponding to HH $\cdots$ H. In all three situations, any gap start-ups in the run of H's can be counted exactly, by determining whether a given character b_{ij} in B is the first non-null character in its row encountered during the run. Incorporating this idea results in the following system of recurrences for exact gap counts.

For $i > 0$ and $j > 0$,

$$H(i,j) = \lambda f_1(j) + \min \begin{cases} H_H(i,j-1), \\ V(i,j-1) + \gamma f_1(j), \\ D(i,j-1) + \gamma f_1(j) \end{cases} \tag{28}$$

$$V(i,j) = \lambda k + \min \begin{cases} H_V(i-1,j), \\ V(i-1,j), \\ D(i-1,j) + \gamma f_1(j) \end{cases} \tag{29}$$

$$D(i,j) = \sum_{b \in \Sigma \cup \{\epsilon\}} f_b(j)\, \delta(a_i, b) + \min \begin{cases} H_D(i-1,j-1), \\ V(i-1,j-1), \\ D(i-1,j-1) + \gamma f_{10}(j) \end{cases} \tag{30}$$

$$H_H(i,j) = \begin{cases} \displaystyle\min_{0 \le j' < j} \left\{ \min \begin{Bmatrix} V(i,j'), \\ D(i,j') \end{Bmatrix} + G_H(j',j) \right\}, & j > 0; \\ H(i,0), & j = 0. \end{cases} \tag{31}$$

$$H_V(i,j) = \begin{cases} \min \begin{Bmatrix} \displaystyle\min_{0 \le j' < j} \{V(i,j') + G_{VV}(j',j)\}, \\ \displaystyle\min_{0 \le j' < j} \{D(i,j') + G_{DV}(j',j)\} \end{Bmatrix}, & i > 0; \\ H(0,j) + \gamma k, & i = 0. \end{cases} \tag{32}$$

$$H_D(i,j) = \begin{cases} \min \left\{ \begin{array}{l} \min\limits_{0 \le j' < j} \{V(i,j') + G_{VD}(j',j)\}, \\ \min\limits_{0 \le j' < j} \{D(i,j') + G_{DD}(j',j)\} \end{array} \right\}, & i > 0 \text{ and } j > 0; \\ V(i,0), & i > 0 \text{ and } j = 0; \\ H(0,j) + \gamma f_0(j+1), & i = 0. \end{cases} \tag{33}$$

$$G_H(j',j) = F(j',j) + g_H(j, j-j'), \tag{34}$$

$$G_{VV}(j',j) = F(j',j) + g_V(j, j-j'), \tag{35}$$

$$G_{DV}(j',j) = F(j',j) + g_V(j, j-j'+1), \tag{36}$$

$$G_{VD}(j',j) = F(j',j) + g_D(j, j-j'), \tag{37}$$

$$G_{DD}(j',j) = F(j',j) + g_D(j, j-j'+1), \tag{38}$$

$$F(j',j) = \left\{ \begin{array}{ll} F(j',j-1) + \lambda f_1(j) + g_H(j-1, j-j'-1), & j' < j; \\ 0, & j' = j. \end{array} \right\}, \tag{39}$$

$$L(i,j) = \left\{ \begin{array}{ll} L(i,j-1) + 1, & j > 0 \text{ and } b_{ij} = \epsilon; \\ 0, & \text{otherwise.} \end{array} \right\}, \tag{40}$$

where we define

$$g_H(j,\ell) := \gamma \sum_{1 \le r \le k} \begin{cases} 1, & b_{r\,j+1} \ne \epsilon \text{ and } L(r,j) \ge \ell; \\ 0, & \text{otherwise.} \end{cases}$$

$$g_V(j,\ell) := \gamma \sum_{1 \le r \le k} \begin{cases} 1, & L(r,j) < \ell; \\ 0, & \text{otherwise.} \end{cases}$$

$$g_D(j,\ell) := \gamma \sum_{1 \le r \le k} \begin{cases} 1, & b_{r\,j+1} = \epsilon \text{ and } L(r,j) < \ell; \\ 0, & \text{otherwise.} \end{cases}$$

In the above, $H_H(i,j)$, $H_V(i,j)$, and $H_D(i,j)$ are the cost of an optimal alignment of the first i characters of A versus the first j columns of B that ends by inserting the jth column of B, *plus* the cost of any gaps started by following with an insertion, deletion, or substitution, respectively. Quantity $F(j',j)$ is the cost of insertions plus gap startups for inserting columns $j'+1$ through j of B, while $L(i,j)$ is the length of the gap in row i of B measured starting from column j of B and extending to the left. Quantities $g_H(j,\ell)$, $g_V(j,\ell)$, and $g_D(j,\ell)$ give the cost of gap startups for following a run of insertions up through column j of B with an insertion, deletion, or substitution, respectively, where ℓ varies depending on the length of the run and whether the run is preceded by a deletion or a substitution.

The boundary conditions for the recurrences are the same as equations (19) through (27) for optimistic-pessimistic gap counts, except that equation (24) is replaced by

$$D(i,0) := V(i,0). \tag{41}$$

Straightforwardly evaluating the recurrences, using 9 tables to store V, D, F, g_H, g_V, g_D, L, f_1, f_{10}, and $|\Sigma|+1$ additional tables to hold the f_c counts for

$c \in \Sigma \cup \{\epsilon\}$, and tracing back to recover the optimal alignment, takes time

$$O(kn^2 + \min\{s,k\}mn + mn^2),$$

where m is the length of A, n is the length of B, k is the number of sequences in B, and s is the size of Σ. The first term is the time to precompute g_H, g_V, g_D, and F, the second term is the time to evaluate H, V, and D assuming H_H, H_V, and H_D are known, and the third term is the time to evaluate H_H, H_V, and H_D directly from their recurrences.

When m and n are large compared to k, which corresponds to the common situation of aligning a few long sequences, the final $O(mn^2)$ term dominates the running time, which is then cubic in the length of the inputs. We now show that by applying the candidate-list technique for sequence alignment with convex gap-costs pioneered independently by Miller and Myers [16] and Galil and Giancarlo [6] (and suggested by Waterman [23]), we can reduce the time to compute the 1-dimensional minimizations for H_H, H_V, and H_D to $O(mn \log n)$ time in total, which is close to quadratic in the length of the inputs.

Applying the candidate-list technique

Examining equations (31) through (33), we can abstract the five 1-dimensional minimization problems for H_H, H_V, and H_D into the problem of computing a function $h(j)$ for $j = 1, 2, \ldots, n$, where

$$h(j) \;:=\; \min_{0 \le i < j} \Big\{ h_i(j) \Big\},$$

and each function $h_i(j)$ has the form

$$h_i(j) \;:=\; a(i) + b(i,j).$$

In our context the function $a(i)$, for a fixed row r of $H_H(r,\cdot)$, $H_V(r,\cdot)$, or $H_D(r,\cdot)$, is either $V(r,i)$ or $D(r,i)$ or the minimum of the two, and the function $b(i,j)$ is either $g_H(j,j-i)$, $g_V(j,j-i)$, $g_D(j,j-i)$, or a slight variant. From this simplified perspective, computing function h is the problem of determining the minimum envelope of a family of curves, where curve h_i of the family starts at position $i+1$ in the domain.

If it can be shown that any two curves in the family satisfy a *crossing-once* property in the sense that once two such curves cross they never cross again, then the minimum envelope of the family can be found very efficiently [16, 6]. The key to proving the crossing-once property is to consider forward differences of the functions. Let $\Delta h_i(j)$ denote $h_i(j) - h_i(j-1)$.

Lemma 1 *Let functions $h_i(j)$ be defined as above for H_H, H_V, and H_D, and suppose $0 \le i \le i'$. Then for all $j > i'$,*

$$\Delta h_i(j) \;\le\; \Delta h_{i'}(j).$$

Proof sketch We can prove separately that the inequality holds when the functions h_i are defined as appropriate for H_H, H_V, or H_D.

First consider the functions h_i defined for H_H. By straightforward manipulation,

$$\Delta h_{i'}(j) - \Delta h_i(j) = g_H(j, j-i') - g_H(j, j-i).$$

Since $g_H(j, \ell)$ is monotonically decreasing in ℓ, this difference is at least zero.

Next consider the functions h_i defined for H_V. It is not hard to show that

$$\Delta h_{i'}(j) - \Delta h_i(j) = 2\Big(g_H(j-1, j-i'-1) - g_H(j-1, j-i-1)\Big),$$

which as argued above is at least zero.

Finally considering functions h_i defined for H_D, it can be shown that

$$\begin{aligned}\Delta h_{i'}(j) - \Delta h_i(j) &= \Big(g_H(j-1, j-i'-1) - g_H(j-1, j-i-1)\Big) \\ &\quad + \Big(g_H(j, j-i') - g_H(j, j-i)\Big),\end{aligned}$$

which again is at least zero. □

An easy consequence of Lemma 1 is the following.

Corollary 1 (Crossing-Once Property) *Let functions $h_i(j)$ be defined as for H_H, H_V, and H_D, and suppose $0 \le i \le i' < j$. If*

$$h_i(j) \le h_{i'}(j),$$

then for all $j' \ge j$,

$$h_i(j') \le h_{i'}(j').$$

Given Corollary 1 we can determine $h(1), h(2), \ldots, h(n)$ in $O(n \log n)$ total time. The idea is to incrementally update, as successive curves are considered, a partition of the domain of h into maximal intervals in which one curve of the family gives the minimum envelope. The extents of the intervals in the partition are determined by computing crossover points between pairs of curves in $O(\log n)$ time using binary search. (For a very readable exposition of this idea in the context of sequence alignment with convex gap-costs, see [12, pp. 293–302].) We summarize this result in the following theorem.

Theorem 3 *An optimal alignment of a sequence versus an alignment under the sum-of-pairs objective with exact gap counts can be computed in time*

$$O(kn^2 + \min\{s, k\}mn + mn \log n),$$

where m is the length of the sequence, n is the length of the input alignment, k is the number of sequences in the input alignment, and s is the size of the alphabet.

4 Alignment vs. alignment

We can generalize further to aligning two multiple alignments, which arises in many multiple alignment heuristics [18]. Now A is an alignment, denoted by a $k \times m$ matrix (a_{ij}), and B is an alignment, denoted by an $\ell \times n$ matrix (b_{ij}). The cost of an alignment of the columns of A against the columns of B is the sum of the costs of all pairwise alignments induced by choosing a row from A and a row from B. We first give recurrences for optimistic-pessimistic gap counts, and then discuss the situation with exact gap counts.

4.1 Optimistic and pessimistic gap counts

Recurrences for the alignment vs. alignment variation with optimistic or pessimistic gap counts can be derived in a similar manner as for the sequence vs. alignment case by summing over all pairs of rows from A and B and counting gap startups using table Γ. The recurrences can then be simplified by rewriting them in terms of functions f^A_{00}, f^B_{00}, $\ldots$, f^A_{11}, f^B_{11}, and $f^A_{\perp 1}$, $f^B_{\perp 1}$, which count the occurrences of patterns of null and non-null characters in the columns of A and B, defined as for the sequence vs. alignment case.

We give the alignment vs. alignment optimistic-pessimistic recurrences in Section A of the Appendix, and state the final result here.

Theorem 4 *An optimal alignment of two alignments under the sum-of-pairs objective with optimistic or pessimistic gap counts can be computed in time*

$$O(km + \ell n + \min\{s^2, k\ell\}mn),$$

where m and n are the lengths of the two alignments, k and ℓ are the corresponding numbers of sequences in the alignments, and s is the size of the alphabet.

4.2 Exact gap counts

In the alignment vs. alignment variation, the idea used to derive recurrences for exact gap counts for the sequence vs. alignment case no longer works. Since A is an alignment as well, preceding a run of insertions by a deletion or a substitution can still give rise to a pair of rows containing $\genfrac{}{}{0pt}{}{-}{-}$, in which case gap startups still cannot be resolved. In fact, we have been unable to find a representation of an alignment that allows us to count gap startups exactly and gives rise to a polynomial number of cases in the dynamic program. (In the sequence vs. alignment case, there is such a representation, namely an integer describing the length of a terminal run of insertions together with a bit describing whether the run was preceded by a deletion or a substitution.) This suggests that it may not be possible to find an optimal solution for the alignment vs. alignment case with exact gap counts in polynomial time.

Gotoh [8] suggests a procedure for the alignment vs. alignment variation that considers a set of candidate alignments at each (i, j)-subproblem in the dynamic program. The set of candidates is apparently potentially all alignments, which is reduced by testing certain inequalities. No worst-case time bound for the procedure is given, but the paper states "the total arithmetic operations used ... is close to [quadratic] in typical cases."

5 Sequence vs. profile

In the sequence vs. profile variation, B is now a profile [10], which is essentially a sequence of distributions over the alphabet, rather than a sequence of characters, obtained by reducing the columns of a multiple alignment to character counts. Many protein families currently are characterized by profiles, thus it is worthwhile developing an algorithm for aligning profiles; the scoring of such alignments with linear gap costs can be put on an objective basis within the optimistic-pessimistic framework.

We describe profile B by a sequence $f_c(1)$, $f_c(2)$, ..., $f_c(n)$, where n is the length of B, $c \in \Sigma \cup \{\epsilon\}$ is a character, and $f_c(i)$ gives the number of occurrences of character c at position i in B. We use k to denote the total number of characters (including the null character) at a position of B, which is constant across positions.

Given that the gapping structure of the individual sequences on which a profile is based is lost, it is not possible to define exact gap counts, but it is possible to define optimistic and pessimistic gap counts. We define *optimistic* gap counts for a sequence-profile alignment to be the minimum, over all multiple alignments obtained by arranging the non-null characters at each position in the profile, of the cost of the resulting sequence-alignment alignment under optimistic gap counts. Similarly, we define *pessimistic* gap counts for a sequence-profile alignment to be the maximum, over all multiple alignments obtained by arranging the non-null characters in the profile, of the cost of the resulting sequence-alignment alignment under pessimistic gap counts.

We give the recurrences for sequence-profile alignment with optimistic or pessimistic gap counts in Section B of the Appendix, and state the final result here.

Theorem 5 *An optimal alignment of a sequence versus a profile under the sum-of-pairs objective with optimistic or pessimistic gap counts can be computed in time $O(smn)$, where m is the length of the sequence, n is the length of the profile, and s is the size of the alphabet.*

6 Profile vs. profile

We can generalize to profile-profile alignment using the same notion of optimistic and pessimistic gap counts developed for sequence-profile alignment. We give the recurrences in Section C of the Appendix, and state the final result here.

Theorem 6 *An optimal alignment of two profiles under the sum-of-pairs objective with optimistic or pessimistic gap counts can be computed in time*

$O(s^2mn)$, where m and n are the lengths of the profiles, and s is the size of the alphabet.

7 Conclusions

We have presented recurrences with complete boundary conditions for the sequence vs. alignment, alignment vs. alignment, sequence vs. profile, and profile vs. profile variations under the sum-of-pairs objective with linear gap costs and optimistic or pessimistic gap counts, and recurrences for the sequence vs. alignment variation with exact gap counts. To our knowledge this is the first presentation in the literature of recurrences for each of these forms of the problem that permits direct implementation. Evaluating the recurrences yields $O(mn)$ time algorithms for all variations with optimistic or pessimistic gap counts, and by applying the candidate-list technique, an $O(mn \log n)$ time algorithm for the sequence vs. alignment variation with exact gap counts, where time is measured in terms of the lengths m and n of the input.

It may be worth noting one interesting observation arising from the sequence vs. alignment and alignment vs. alignment recurrences: With respect to sum-of-pairs alignment with optimistic or pessimistic gap counts, by storing just four additional integers at each position in a profile (namely the counts f_{00}, f_{01}, f_{10}, and f_{11}), profiles become equivalent in descriptive power to the original multiple alignment of the family.

Further research

The main unresolved theoretical problem is the alignment vs. alignment variation with exact gap counts. We suspect one can construct an input that forces Gotoh's procedure [8] for the alignment vs. alignment variation to take exponential time, and we conjecture that the problem in general is NP-complete. An intractability result for this problem would clarify why it is so hard to incorporate exact gap counts into many already NP-complete multiple alignment problems (see for instance [11]).

Finally from a practical point of view, it would be interesting to perform a computational study comparing optimistic versus pessimistic versus exact gap counts, and full alignments versus profiles, now that complete recurrences are available for the problem variations.

References

[1] Altschul, S.F. "Gap costs for multiple sequence alignment." *Journal of Theoretical Biology* 138, 297–309, 1989.

[2] Anson, E.L. and E.W. Myers. "ReAligner: a program for refining DNA sequence multi-alignments." Proceedings of the 1st ACM Conference on *Computational Molecular Biology*, 9–13, 1997.

[3] Carrillo, H. and D. Lipman. "The multiple sequence alignment problem in biology." *SIAM Journal on Applied Mathematics* 48, 1073–1082, 1988.

[4] Dayhoff, M.O., R.M. Schwartz and B.C. Orcutt. "A model of evolutionary change in proteins." In *Atlas of Protein Sequence and Structure* 5:3, M.O. Dayhoff editor, 345–352, 1978.

[5] Fredman, M.L. "Algorithms for computing evolutionary similarity measures with length independent gap penalties." *Bulletin of Mathematical Biology* 46:4, 553–566, 1984.

[6] Galil, Z. and R. Giancarlo. "Speeding up dynamic programming with applications to molecular biology." *Theoretical Computer Science* 64, 107–118, 1989.

[7] Gotoh, O. "An improved algorithm for matching biological sequences." *Journal of Molecular Biology* 162, 705–708, 1982.

[8] Gotoh, O. "Optimal alignment between groups of sequences and its application to multiple sequence alignment." *Computer Applications in the Biosciences* 9:3, 361–370, 1993.

[9] Gotoh, O. "Further improvement in methods of group-to-group sequence alignment with generalized profile operations." *Computer Applications in the Biosciences* 10:4, 379–387, 1994.

[10] Gribskov, M., A.D. McLachlan, and D. Eisenberg. "Profile analysis: detection of distantly related proteins." *Proceedings of the National Academy of Sciences USA* 84, 4355–4358, 1987.

[11] Gupta, S., J. Kececioglu and A. Schäffer. "Improving the practical space and time efficiency of the shortest-paths approach to sum-of-pairs multiple sequence alignment." *Journal of Computational Biology* 2:3, 459-472, 1995.

[12] Gusfield, D. *Algorithms on Strings, Trees, and Sequences: Computer Science and Computational Biology.* Cambridge University Press, New York, 1997.

[13] Henikoff, S. and J.G. Henikoff. "Amino acid substitution matrices from protein blocks." *Proceedings of the National Academy of Sciences USA* 89, 10915–10919, 1992.

[14] Hirschberg, D.S. "A linear space algorithm for computing longest common subsequences." *Communications of the ACM* 18, 341–343, 1975.

[15] Lipman, D.G., S.F. Altschul and J.D. Kececioglu. "A tool for multiple sequence alignment." *Proceedings of the National Academy of Sciences USA* 86, 4412-4415, 1989.

[16] Miller, W. and E.W. Myers. "Sequence comparison with concave weighting functions." *Bulletin of Mathematical Biology* 50, 97–120, 1988.

[17] Myers, E.W. and W. Miller. "Optimal alignments in linear space." *Computer Applications in the Biosciences* 4:1, 11–17, 1988.

[18] Myers, G., S. Selznick, Z. Zhang and W. Miller. "Progressive multiple alignment with constraints." Proceedings of the 1st ACM Conference on *Computational Molecular Biology*, 220–225, 1997.

[19] Sankoff, D. and J.B. Kruskal, editors. *Time Warps, String Edits, and Macromolecules: The Theory and Practice of Sequence Comparison.* Addison-Wesley, Reading, MA, 1983.

[20] Setubal, J. and J. Meidanis. *Introduction to Computational Molecular Biology.* PWS Publishing Company, Boston, 1997.

[21] Taylor, E.W., A. Bhat, R. Nadimpalli, W. Zhang and J.D. Kececioglu. "HIV-1 encodes a sequence overlapping env gp41 with highly significant similarity to selenium dependent glutathione peroxidases." *Journal of Acquired Immune Deficiency Syndromes and Human Retrovirology* 15:5, 393–394, 1997.

[22] Wang, L. and T. Jiang. "On the complexity of multiple sequence alignment." *Journal of Computational Biology* 1:4, 337–348, 1994.

[23] Waterman, M.S. "Efficient sequence alignment algorithms." *Journal of Theoretical Biology* 108, 333–337, 1984.

[24] Waterman, M.S. *Introduction to Computational Biology: Maps, Sequences, and Genomes.* Chapman and Hall, London, 1995.

[25] Zhang, W., J.D. Kececioglu and E.W. Taylor. "Assessing distant homology between an aligned protein family and a proposed member through accurate sequence alignment." Technical Report 97-3, Department of Computer Science, The University of Georgia, October 1997. Submitted to *Journal of Molecular Biology.*

A Alignment vs. alignment

The recurrences for the alignment vs. alignment variation with optimistic or pessimistic gap counts are as follows. For $i > 0$ and $j > 0$,

$$
\begin{aligned}
H(i,j) \;&=\; \lambda\, k\, f_1^B(j) \\
&\quad + \min \begin{cases} H(i,j-1) \;+\; \gamma \begin{cases} 0, & \text{opt;} \\ k\, f_{01}^B(j), & \text{pess.} \end{cases} \\ V(i,j-1) \;+\; \gamma \begin{cases} f_1^A(i)\, f_1^B(j), & \text{opt;} \\ k\, f_1^B(j), & \text{pess.} \end{cases} \\ D(i,j-1) \;+\; \gamma \begin{cases} f_1^A(i)\, f_1^B(j), & \text{opt;} \\ f_1^A(i)\, f_1^B(j) \;+\; f_0^A(i)\, f_{01}^B(j), & \text{pess.} \end{cases} \end{cases} \qquad (42) \\
V(i,j) \;&=\; \lambda\, \ell\, f_1^A(i)
\end{aligned}
$$

$$+\min\begin{cases} H(i-1,j) + \gamma\begin{cases} f_1^A(i)\, f_1^B(j), & \text{opt;}\\ \ell\, f_1^A(i), & \text{pess.}\end{cases}\\ V(i-1,j) + \gamma\begin{cases} 0, & \text{opt;}\\ \ell\, f_{01}^A(i), & \text{pess.}\end{cases}\\ D(i-1,j) + \gamma\begin{cases} f_1^A(i)\, f_1^B(j), & \text{opt;}\\ f_1^A(i)\, f_1^B(j) + f_{01}^A(i)\, f_0^B(j), & \text{pess.}\end{cases}\end{cases} \tag{43}$$

$$D(i,j) = \sum_{a,b\,\in\,\Sigma\cup\{\epsilon\}} f_a^A(i)\, f_b^B(j)\, \delta(a,b)$$

$$+\min\begin{cases} H(i-1,j-1) + \gamma\begin{cases} f_1^A(i)\, f_{10}^B(j), & \text{opt;}\\ f_1^A(i)\, f_0^B(j) + f_0^A(i)\, f_{01}^B(j), & \text{pess.}\end{cases}\\ V(i-1,j-1) + \gamma\begin{cases} f_{10}^A(i)\, f_1^B(j), & \text{opt;}\\ f_0^A(i)\, f_1^B(j) + f_{01}^A(i)\, f_0^B(j), & \text{pess.}\end{cases}\\ D(i-1,j-1) + \gamma\begin{cases} f_1^A(i)\, f_{10}^B(j) + f_{10}^A(i)\, f_1^B(j), & \text{opt;}\\ f_1^A(i)\, f_{10}^B(j) + f_{10}^A(i)\, f_1^B(j) \\ \quad + f_{00}^A(i)\, f_{01}^B(j) + f_{01}^A(i)\, f_{00}^B(j), & \text{pess.}\end{cases}\end{cases} \tag{44}$$

For $i = 0$ and $j > 0$,

$$H(0,j) = H(0,j-1) + \lambda\, k\, f_1^B(j) + \gamma\, k\, f_{\perp 1}^B(j), \tag{45}$$
$$V(0,j) = H(0,j) + \gamma\, \ell\, f_{\perp 1}^A(1), \tag{46}$$
$$D(0,j) = V(0,j). \tag{47}$$

For $i > 0$ and $j = 0$,

$$V(i,0) = V(i-1,0) + \lambda\, \ell\, f_1^A(i) + \gamma\, \ell\, f_{\perp 1}^A(i), \tag{48}$$
$$H(i,0) = V(i,0) + \gamma\, k\, f_{\perp 1}^B(1), \tag{49}$$
$$D(i,0) = H(i,0). \tag{50}$$

For $i = 0$ and $j = 0$,

$$V(0,0) = 0, \tag{51}$$
$$H(0,0) = 0, \tag{52}$$
$$D(0,0) = 0. \tag{53}$$

B Sequence vs. profile

The recurrences for the sequence vs. profile variation with optimistic or pessimistic gap counts are as follows. For $i > 0$ and $j > 0$,

$$H(i,j) = \lambda f_1(j) + \min\begin{cases} H(i,j-1) + \gamma\begin{cases} 0, & \text{opt;}\\ \min\{f_1(j),\ f_0(j-1)\}, & \text{pess.}\end{cases}\\ V(i,j-1) + \gamma f_1(j),\\ D(i,j-1) + \gamma f_1(j)\end{cases} \tag{54}$$

$$V(i,j) \;=\; \lambda k + \min \begin{cases} H(i-1,j) + \gamma \begin{cases} f_1(j), & \text{opt;} \\ k, & \text{pess.} \end{cases} \\ V(i-1,j), \\ D(i-1,j) + \gamma f_1(j) \end{cases} \tag{55}$$

$$\begin{aligned} D(i,j) \;=\; & \sum_{b \,\in\, \Sigma \cup \{\epsilon\}} f_b(j)\,\delta(a_i,b) \\ & + \min \begin{cases} H(i-1,j-1) + \gamma \begin{cases} \max\{f_0(j) - f_0(j-1),\, 0\}, & \text{opt;} \\ f_0(j), & \text{pess.} \end{cases} \\ V(i-1,j-1), \\ D(i-1,j-1) + \gamma \begin{cases} \max\{f_0(j) - f_0(j-1),\, 0\}, & \text{opt;} \\ \min\{f_0(j),\, f_1(j-1)\}, & \text{pess.} \end{cases} \end{cases} \end{aligned} \tag{56}$$

The boundary conditions are the same as for the sequence vs. alignment variation, except we now define

$$f_{\perp 1}(j) \;:=\; F_{\perp 1}(j) - F_{\perp 1}(j-1)$$

for $1 \le j \le n$, where

$$F_{\perp 1}(j) \;=\; \begin{cases} \left\{ \begin{array}{ll} \max\{F_{\perp 1}(j-1),\, f_1(j)\}, & \text{opt;} \\ \min\{F_{\perp 1}(j-1) + f_1(j),\, k\}, & \text{pess.} \end{array} \right\}, & j > 0; \\ 0, & j = 0. \end{cases} \tag{57}$$

and we also define $f_0(0) := 0$ and $f_1(0) := k$.

C Profile vs. profile

The recurrences for the profile vs. profile variation with optimistic or pessimistic gap counts are as follows, where for $S \in \{A, B\}$ and $x, y \in \{0, 1\}$, we define

$$\begin{aligned} g^S_{xy}(i) &:= \min\{f^S_y(i),\, f^S_x(i-1)\}, \\ h^S_{xy}(i) &:= \max\{f^S_y(i) - f^S_x(i-1),\, 0\}. \end{aligned}$$

For $i > 0$ and $j > 0$,

$$H(i,j) \;=\; \lambda k\, f^B_1(j) + \min \begin{cases} H(i,j-1) + \gamma \begin{cases} 0, & \text{opt;} \\ k\, g^B_{01}(j), & \text{pess.} \end{cases} \\ V(i,j-1) + \gamma \begin{cases} f^A_1(i)\, f^B_1(j), & \text{opt;} \\ k\, f^B_1(j), & \text{pess.} \end{cases} \\ D(i,j-1) + \gamma \begin{cases} f^A_1(i)\, f^B_1(j), & \text{opt;} \\ k\, g^B_{01}(j) + f^A_1(i)\, h^B_{01}(j), & \text{pess.} \end{cases} \end{cases} \tag{58}$$

$$V(i,j) = \lambda \ell f_1^A(i) + \min \begin{cases} H(i,j-1) + \gamma \begin{cases} f_1^A(i)\, f_1^B(j), & \text{opt;} \\ \ell f_1^A(i), & \text{pess.} \end{cases} \\ V(i,j-1) + \gamma \begin{cases} 0, & \text{opt;} \\ \ell g_{01}^A(i), & \text{pess.} \end{cases} \\ D(i,j-1) + \gamma \begin{cases} f_1^A(i)\, f_1^B(j), & \text{opt;} \\ \ell g_{01}^A(i) + f_1^B(j)\, h_{01}^A(i), & \text{pess.} \end{cases} \end{cases} \tag{59}$$

$$\begin{aligned} D(i,j) &= \sum_{a,b \,\in\, \Sigma \cup \{\epsilon\}} f_a^A(i)\, f_b^B(j)\, \delta(a,b) \\ &+ \min \begin{cases} H(i-1,j-1) + \gamma \begin{cases} f_1^A(i)\, h_{00}^B(j), & \text{opt;} \\ f_1^A(i)\, f_0^B(j) + f_0^A(i)\, g_{01}^B(j), & \text{pess.} \end{cases} \\ V(i-1,j-1) + \gamma \begin{cases} f_1^B(j)\, h_{00}^A(i), & \text{opt;} \\ f_1^B(j)\, f_0^A(i) + f_0^B(j)\, g_{01}^A(i), & \text{pess.} \end{cases} \\ D(i-1,j-1) + \gamma \begin{cases} g_{11}^A(i)\, h_{00}^B(j) + h_{00}^A(i)\, g_{11}^B(j) \\ \quad + h_{00}^A(i)\, h_{11}^B(j) + h_{11}^A(i)\, h_{00}^B(j), & \text{opt;} \\ h_{01}^A(i)\, g_{10}^B(j) + g_{10}^A(i)\, h_{01}^B(j) \\ \quad + g_{10}^A(i)\, g_{01}^B(j) + g_{01}^A(i)\, g_{10}^B(j) \\ \quad + g_{01}^A(i)\, h_{10}^B(j) + h_{10}^A(i)\, g_{01}^B(j), & \text{pess.} \end{cases} \end{cases} \end{aligned} \tag{60}$$

The boundary conditions for the recurrences are the same as for the alignment vs. alignment variation with optimistic or pessimistic gap counts, except $f_{\perp 1}^A(i)$ and $f_{\perp 1}^B(j)$ are now defined for the two profiles A and B as in the sequence vs. profile variation.

Efficient Special Cases of Pattern Matching with Swaps

Amihood Amir* — Georgia Tech and Bar-Ilan University
Gad M. Landau** — Polytechnic University and Haifa University
Moshe Lewenstein*** — Bar-Ilan University
Noa Lewenstein† — Bar-Ilan University

Abstract. Let a text string T of n symbols and a pattern string P of m symbols from alphabet Σ be given. A *swapped version* T' of T is a length n string derived from T by a series of *local swaps*, (i.e. $t'_\ell \leftarrow t_{\ell+1}$ and $t'_{\ell+1} \leftarrow t_\ell$) where each element can participate in *no more than one swap*.

The *Pattern Matching with Swaps* problem is that of finding all locations i for which there exists a swapped version T' of T where there is an exact matching of P in location i of T'.

It was recently shown that the *Pattern Matching with Swaps* problem has a solution in time $O(nm^{1/3} \log m \log^2 \sigma)$, where $\sigma = min(|\Sigma|, m)$.

We consider some interesting special cases of patterns, namely, patterns where there is no *length-one run*, i.e. there are no $a, b, c \in \Sigma$ where $b \neq a$ and $b \neq c$ and where the substring abc appears in the pattern. We show that for such patterns the pattern matching with swaps problem can be solved in time $O(n \log^2 m)$.

Key Words: Design and analysis of algorithms, combinatorial algorithms on words, pattern matching, pattern matching with swaps, approximate pattern matching, generalized pattern matching.

1 Introduction

The *exact string matching* problem of seeking all locations in a text string where a pattern string matches, is a classical problem in computer science. Much effort

* College of Computing, Georgia Institute of Technology, Atlanta, GA 30332-0280, (404)894-5224; Partially supported by NSF grant CCR-96-10170 and the Israel Ministry of Science and the Arts grants 6297 and 8560.

** Department of Computer Science, Haifa University, Haifa 31905, Israel (972-4)824-0103; landau@poly.edu; partially supported by NSF grants CCR-9305873 and CCR-9610238.

*** Department of Mathematics and Computer Science, Bar-Ilan University, 52900 Ramat-Gan, Israel, (972-3)531-8407; moshe@cs.biu.ac.il.

† Department of Mathematics and Computer Science, Bar-Ilan University, 52900 Ramat-Gan, Israel, (972-3)531-8407; noa@cs.biu.ac.il; Partially supported by the Israel Ministry of Science and the Arts grant 8560.

was devoted to its solution in the early 70's (e.g. [4, 5, 7, 6, 14]) which provided algorithms, data structures, and techniques that became textbook materials.

The recent explosion in multimedia, digital libraries and the world wide web, has spurred renewed efforts in the pattern matching area [12]. The new research directions are *generalized matching* and *approximate matching.*

In *generalized matching,* one still seeks all exact occurrences of the pattern in the text, but a *match* is not *equality.* Rather, a more general matching relation is defined. Examples are string matching with "don't cares" [5, 13], less-than matching [3], or a matching relation defined by a graph on the alphabet [11].

In *approximate matching,* a distance metric between strings is defined. One then seeks all text locations where the pattern matches the text within the allowed distance under the given metric. An approximation can clearly be defined on a generalized matching relation, however the earliest distance metrics were defined on exact matching. Levenshtein used the *edit distance* of *insertions, deletions* and *mismatches* [9]. Lowrance and Wagner [10] added to the distance metric the *swap* as a unit cost error. The *swap,* where the order of two consecutive symbols is reversed, is one of the most typical typing errors. A much more complex version of swaps occurs in gene mutations.

The *approximate matching problem* is that of finding the smallest distance between every text location and the pattern. In the case of the Lowrance and Wagner edit distance, we want the smallest number of edit operations by which every text location will be brought to match the pattern. The fastest known deterministic worst-case solution to this problem takes time $O(nm)$, where n is the text length and m is the pattern length. This solution uses dynamic programming. Despite numerous atempts to "break" the $O(nm)$ barrier, there is still no known algorithm that solves the approximate pattern matching problem with the above four edit operations in time $o(nm)$.

Attempts were made in the past to limit the edit operations, and see if more efficient algorithms result. It was shown [1, 8] that if only mismatches are used, then one can indeed solve the approximate pattern matching problem in time $O(n\sqrt{m \log m})$.

In a recent paper [2], the swap edit operation was isolated. Both the approximate matching problem with the number of swaps as the error metric, and the exact swapped matching problem, were considered. In the latter, a generalized matching relation - *the swap match* is defined.

A pattern P *swap matches* a text T at location i if an unrestricted number of local swaps can bring the text to exactly match the pattern, provided only adjacent symbols are swapped, and no symbol is swapped more than once. In [2] it was shown that the problem of pattern matching with swaps can be solved in time $O(nm^{1/3} \log m \log^2 \sigma)$, where $\sigma = min(|\Sigma|, m)$. It was conjectured there that the problem can be solved in time $O(n \text{ polylog } m)$

In this paper we consider several special cases of patterns. If every time a new symbol appears in the pattern, there is a *run* of at least *four* consecutive appearances of the symbol, then we show that the problem can be solved in linear time for general alphabets.

If the length of the minimum run is limited to *two* rather than four, we provide an $O(n \log^2 m)$ algorithm for general alphabets, thus giving evidence in support of the conjecture of [2].

2 Problem Definition – Pattern Matching with Swaps

Definition: Let $T = t_1, \ldots, t_n$ be a *text* string over alphabet Σ and $P = p_1, \ldots, p_m$ be a *pattern* string over Σ. The string $\overline{T}$ is *derived from* T *via a swap* if $\exists \ell$, $1 \leq \ell \leq n-1$, such that $\overline{t}_\ell = t_{\ell+1}$, $\overline{t}_{\ell+1} = t_\ell$, and for every j, $j \neq \ell, \ell+1$, $\overline{t}_j = t_j$. T' is a *swapped version* of T if there is a sequence $T = T^1, \cdots, T^s = T'$ where T^{x+1} is derived from T^x via a swap and where each element t_ℓ is involved in no more than one swap derivation in the sequence.

The *Pattern Matching with Swaps Problem* is the following:
INPUT: Text string $T = t_1, \ldots, t_n$ and pattern string $P = p_1, \ldots, p_m$ over alphabet Σ.
OUTPUT: All locations i for which there exists a swapped version T' of T where there is an exact occurrence of P in T' starting from i, i.e. where $t'_i = p_1, t'_{i+1} = p_2, \ldots, t'_{i+m-1} = p_m$.

For ease of the exposition we will also consider the following restriction on the matching.

Restriction: When matching a pattern P with swapped version T' in location i we do not allow the symbol t'_i to have participated in any swap. We call this the *Swapped Pattern Matching with Fixed Starting Symbol* problem.

The following lemma, proven in [2], assures us that this restriction does not limit generality.

Lemma 1. *The Swapped Pattern Matching with Fixed Starting Symbol can be solved in time $O(f(n,m))$ iff the Pattern Matching with Swaps problem can be solved in time $O(f(n,m))$.*

Further Restriction: As we did in the Swapped Pattern Matching with Fixed Starting Point, we may in addition restrict the Swapped Matching so that the ending point is not allowed to participate in a swap. We get the *Swapped Matching with Fixed Endpoints* problem.

In a similar manner to Lemma 1, one can show that any solution of time complexity $O(f(n,m))$ for the Swapped Pattern Matching with Fixed Endpoints problem, can be used to derive a solution for the general Swapped Matching problem in time $O(f(n,m))$. From now on, in all our algorithms we use any of the fixed endpoints version of the problem as convenience dictates.

3 Special Patterns - Linear Time Algorithm

In the next sections we will analyze special properties of the pattern and show that under many circumstances swapped pattern matching can be done significantly faster than our general algorithm. There are some special cases where the swapped matching problem can be solved by an extremely simple method in linear time. One of them is the case where the pattern is composed of runs whose length is not smaller than 4. We start with this case because of its simplicity. We later relax the restrictions on the minimum run-length and still provide an efficient algorithm.

Definitions:

1. Let $S = s_1 \ldots s_m$ be a string. We say that a *run of length l* starts in position i if $s_{i-1} \neq s_i$, $s_i = s_{i+1} = \cdots = s_{i+l-1}$, and $s_{i+l-1} \neq s_{i+l}$. We denote such a run with s_i^l.
   The run lengths can be no less than 1 and no longer than m.
2. The *run-length notation* of string S is $a_1^{\ell_1} a_2^{\ell_2} \cdots a_r^{\ell_r}$, where $a_i \neq a_{i+1}$, $i = 1, \ldots, r-1$.

Note that if every run length in the pattern is at least 4, a swapped version of the text that matches the pattern has a very distinct look. Clearly swapping elements within a run does not change the text's appearance. The only time that swaps can have an effect is on a seam between runs. However, because the pattern has no runs of length smaller than 4, a swap would force a situation of the type $aababb$, i.e. two length 1 runs surrounded by runs of length 2 or more. In fact, this is the only possibility of a swapped text matching the pattern. This suggests the following algorithm.

Algorithm for Swapped Matching with Long Runs (Fixed Endpoints)

1. Scan the text T and construct text T' where every situation of the type $xxyxyy$ is converted to $xxxyyy$
2. Run any linear time exact matching algorithm on text T' and pattern P.
3. For every match found in location i declare it a swapped match if $p_1 = t_i$ and if $p_m = t_{i+m-1}$ in T.

End Algorithm

Time: $O(n)$.

Correctness: It follows from the discussion that every swapped match of P in the text is found in step 2. We still need to make sure that we didn't declare a match in step 2 for a location that doesn't have a swapped match. Once again, from the discussion it is clear that the only way that this can happen is on the endpoints of the pattern. In step 3 we verify which of those found have fixed endpoints.

4 Special Patterns - Runs of Length 3

Assumption: Let us further reduce the restriction on the maximum run-length in the pattern, and assume it is 3.

The following observation will be the key to a more efficient algorithm:

Observation 1 *Let $a_1^{\ell_1} a_2^{\ell_2} \cdots a_r^{\ell_r}$ be the run length notation of P, and let i be a text location where there is a swapped occurrence of P. If $\ell_j \geq 3$, $j = 1, \ldots, r$ then all the following locations do not swap (i.e. the pattern symbol exactly matches the text symbol in that location):*

$t_{i+1}, \ldots, t_{i+\ell_1-2}$
$t_{i+\ell_1+1}, \ldots, t_{i+\ell_1+\ell_2-2}$

.

.

.

$t_{i+\ell_1+\ell_2+\cdots\ell_{r-1}+1}, \ldots, t_{i+\ell_1+\ell_2+\cdots\ell_r-2}$.

The above observation is true since there is no point in swapping a pair of equal symbols. It means that in every run, only the two symbols at the edge may be swapped. All the other symbols appear in the text and pattern in exactly the same location, call them the *fixed part*. Call the leftmost (rightmost) symbol in the fixed part in a run the *left (right) fixed edge*.

Example: Let $P = aaaabbbbcccaaabbbb = a^4b^4c^3a^3b^4$. Consider a swapped matching starting at text location i (we use T_i to denote the text substring $t_i \cdots t_{i+|P|-1}$). The connected locations have no swaps.

$$
\begin{array}{lcccccccccccccccccc}
P = & a & a & a & a & b & b & b & b & c & c & c & a & a & a & b & b & b & b \\
 & & | & | & & & | & | & & & | & & & | & & & | & | & \\
T_i = & * & a & a & * & * & b & b & * & * & c & * & * & a & * & * & b & b & *
\end{array}
$$

There are five fixed parts and accordingly five left (and right) fixed edges. a in the second pattern location is an example of a left fixed edge and b in the seventh location is an example of a right fixed edge.

This leads to the following idea. Let us first find all locations where there is an *exact matching* between the part of the pattern that does not swap and the respective text locations. We mark all starts of such matchings as *potential starts*. The second step is verifying that the edges are legal. Fortunately, the legality of the edge is a local property, as can be seen from the following lemma.

Lemma 2. (Fixed Edge Local Property) *Let P be a pattern whose minimal run has length at least 3. Then there is a swapped occurrence of P at location i in T iff there is an exact matching of the pattern fixed part and the corresponding text location, and for every left fixed edge t_j either* (i) $(t_j = t_{j-1})$ *or* (ii) $(t_j = t_{j-2})$ *and for every right fixed edge t_j either* (i) $(t_j = t_{j+1})$ *or* (ii) $(t_j = t_{j+2})$.

Proof: Note that condition (i) means there is no swap on the edge and condition (ii) means there is a swap. Also note that for every two consecutive runs, condition (i) of the rightmost fixed edge holds for a run iff condition (i) of the leftmost fixed edge holds for the following run. With this in mind the lemma is obvious, since it states that there is a swapped occurrence iff the symbols of the fixed part are not swapped, and for every edge, either there is a swap or there isn't one. □

The importance of the above lemma is that it allows us to check whether a text pair is legal or not, *independently of the pattern.*

Before proceeding with our algorithm, we define one more tool that will be required.

Definition: Let $T = t_1 \cdots t_n$ and $P = p_1 \cdots p_m$ be a text and pattern over alphabet $\Sigma \cup \{\phi\}$, where $\phi \notin \Sigma$ represent a "don't care" symbol. The *String Matching with Don't Cares Problem* is that of finding all locations i in T where $\forall j,\ j = 0, \ldots, m-1$ either $t_{i+j} = p_{1+j}$ or $t_{i+j} = \phi$ or $p_{1+j} = \phi$. In other words, this is the exact matching problem but there is a wildcard symbol that is permitted to match every other alphabet symbol.

Fischer and Paterson [5] showed a convolution-based algorithm for the string matching with don't cares problem whose time complexity is $O(n \log m)$ for alphabets of finite fixed size, and $O(n \log^2 m)$ for general alphabets (in a computation model where a word has at least $\log m$ bits).

Our algorithm is now complete. Let $P = a_1^{\ell_1} a_2^{\ell_2} \cdots a_r^{\ell_r}$, and let $\ell_j \geq 3,\ j = 1, \ldots, r$. Let $\phi \notin \Sigma$ represent the "don't care" symbol.

Algorithm for Swapped Matching with Runs of length at least 3

1. **The fixed part:** Construct $P_{fixed} = \phi a_1^{\ell_1 - 2} \phi\phi a_2^{\ell_2 - 2} \phi\phi \cdots \phi\phi a_r^{\ell_r - 2} \phi$. Do a string matching with don't cares of P_{fixed} in T. Declare every match a potential start of a swapped occurrence of P in T.

2. **The leftmost fixed edges:**
   (a) Construct new text T' where

$$t'_j = \begin{cases} 1, & \text{if } j{=}2 \text{ and } t_1 = t_2 \\ & \text{or if } j \geq 3 \text{ and } [(t_j = t_{j-1}) \vee (t_j = t_{j-2})]; \\ 0, & \text{otherwise.} \end{cases}$$

   (b) Construct $P_{left-edge} = \phi 1 \phi^{\ell_1 - 1} 1 \phi^{\ell_2 - 1} 1 \cdots \phi^{\ell_{r-1}-1} 1$.
   (c) Do a string matching with don't cares of $P_{left-edge}$ in T'. For every match j, declare j a potential start of a swapped occurrence of P in T.
3. **The rightmost fixed edges:**
   (a) Construct new text T' where

$$t'_j = \begin{cases} 1, & \text{if } j{=}n-1 \text{ and } t_{n-1} = t_n \\ & \text{or if } j \leq n-2 \text{ and } [(t_j = t_{j+1}) \vee (t_j = t_{j+2})]; \\ 0, & \text{otherwise.} \end{cases}$$

   (b) Construct $P_{right-edge} = \phi^{\ell_1 - 2} 1 \phi^{\ell_2 - 1} 1 \cdots \phi^{\ell_{r-1}-1} 1 \phi^{\ell_r - 1} 1$.
   (c) Do a string matching with don't cares of $P_{right-edge}$ in T'. For every match j, declare j a potential start of a swapped occurrence of P in T.
4. The intersection of the potential starts in the fixed part step, the leftmost edges step and the rightmost edges step is exactly the set of starts of all swapped matchings of P in T.

End Algorithm

Correctness: Immediate from the fixed edge local property lemma.

Time: $O(n \log m)$ for the edge pairs step and $O(n)$ for the intersection. For constant size Σ, the fixed part step is also done in time $O(n \log m)$. For general alphabets the fixed part step takes time $O(n \log^2 m)$. We conclude:

Total Algorithm Time: $O(n \log m)$ for finite fixed size alphabets. $O(n \log^2 m)$ for general alphabets.

Note: The algorithm can be easily modified to find all swapped matches of a pattern with don't care symbols as long as all runs of alphabet symbols ($\neq \phi$) are of length at least three and the runs of don't care symbols are at least of length two.

5 Special Patterns - Runs of Length 2

We further reduce our restriction to patterns where the minimum run has length 2. At this point we no longer have any fixed part to the pattern. For example, if all runs have length 2, every element may swap. However, in such a dire situation we take advantage of another pattern property.

The following lemma establishes conditions under which it is possible to concatenate two strings and assume that if there are consecutive swapped appearances of each, there is a swapped appearance of the concatenated string. This is the local property we will need for patterns with runs of length two.

Lemma 3. (Concatenating Patterns Property) *Let $S = s_1 \cdots s_m$, $S' = s'_1 \cdots s'_{m'}$, be two strings where $s_{m-1} = s_m = \sigma$ and $s'_1 = s'_2 = \rho$. Assume that there is a swapped matching of S starting at location i of text T, and that there is a swapped matching of S' starting at location $i+m$ of the text.*
Let $P = SS'$, i.e. P is the concatenation of S and S'. Then there exists a swapped matching of P starting at location i of T.

Proof: We assume that there is a sequence of text swaps that allows matching S starting at location i, and *another* sequence of text swaps that allows matching S' starting at location $i+m$. The lemma says that these two sequences do not conflict. In fact, we will show that both S and S' have to use exactly the same text swaps on the seam (locations $i+m-1$ and $i+m$). If this is the case, we are done since the other swaps have no affect upon each other.

Since $s_{m-1} = s_m$ there can not be a swap between t_{i+m-2} and t_{i+m-1}. For the same reason t_{i+m} and t_{i+m+1} can not swap. In fact t_{i+m-1} and t_{i+m} can only swap with each other.
Therefore, if $t_{i+m-1} = \sigma$ then t_{i+m} must equal ρ and there will be no external swaps in the set of swaps of the matchings of S and S'. Otherwise if $t_{i+m-1} \neq \sigma$ it forces t_{i+m} to be σ and t_{i+m-1} to be ρ. In this case there must be external swaps in both S and S'. This swap serves P. □

This leads us to an algorithm for patterns with run-length 2 only. The idea is as follows: for each $\sigma \in \Sigma$ scan the text and mark each location that has a swapped matching of $\sigma\sigma$ with 1 and every other location 0. Mark each pattern location in which $\sigma\sigma$ begins with 1 and every other pattern location with ϕ. Now simply do pattern matching with don't cares. Merging this with the algorithm of the previous section yields an algorithm with time complexity $O(|\Sigma| n \log m)$ and correctness following from the Concatenating Patterns Property. We now show a method that eliminates the $|\Sigma|$ factor from the complexity.

If at each text location there is at most one type of pair that has a swapped matching then we can adapt the idea we just displayed by building a new text containing at each location σ if $\sigma\sigma$ matches with swaps at this location and apply don't care pattern matching with pattern $p_1 \phi p_3 \cdots p_{n-1} \phi$.
In general there may be more than one type of pair with a swapped matching at any text location but clearly not more than two. If both $\sigma\sigma$ and $\rho\rho$ have a swapped matching at text location i then the text beginning at text location $i-1$ must be of the form $\sigma\rho\rho\sigma$, $\sigma\rho\sigma\rho$, $\rho\sigma\rho\sigma$ or $\rho\sigma\sigma\rho$. This leads us to the following lemma:

Lemma 4. (Concatenating Sandwiched Pairs Property) *Let $S = s_1 \cdots s_m$, $S' = s'_1 \cdots s'_{m'}$ and $U = u_0^2 u_1^2 \ldots u_k^2$, such that $s_{m-1} = s_m \neq u_0$ and $u_k \neq s'_1 = s'_2$*

(for notation, we assume $s_m = u_{-1}$ *and* $s'_1 = u_{k+1}$*). Assume that there is a swapped matching of* S *starting at location* i *of text* T*, and that there is a swapped matching of* S' *starting at location* $i + m + 2k + 2$ *of the text.*
Let $P = SUS'$ *then* P *has a swapped matching at text location* i *iff for every* $0 \leq j \leq k$

a. *if* $u_{j-1} = u_{j+1}$ *then both, and only, the pairs* u_j^2 *and* u_{j-1}^2 *have a swapped matching at text location* $i + m + 2j$*.*

b. *if* $u_{j-1} \neq u_{j+1}$ *then* u_j^2 *is the only pair that has a swapped matching at text location* $i + m + 2j$*.*

Proof: The 'only if' side follows from induction on the concatenating patterns property.
Therefore, let us assume that $P = SUS'$ has a swapped matching at text location i. Let $0 \leq j \leq k$. Clearly u_j^2 must match with swaps at text location $i + m + 2j$. But since $p_{m+2j+1} = p_{m+2j+2}$ $(= u_j)$, t_{i+m+2j} can not swap with $t_{i+m+2j+1}$. Following the same reasoning $t_{i+m+2j+2}$ and $t_{i+m+2j+3}$ cannot swap and likewise for $t_{i+m+2j-2}$ and $t_{i+m+2j-1}$. The only swaps possible between these elements is a swap between $t_{i+m+2j-1}$ and t_{i+m+2j} and a swap between $t_{i+m+2j+1}$ and $t_{i+m+2j+2}$. This means that the only other pair that can match with swaps at location $i + m + 2j$ is u_{j-1}^2 and this happens only if $u_{j-1} = u_{j+1}$. □

We are now ready for our algorithm. The *concatenating patterns property* assures us that text locations that have a match with swaps for all runs of length at least 3 and have a match with swaps for runs of length 2 separately are exactly the text locations where the whole pattern has a swapped matching.
Checking for locations where the runs of length 3 match with swaps is done using the algorithm from the previous section. For the runs of length 2 we use the *concatenating sandwiched pairs property.* The idea is to create a new text with symbols corresponding to the letters of the pairs that have a swapped matching at that text location, more precisely if at text location i no pair matches with swaps then t'_i will be 0, if only one pair matches with swaps, e.g. 'aa', t'_i will be 'a' and if two pairs have a swapped matching, e.g. 'aa' and 'cc', t'_i will be the new symbol $< a, c >$ which is formed from the corresponding letters in lexicographic order.
We also create a new pattern with a symbol at the beginning of every pair depending on the neighboring pairs. If the neighboring pairs are of different types then we leave the letter already there, e.g. for 'bb aa cc' the first 'a' remains. If, on the other hand, the neighboring pairs are of the same type we replace it with a new symbol consisting of a letter from each type, e.g. for 'cc aa cc' the first 'a' will be replaced by $< a, c >$. The second symbol is always replaced by a don't care symbol. The matches of the new pattern in the new text capture the locations at which the length 2 runs have a swapped matching. Note that even though we have created a special pattern we still allow swaps only in the text.

The pattern we are handling is of the form $P = a_1^{\ell_1} B_1 a_2^{\ell_2} B_2 \cdots a_{r-1}^{\ell_{r-1}} B_{r-1} a_r^{\ell_r}$ where
$a_i \in \Sigma, \quad i = 1, \ldots, r.$
$\ell_i \geq 3, \quad i = 2, \ldots, r-1.$
$\ell_i = 0$ or $\ell_i \geq 3, \quad i = 1, r.$
$B_i =$ substrings over Σ where all runs are of length *exactly* 2. The first symbol of B_i is not equal to a_i, and the last symbol of B_i is not equal to a_{i+1}. The length of B_i is $b_i, \quad i = 1, \ldots, r-1$. Note that B_i may be empty, i.e. $b_i = 0$. In this case $a_i \neq a_{i+1}$.

For the sake of simplicity we also assume $\ell_1 \neq 0$ and $\ell_r \neq 0$, since this is the case in the concatenated sandwiched pair property. The only addition to the algorithm below for cases where the pattern starts (ends) with a run of length two, is the following. Create pattern P_d by deleting the first (last) two letters of the pattern P. Run the algorithm below for finding all locations of P_d and then, for each location found, check the leading (trailing) pair of characters. Retain only the locations where the leading (trailing) pair swap-match as well.

Example for the case of $\ell_1 \neq 0$ and $\ell_r \neq 0$:

$T =$	c	c	a	c	b	a	b	a	c	a	a	c	c	d	e	$\cdots$
$T' =$	c	c	c	a	b	$<a,b>$	a	a	a	$<a,c>$	0	c	0	$\cdot$	$\cdot$	$\cdots$
$P =$	c	c	c	a	a	b	b	a	a	c	c	a	a	a		
$P' =$				a	ϕ	$<a,b>$	ϕ	a	ϕ	$<a,c>$	ϕ					

As can be seen, there is an exact matching with "don't cares" of P' starting at location 4 of T'. Note that P' represent only the run-length-two part of P.

Algorithm for Swapped Matching with Runs of length at least 2

1. **Runs of length at least 3:** Construct $P' = a_1^{\ell_1} \phi^{b_1} a_2^{\ell_2} \phi^{b_2} \cdots \phi^{b_{r-1}} a_r^{\ell_r}$. Run the algorithm for swapped matching with runs of length at least 3 and find all text locations where there is a swapped match of P'.
2. **Runs of length 2 exactly:**
   (a) Construct new text T' where

$$t'_j = \begin{cases} 0, & \text{if there is no swapped match of } \sigma\sigma \text{ for any } \sigma \in \Sigma \\ & \text{starting at location } j; \\ \sigma, & \text{if } \sigma \text{ is the only letter for which there exists a swapped} \\ & \text{match of } \sigma\sigma \text{ starting at location } j; \\ <\sigma, \rho>, & \text{if } \sigma < \rho \text{ and } \sigma\sigma \text{ and } \rho\rho \text{ both have a swapped matching} \\ & \text{at text location } j. \end{cases}$$

   (b) Construct pattern P' where
   $p'_j = \phi$ for all p_j that are not in the B_i parts of the pattern, $i = 1, \ldots, r-1$. For the elements of $B_i, \quad i = 1, \ldots, r-1$, with first

element p_{k_i+2j}, we set $p'_{k_i+2j+1} = \phi$, and

$$p'_{k_i+2j} = \begin{cases} < p_{k_i+2j-1}, p_{k_i+2j} >, & \text{if } p_{k_i+2j-1} = p_{k_i+2j+2} \text{ and } p_{k_i+2j-1} < p_{k_i+2j}; \\ < p_{k_i+2j}, p_{k_i+2j-1} >, & \text{if } p_{k_i+2j-1} = p_{k_i+2j+2} \text{ and } p_{k_i+2j-1} > p_{k_i+2j}; \\ p_{k_i+2j}, & \text{otherwise.} \end{cases}$$

(c) Do a string matching with don't cares of P' in T'.

3. The intersection of the potential starts in the length 3 runs step and the potential starts in the length 2 runs step is exactly the desired set of starts of all swapped matchings of P in T.

End Algorithm

Correctness: Follows from the concatenating sandwiched pairs property and the concatenating patterns property.

Time: $O(n \log m)$ for finite fixed size alphabets, $O(n \log^2 m)$ for general alphabets.

6 Conclusions and Open Problem

We have shown that for all cases where the minimum run in the pattern is at least 2 there is a $O(nlog^2m)$ algorithm. Yet, for the general case, the best known algorithm is $O(nm^{1/3}$ polylog $m)$. We believe that the general swapped pattern matching problem can be solved in time $O(n$ polylog $m)$.

References

1. K. Abrahamson. Generalized string matching. *SIAM J. Computing*, 16(6):1039–1051, 1987.
2. A. Amir, Y. Aumann, G. Landau, M. Lewenstein, and N. Lewenstein. Pattern matching with swaps. *Proc. 38th IEEE FOCS*, pages 144–153, 1997.
3. A. Amir and M. Farach. Efficient 2-dimensional approximate matching of half-rectangular figures. *Information and Computation*, 118(1):1–11, April 1995.
4. R.S. Boyer and J.S. Moore. A fast string searching algorithm. *Comm. ACM*, 20:762–772, 1977.
5. M.J. Fischer and M.S. Paterson. String matching and other products. *Complexity of Computation, R.M. Karp (editor), SIAM-AMS Proceedings*, 7:113–125, 1974.
6. R. Karp, R. Miller, and A. Rosenberg. Rapid identification of repeated patterns in strings, arrays and trees. *Symposium on the Theory of Computing*, 4:125–136, 1972.
7. D.E. Knuth, J.H. Morris, and V.R. Pratt. Fast pattern matching in strings. *SIAM J. Computing*, 6:323–350, 1977.
8. S. Rao Kosaraju. Efficient string matching. Manuscript, 1987.

9. V. I. Levenshtein. Binary codes capable of correcting, deletions, insertions and reversals. *Soviet Phys. Dokl.*, 10:707–710, 1966.
10. R. Lowrance and R. A. Wagner. An extension of the string-to-string correction problem. *J. of the ACM*, pages 177–183, 1975.
11. S. Muthukrishnan and H. Ramesh. String matching under a general matching relation. *Information and Computation*, 122(1):140–148, 1995.
12. A. Pentland. Invited talk. NSF Institutional Infrastructure Workshop, 1992.
13. R. Y. Pinter. Efficient string matching with don't care patterns. In Z. Galil A. Apostolico, editor, *Combinatorial Algorithms on Words*, volume 12, pages 11–29. NATO ASI Series F, 1985.
14. P. Weiner. Linear pattern matching algorithm. *Proc. 14 IEEE Symposium on Switching and Automata Theory*, pages 1–11, 1973.

Aligning DNA Sequences to Minimize the Change in Protein

(Extended Abstract)

Yufang Hua[1], Tao Jiang[2*], Bin Wu[3**]

[1] Dept 659, IBM Canada, Toronto, Ont. M3C 1W3, Canada. E-mail: yhua@VNET.IBM.COM

[2] Department of Computer Science, McMaster University, Hamilton, Ont. L8S 4K1, Canada. E-mail: jiang@maccs.mcmaster.ca

[3] Department of Computer Science, McMaster University, Hamilton, Ont. L8S 4K1, Canada. E-mail: binwu@maccs.mcmaster.ca

Abstract. We study an alignment model for coding DNA sequences recently proposed by J. Hein that takes into account both DNA and protein information, and attempts to minimize the total amount of evolution at both DNA and protein levels. Assuming that the gap penalty function is affine, we design a quadratic time dynamic programming algorithm for the model. Although the algorithm theoretically solves an open question of Hein, its running time is impractical because of the large constant factor embedded in the quadratic time complexity function. We therefore consider a mild simplification of Hein's model and present a much more efficient algorithm for the simplified model. The algorithms have been implemented and tested on both real and simulated sequences, and it is found that they produce almost identical alignments in most cases.

1 Introduction

Sequence alignment is a model of comparing DNA or protein sequences under the assumptions that (i) insertion, deletion, and mutation are the elementary evolutionary events and (ii) evolution usually takes the most economic course. Classical alignment algorithms either align DNA sequences based on DNA evolution or align protein sequences based on protein evolution [8, 10, 12]. It is well known that protein evolves slower than its coding DNA, and hence alignments of protein are usually more reliable than that of the underlying DNA.

We are interested in the alignment of coding DNA sequences. It is clearly desirable that an alignment of coding DNA sequences incorporate the information from their protein sequences. A straightforward method is to align the protein sequences first and then back-translate the alignment into DNA. The method has several shortfalls including (i) it forces insertions and deletions (abbreviated

* Supported in part by NSERC Operating Grant OGP0046613 and Canadian Genome Analysis and Technology Grant GO-12278.

** Supported in part by NSERC Operating Grant OGP0046613.

as *indels*) to occur at codon boundaries and (ii) it ignores homologies at the DNA level. Recently, Hein proposed a model of DNA sequence alignment where evolutionary changes at both the DNA and protein levels are dealt with simultaneously [4]. The basic idea is that in computing an alignment, we consider each nucleotide mutation and indel, and penalize it appropriately taking into account any amino acid change it might induce. The model allows indels to occur within codons and assumes that each indel involves a multiple of three nucleotides so that the reading frame never changes during the evolution. A gap (*i.e.* a block of consecutive spaces; representing an indel) of length i is penalized with a cost $g(i)$, where g is any positive function satisfying $g(i) + g(j) \geq g(i+j)$. A dynamic programming algorithm is demonstrated in [4] for computing optimal alignments in this model that runs in $O(m^2n^2)$ time, where m and n are the lengths of the two DNA sequences aligned. The algorithm is too slow to be useful in practice even for moderate m and n. It is left as an open question in [4] whether the time complexity can be improved to $O(mn)$ when the gap penalty function is affine, *i.e.* $g(i) = g_{open} + i * g_{ext}$ for some constants g_{open} and g_{ext}. Affine functions are perhaps the most popular among gap functions. A fast heuristic algorithm for the problem, assuming affine gaps, is proposed in [5, 6] which does not guarantee an optimal alignment.

In this paper we first design a dynamic programming algorithm that computes an optimal alignment in $O(mn)$ time, assuming affine gaps. However, the algorithm is impractical because of the large constant factor embedded in its time complexity function. The large constant factor comes from the fact that the algorithm has to compute $16644mn$ table entries. Since such large constants seem to be inherent in all quadratic time algorithms for Hein's model, we simplify the model slightly by considering indels primarily at the DNA level. [1] A much more efficient quadratic time algorithm is presented for the simplified model which needs only to compute $292mn$ table entries, again assuming affine gaps. Although the framework of the algorithm is still dynamic programming, the crux of this algorithm is a careful partition of the state space in order to minimize the total number of table entries that it has to compute. Both algorithms have been implemented in GNU C and tested on real and simulated data. Our algorithm for the simplified model runs reasonably fast on a SPARC Ultra II Model 1300, and constructs almost identical alignments as the other algorithm in most cases.

The rest of the paper is organized as follows. In the next section, we describe Hein's model of coding DNA sequence alignment and the notion of a codon alignment. Our quadratic time algorithm for this model is presented in Section 3. We then simplify the model and give a faster algorithm in Section 4. Section 5 discusses some issues arising in the implementation of these algorothms and also gives some test results.

---

[1] In Hein's model, a nucleotide indel is also penalized at the protein level. At this level, the indel generally corresponds to an amino acid substitution and an amino acid indel as illustrated in Figure 2. Our simplified model will disregard the amino acid substitution.

2 Hein's Model and Codon Alignment

Let $A = a_1a_2a_3 \cdots a_{3m-2}a_{3m-1}a_{3m}$ and $B = b_1b_2b_3 \cdots b_{3n-2}b_{3n-1}b_{3n}$ be two coding DNA sequences consisting of m and n codons respectively. Each sequence has a fixed reading frame starting at the first base. An *alignment* of A and B is a correspondence between the bases in A and B, and postulates a possible evolution from A and B in terms of single nucleotide mutations and indels of blocks of nucleotides. An alignment can also be conveniently expressed as a path in a grid graph. Figure 1 demonstrates an alignment of TTG and TTGCTC and the corresponding path. It postulates that a mutation $G \to C$ and an insertion of TGC have happened in the evolution from TTG to TTGCTC. Since indels of length other than a multiple of three change the reading frame and hence the entire protein, for simplicity, we will assume that all indels have lengths divisible by three as in [4]. In other words, we will only consider alignments of A and B with gaps of lengths divisible by three.

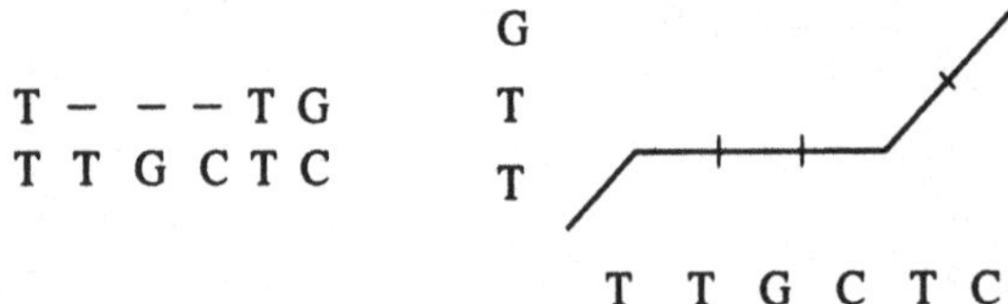

Fig. 1. An alignment and its corresponding path representation.

The cost of an alignment between A and B is decided by both the evolutionary events of the nucleotides postulated by the alignment and the evolutionary changes at the protein level. We will look at the three events mutation, insertion and deletion separately. For each pair of nucleotides a and b, let $c_d(a, b)$ denote the cost of substituting b for a, without worrying about the effect of this change at the protein level. For each pair of codons $e_1e_2e_3$ and $f_1f_2f_3$, let $c_p(e_1e_2e_3, f_1f_2f_3)$ denote the cost of substituting the amino acid coded by $b_1b_2b_3$ for the amino acid coded by $e_1e_2e_3$. For any integer i, functions $g_d(i)$ and $g_p(i)$ denote the costs of inserting (or deleting) a block of i nucleotides and a block of i amino acids, respectively. For convenience, let $g(i) = g_d(3i) + g_p(i)$.

• Mutation. The *combined cost* of a nucleotide mutation $e_1 \to f_1$ in codon $e_1e_2e_3$ is $c_d(e_1, f_1) + c_p(e_1e_2e_3, f_1e_2e_3)$. The combined costs of mutations at the second or third positions of a codon are defined in a similar way.
• Insertion. Consider the event of inserting $3i$ nucleotides $f_1 \cdots f_{3i}$ in the codon $e_1e_2e_3$. If the insertion happens to the immediate left of e_1 or the immediate right of e_3, its combined cost is simply $g(i)$. Otherwise suppose that the string $f_1 \cdots f_{3i}$ is inserted between the nucleotides e_1 and e_2. Then the combined cost

of the insertion is

$$g(i) + \min\{c_p(e_1e_2e_3, e_1f_1f_2), c_p(e_1e_2e_3, f_{3i}e_2e_3)\}$$

The case when the insertion happens between the nucleotides e_2 and e_3 is handled similarly.

• Deletion. This is symmetric to insertion. Consider the event of deleting $3i$ nucleotides from a sequence of $i+1$ codons $e_1e_2e_3\cdots e_{3i+1}e_{3i+2}e_{3i+3}$. If the deletion happens at e_1 or e_4, its combined cost is simply $g(i)$. Otherwise suppose that the string $e_2\cdots e_{3i+1}$ is deleted. Then the combined cost of the deletion is

$$g(i) + \min\{c_p(e_1e_2e_3, e_1e_{3i+2}e_{3i+3}), c_p(e_{3i+1}e_{3i+2}e_{3i+3}, e_1e_{3i+2}e_{3i+3})\}.$$

The case when $e_3\cdots e_{3i+2}$ is deleted can be handled similarly.

Although an alignment of A and B postulates a set of evolutionary events that transform A into B, it does not specify the order that the events should happen. In fact, all permutations of the events are possible. However, different permutations may yield different overall combined costs. For example, in Figure 2, the overall combined cost is $g(1) + c_d(G, C)$ if the insertion happens first or $c_d(G,C) + c_p(TTG, TTC) + g(1) + c_p(TTC, TTG)$ if the mutation happens first. In other words, the evolutionary events are no longer independent when it comes to computing the combined cost. An event may influence the cost of other events. Therefore, we define the cost of an alignment of A and B as the minimum overall combined cost among all possible permutations of the evolutionary events postulated by the alignment. An *optimal* alignment is one with the minimum cost.

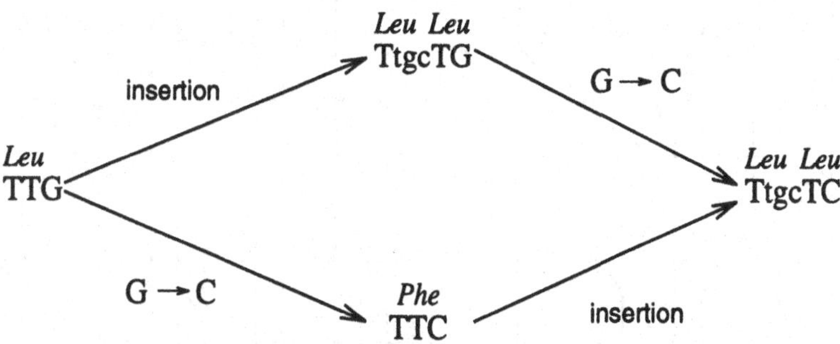

Fig. 2. Different orders yield different costs.

Computing an optimal alignment of A and B is not an easy task due to the influence between events. The notion of a *codon alignment* introduced in [4] will help simplify the matter. An alignment of $A = a_1a_2a_3\cdots a_{3m-2}a_{3m-1}a_{3m}$ and $B = b_1b_2b_3\cdots b_{3n-2}b_{3n-1}b_{3n}$ is called a codon alignment if

1. $m = 0$ or
2. $n = 0$ or
3. There do not exist i and j, $1 \leq i \leq 3m$ and $1 \leq j \leq 3n$, such that a_i is aligned with b_j, and (i) $i \bmod 3 = \mathrm{j} \bmod 3 = 1$ and $i + j > 2$ or (ii) $i \bmod 3 = \mathrm{j} \bmod 3 = 0$ and $i + j < 3m + 3n$.

In other words, except in the first and last columns, a codon alignment does not align a base at some codon boundary of A with a base at any codon boundary of B. For example, the alignment in Figure 1 is in fact a codon alignment. The cost of a codon alignment is defined the same way as for an alignment.

It is known [4] that there are 11 distinct types of codon alignment, as depicted in Figure 3. Observe that each codon alignment can involve at most 5 evolutionary events. Hence, the cost of a codon alignment, which is the minimum total combined cost over all possible permutations of the events postulated by the alignment, can be computed in linear time.

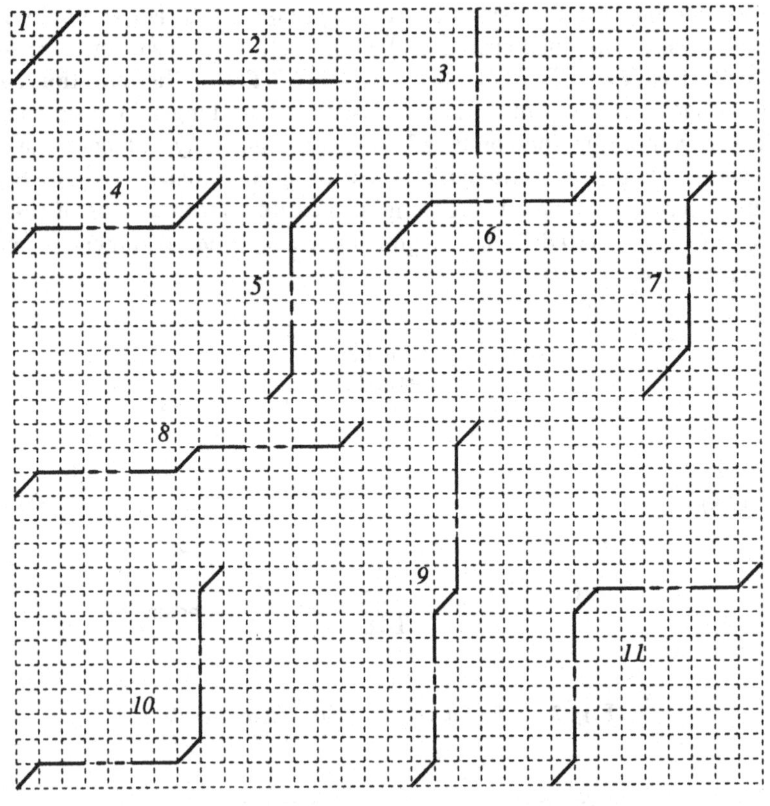

Fig. 3. The 11 types of codon alignment.

We can always decompose an alignment of A and B uniquely into a sequence of maximal codon alignments, as illustrated in Figure 4. Although the evolutionary events in a same codon alignment may influence each other's cost, events in different codon alignments are independent. This gives rise to a straightforward dynamic programming algorithm for computing an optimal alignment of A and B in $O(m^2n^2)$ time, as described in [4]. All of our constructions will also be based on such decompositions into codon alignments.

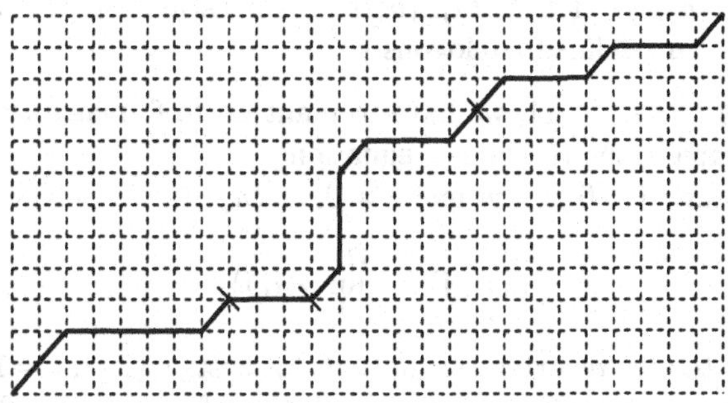

Fig. 4. Decomposing an alignment into codon alignments.

3 A Quadratic Time Algorithm Assuming Affine Gaps

From now on, we assume that the gap penalty function $g(i)$ is affine, *i.e.* $g(i) = g_{open} + i * g_{ext}$ for some fixed non-negative constants g_{open} and g_{ext}. Again, consider sequences $A = a_1a_2a_3 \cdots a_{3m-2}a_{3m-1}a_{3m}$ and $B = b_1b_2b_3 \cdots b_{3n-2}b_{3n-1}b_{3n}$. For any indices $i = 0, \ldots, m$ and $j = 0, \ldots, n$, let $A(i) = a_1a_2a_3 \cdots a_{3i-2}a_{3i-1}a_{3i}$, $B(j) = b_1b_2b_3 \cdots b_{3j-2}b_{3j-1}b_{3j}$, and $c(i,j)$ denote the cost of an optimal alignment between the prefix $A(i)$ and prefix $B(j)$. In order to derive a recurrence equation for $c(i,j)$, we need introduce the following notations.

We will classify an alignment into 11 classes according to the type of its last codon alignment.

1. For type $t = 1,2,3$, let $c_t(i,j)$ denote the cost of an optimal alignment between $A(i)$ and $B(j)$ whose last codon alignment is of type t.
2. For $t = 4,6$ and any nucleotides $x_1, x_2, x_3 \in \{A,C,G,T\}$, $c_t(i,j,x_1x_2x_3)$ denotes the cost of an optimal alignment between $A(i)$ and $B(j)x_1x_2x_3$ ending with a codon alignment of type t. Also define $c_t(i,j) = c_t(i,j-1, b_{3j-2}b_{3j-1}b_{3j})$.
3. For $t = 5,7$ and any nucleotides $x_1, x_2, x_3 \in \{A,C,G,T\}$, $c_t(i,x_1x_2x_3,j)$ denotes the cost of an optimal alignment between $A(i)x_1x_2x_3$ and $B(j)$ ending with a codon alignment of type t. Also define $c_t(i,j) = c_t(i-1, a_{3i-2}a_{3i-1}a_{3i}, j)$.
4. For any nucleotides $x_1, x_2, x_3, x_4, x_5, x_6 \in \{A,C,G,T\}$, $c_8(i,j,x_1x_2x_3x_4x_5x_6)$ denotes the cost of an optimal alignment between $A(i)$ and $B(j)x_1x_2x_3x_4x_5x_6$

ending with a codon alignment of type 8. Also define $c_8(i,j) = c_8(i, j-1, b_{3j-5}b_{3j-4}b_{3j-3}b_{3j-2}b_{3j-1}b_{3j})$. The notations $c_9(i, x_1x_2x_3x_4x_5x_6, j)$ and $c_9(i,j)$ are defined analogously.

5. For any nucleotides $x_1, x_2, x_3, y_1, y_2, y_3 \in \{A, C, G, T\}$, $c_{10}(i, x_1x_2x_3, j, y_1y_2y_3)$ denotes the cost of an optimal alignment between $A(i)x_1x_2x_3$ and $B(j)y_1y_2y_3$ ending with a codon alignment of type 10. Also define $c_{10}(i,j) = c_{10}(i-1, a_{3i-2}a_{3i-1}a_{3i}, j-1, b_{3j-2}b_{3j-1}b_{3j})$. The notations $c_{11}(i, x_1x_2x_3, j, y_1y_2y_3)$ and $c_{11}(i,j)$ are defined analogously.

Here, for types $t = 4, \ldots, 11$, we have to plant up to 6 imaginary trailing bases in order to complete the recurrence equations.

Clearly, for any $i = 0, \ldots, m$ and $j = 0, \ldots, n$,

$$c(i,j) = \min_{t=1}^{11} c_t(i,j)$$

Now it suffices to give recurrence equations for costs $c_t(i,j)$, $t = 1, \ldots, 11$, using $c(i,j)$. We will only give the recurrence equations for $t = 1, 2, 4, 8, 10$. The other cases are highly symmetric to these types. For simplicity, assume that $i > 1$ and $j > 1$. In the following, when there is a *unique* codon alignment between sequences X and Y of type t, we use $ca_t(X, Y)$ to denote the cost of that codon alignment.

$$c_1(i,j) = c(i-1, j-1) + ca_1(a_{3i-2}a_{3i-1}a_{3i}, b_{3j-2}b_{3j-1}b_{3j})$$

$$c_2(i,j) = \min\{c_2(i, j-1) + g_{ext}, c(i, j-1) + g(1)\}$$

$$\begin{aligned} c_4(i, j, x_1x_2x_3) = \min\{ & c_4(i, j-1, x_1x_2x_3) + g_{ext}, \\ & c(i-1, j-1) + \\ & ca_4(a_{3i-2}a_{3i-1}a_{3i}, b_{3j-2}b_{3j-1}b_{3j}x_1x_2x_3)\} \end{aligned}$$

$$\begin{aligned} c_8(i, j, x_1x_2x_3x_4x_5x_6) = \min\{ & c_8(i, j-1, b_{3j-2}b_{3j-1}b_{3j}x_4x_5x_6) + g_{ext}, \\ & c_8(i, j-1, x_1x_2x_3x_4x_5x_6) + g_{ext}, \\ & c(i-1, j-1) + \\ & ca_8(a_{3i-2}a_{3i-1}a_{3i}, b_{3j-2}b_{3j-1}b_{3j}x_1x_2x_3x_4x_5x_6)\} \end{aligned}$$

$$\begin{aligned} c_{10}(i, x_1x_2x_3, j, y_1y_2y_3) = \min\{ & c_{10}(i-1, x_1x_2x_3, j, y_1y_2y_3) + g_{ext}, \\ & c_{10}(i, x_1x_2x_3, j-1, y_1y_2y_3) + g_{ext}, \\ & c(i-1, j-1) + \\ & ca_{10}(a_{3i-2}a_{3i-1}a_{3i}x_1x_2x_3, b_{3j-2}b_{3j-1}b_{3j}y_1y_2y_3)\} \end{aligned}$$

The equations for $t = 1, 2, 4, 10$ are self-explanatory. The equation for $t = 8$ is elaborated in Figure 5. The first term in the minimization corresponds to the case when the second gap (from the left) is longer than one codon (case (a) in the figure), the second term represents the case when the second gap is one codon long and the first gap is longer than one codon (case (b)), and the third term corresponds to the case when both gaps are one codon long (case (c)).

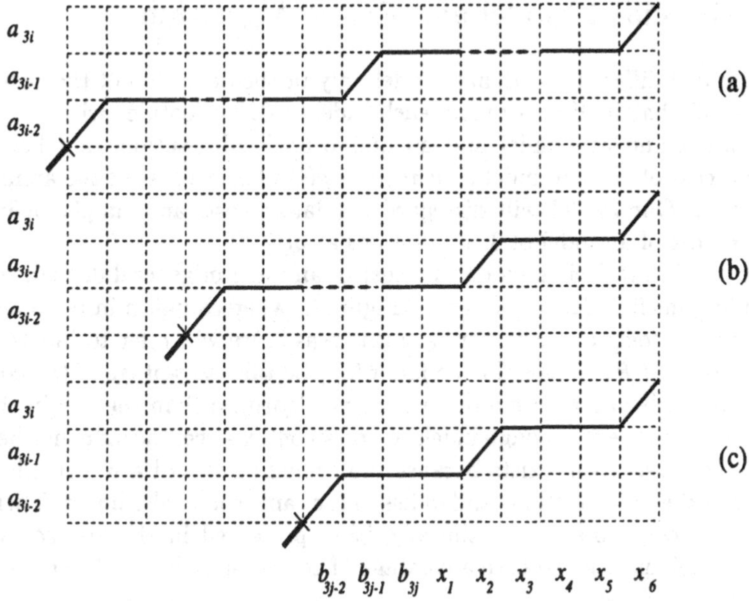

Fig. 5. An illustration for the recurrence equation for $c_8(i, j, x_1x_2x_3x_4x_5x_6)$.

The base cases of the above recurrence equations can be easily formulated. Since the cost $ca_t(X, Y)$ can be computed in $O(1)$ time for any sequences X and$Y$ of lengths at most 9 bases, the recurrence equations obviously imply a dynamic programming algorithm for computing $c(m, n)$ in $O(mn)$ time. This algorithm can be easily expanded to also produce an optimal alignment between A and B, using the standard back-tracing technique [2, 3].

A version of the algorithm has been implemented in GNU C, called *Codon Alignment Tool* (CAT) [7]. To avoid computing the cost $ca_t(X, Y)$ repeatedly for the same short sequences X and Y, we use a table, indexed by X and Y, to store the value $ca_t(X, Y)$ once it is computed so that for each pair X and Y, the cost $ca_t(X, Y)$ is computed at most once. Although this technique greatly improves the time efficiency, the program is still quite slow due to the fact that it has to compute 12 tables for $c(i, j)$ and $c_t(i, j)$, where $t = 1, \ldots, 11$, with a total size of $4 + 4 * 64 + 4 * 4096 = 16644mn$ entries before obtaining the value $c(m, n)$. Clearly, codon alignments of types 8 through 11 are the main reason why such large tables are required. Because of the influence between evolutionary events within a same codon alignment and the events may happen in any of up to 5! different orders, the dynamic programming algorithm has to hypothesize 6 trailing bases for each of these four types, and carry out the computation for each of the 4096 hypotheses. In the next section, we will simplify Hein's model slightly and present a much faster algorithm.

4 A Simplified Model and Faster Algorithm

Our model differs from Hein's model only in the definition of the cost of an indel. Recall that in Hein's model each indel of $3i$ nucleotides within a codon induces an amino acid indel and an amino acid substitution, and hence the combined cost of such an indel is defined as $g(i)$ plus the cost of the amino acid substitution. Our model will disregard the latter cost, and simply define the combined cost of an indel of $3i$ nucleotides as $g(i)$. [2]

Observe that in Hein's model the cost of an indel in general depends on the surrounding nucleotides, as shown in Figure 2, whereas indels in our model do not have such context sensitivity. For this reason we will refer to indels in our model as *context-free* indels. In the following, we take advantage of the context-freeness in indels and devise a more efficient algorithm than the one in the last section. Note that, even though indels are now context-free, the influence between evolutionary events still exists because the combined cost of a substitution may depend on other substitutions and indels in the same codon alignment. Therefore, it does not seem possible for the algorithm presented in the last section (or simple extensions of it) to take advantage of context-free indels. We have to use a different technique.

The framework of our algorithm is still dynamic programming based on codon alignments. We again classify an alignment according to the type of its last codon alignment. The new idea is to refine the classes according to the order of some events in the last codon alignment so we could avoid having to hypothesize (or equivalently, remember) too many nucleotides. This will greatly reduce the total size of the tables required. To demonstrate the idea, we will consider the computation for alignments ending with codon alignments of types $t = 4, 6, 8, 10$. The other cases are either straightforward or symmetric to these types.

Recall that $A = a_1a_2a_3 \cdots a_{3m-2}a_{3m-1}a_{3m}$, $B = b_1b_2b_3 \cdots b_{3n-2}b_{3n-1}b_{3n}$, and for any $i = 0, \dots, m$ and $j = 0, \dots, j$, $c(i,j)$ denotes the cost of an optimal alignment between $A(i)$ and $B(j)$. To define recurrence equations, we will consider *partial* (*i.e.* incomplete) codon alignments consisting a front portion of some codon alignment and *restricted* (partial) codon alignments whose events are required to occur only in some specific orders.

A type 4 codon alignment involves 4 evolutionary events as shown in Figure 6(a). We consider alignments ending with (partial) type 4 codon alignments, and partition them into 8 classes depending on the relative order of events 1 and

[2] It is unclear such a simplification is biologically plausible, although one supporting argument may be that the amino acid substitution is a superficial event. Our tests on real and simulated data in Section 5 will show that optimal alignments for the two models are in fact very similar.

2 and the nucleotide x_1. For any nucleotide $x_1 \in \{A, C, G, T\}$, let $p_4(i, j, x_1, 0)$ denote the cost of an optimal alignment between $A(i)$ and $B(j)$ ending with a restricted partial codon alignment of type 4 consisting of events 1 and 2 with event 1 occurring before event 2. Moreover, event 1 results in the nucleotide x_1. Here, the value $p_4(i, j, x_1, 0)$ includes the combined cost of event 2 but does not include the combined cost of event 1 because it depends on events 3 and 4. The latter cost will be added when the partial codon alignment is completed. So, let $c_4(i, j, x_1, 0)$ denote the cost of an optimal alignment between $A(i)$ and $B(j)$ ending with a restricted codon alignment of type 4 with event 1 preceding event 2. Then it is easy to compute $c_4(i, j, x_1, 0)$ from the values $c(i-1, j-1)$ and $p_4(i, j-1, x_1, 0)$, and the nucleotides $x_1, a_{3i-2}, a_{3i-1}, a_{3i}, b_{3j-2}, b_{3j-1}, b_{3j}$. We only have to consider 12 orders and add the combined costs of events 1, 3 and 4 to $p_4(i, j-1, x_1, 0)$. Similarly, we can define the costs $p_4(i, j, x_1, 1)$ and $c_4(i, j-1, x_1, 1)$ assuming that event 1 occurs after event 2. The only difference is that now the combined cost of event 1 should be included in the value $p_4(i, j, x_1, 1)$ so it does not have to be accounted for at the completion of the codon alignment (which would have been impossible because we do not remember the two nucleotides in B following x_1). Observe that when event 1 is preceded by event 2, the combined cost of event 1 is independent of events 3 and 4. Finally, the cost of an optimal alignment between $A(i)$ and $B(j)$ ending with a type 4 codon alignment, denoted $c_4(i, j)$, is easily computed as

$$c_4(i, j) = \min_{\substack{x_1 \in \{A,C,G,T\} \\ \sigma \in \{0,1\}}} c_4(i, j, x_1, \sigma)$$

Note that, the above equation only requires a table of $8mn$ entries (for storing $p_4(i, j, x_1, \sigma)$) to compute $c_4(i, j)$ instead of a table of $64mn$ entries as required in the last section.

Alignments ending with type 6 codon alignments can be treated in the same spirit. However, instead of "cutting" the codon alignment at event 2 we should not cut it at event 3 (to obtain the partial codon alignment), instead of considering the relative order of events 1 and 2 we consider the order of events 4 and 3, and instead of remembering the nucleotide x_1 we hypothesize the nucleotide x_2 (see Figure 6(b)). Thus, we define $p_6(i, j, x_2, 0)$ assuming that event 4 precedes event 3 and event 4 results in the nucleotide x_2, and define $p_6(i, j, x_2, 1)$ assuming the opposite order. The only tricky point is that $p_6(i, j, x_2, 0)$ should include the combined cost of event 4 while $p_6(i, j, x_2, 1)$ does not. Again, computing the costs $c_6(i, j)$ requires only a table of $8mn$ entries.

The treatment of alignments ending with type 8 codon alignments combines the techniques for both type 4 and type 6 codon alignments, and builds on the information $p_4(i, j, x_1, 0)$ and $p_4(i, j, x_1, 1)$. Define $\sigma_1 = 0$ if event 1 precedes event 2 or $\sigma_1 = 1$ otherwise, and $\sigma_2 = 0$ if event 5 precedes event 4 or $\sigma_2 = 1$ otherwise. For any nucleotides $x_1, x_2 \in \{A, C, G, T\}$ and orders $\sigma_1, \sigma_2 \in \{0, 1\}$, let $p_8(i, j, x_1, \sigma_1, x_2, \sigma_2)$ denote the cost of an optimal alignment between $A(i)$ and $B(j)$ ending with a restricted partial codon alignment of type 8 consisting of events 1 through 4 such that (i) event 1 results in the base x_1, (ii) the relative

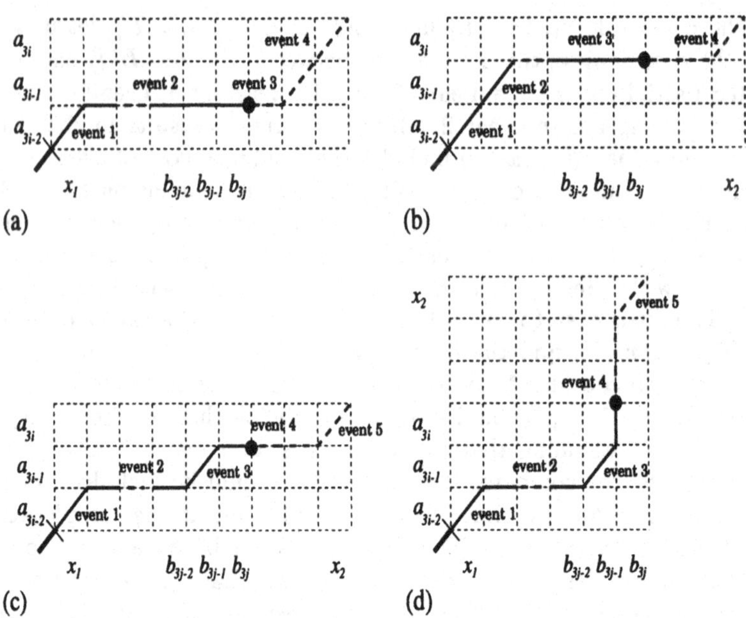

Fig. 6. Dealing with trailing codon alignments of types 4, 6, 8, 10.

order between events 1 and 2 is as prescribed by σ_1, (iii) event 5 results in the base x_2, and (iv) the relative order between events 5 and 4 is as prescribed by σ_2. (See Figure 6(c).) Again, value $p_8(i,j,x_1,\sigma_1,x_2,0)$ should include the combined cost of event 5 while $p_8(i,j,x_1,\sigma_1,x_2,1)$ does not. The cost $p_8(i,j,x_1,\sigma_1,x_2,\sigma_2)$ can be easily computed from the values $p_8(i,j-1,x_1,\sigma_1,x_2,\sigma_2)$ and $p_4(i,j-1,x_1,\sigma_1)$, and the nucleotides $x_1, x_2, a_{3i-2}, a_{3i-1}, a_{3i}, b_{3j-2}, b_{3j-1}, b_{3j}$. Therefore, we can compute the costs $c_8(i,j)$ by using a table of $64mn$ entries for storing $p_8(i,j,x_1,\sigma_1,x_2,\sigma_2)$.

Similarly, we can deal with alignments ending with type 10 codon alignments by combining the techniques for type 4 and type 7 codon alignments, make use of the information $p_4(i,j,x_1,0)$ and $p_4(i,j,x_1,1)$. We still cut the codon alignment at event 4 and consider the order of events 5 and 4; but we hypothesize the nucleotide x_2 instead of b_{3j} (see Figure 6(d)). The cost $p_{10}(i,j,x_1,\sigma_1,x_2,\sigma_2)$ is defined in a straightforward way, and requires ia table of $64mn$ entries to store.

The above discussion yields a quadratic time dynamic programming algorithm which needs to compute 12 tables of a total size of only $(4+4*8+4*64)mn = 292mn$ entries. (The first four tables are used for storing $c(i,j)$, $c_1(i,j)$, $c_2(i,j)$, and $c_3(i,j)$.) The algorithm has been implemented in GNU C, called *Context-free CAT*. In the next section, we compare this and the algorithm presented in the previous section by running the programs CAT and Context-free CAT on both real and simulated data.

5 A Comparison of the Two Algorithms

We have performed tests of the two programs CAT and Context-free CAT on 3 pairs of HIV1 and HIV2 sequences and 13 groups of simulated sequences of length 100 through 1500 bases. The three pairs of real data include (i) HIV1 *gag* gene (bases 790..2304) and HIV2 *gag* gene (bases 548..2113), (ii) HIV1 *vif* gene (bases 5053..5631) and HIV2 *vif* gene (bases 4868..5515), and (iii) HIV1 *nef* gene (bases 8784..9434) and HIV2 *nef* gene (bases 8562..9329). Since we are not sure how to combine cost parameters for amino acids with those of nucleotides, two combinations were considered (a) Dayhoff PAM 40 Matrix for amino acids [1] and DNA PAM 30 Matrix for nucleotides [13] and (b) Dayhoff PAM 40 Matrix for amino acids and DNA PAM 47 Matrix for nucleotides [13]. Overall, CAT and Context-free CAT produced very similar alignments in these tests. The following table summarizes the discrepancy between the alignments produced by the two programs.

Table 1. Discrepancy between the indels produced by CAT and Context-free CAT on the HIV data.

	PAM 40 & DPAM 30		PAM 40 & DPAM 47	
	location	type	location	type
HIV1&2 *gag*	2/14	4/12	1/10	3/9
HIV1&2 *vif*	1/7	1/6	1/7	2/6
HIV1&2 *nef*	1/7	3/6	0/7	4/6

In the table, we first count the number of codon alignments involving indels (*i.e.* any codon alignment except those of type 1) that are placed at different locations by the two programs, and then the number of codon alignments of that are at the same locations but have different types. For example, the entry 2/14 means that out of the 14 codon alignments involving indels, two are placed at different locations by the two programs, and the entry 4/12 means that out of the 12 remaining codon alignments four have different types. In all the cases where indels are placed at different locations, one program merges two adjacent indels produced by the other program. On the other hand, the discrepancy in the types of codon alignments is always because Context-free CAT would sometimes expand a type 2 or 3 codon alignment produced by CAT into a codon alignment of type $4, 5, 6$, or 7 by shifting the indel inside an adajcent codon alignment. It is interesting to note that CAT produces very few codon alignments of types higher than 3 while Context-free CAT produces types $4, 5, 6$, and 7 almost as frequently as types 2 and 3. Also observe that the above discrepancies between CAT and Context-free CAT do not change very much with the two pairs of cost parameters we used.

The 13 groups of simulated sequences were generated randomly on a naive stochastic model using some fixed mutation and indels rates. The amino acid mutation/indel rates are based on Dayhoff PAM 120 Matrix [1] and the nucleotide mutation/indel rates are based on [11]. We ran CAT and Context-free CAT on these groups of data using cost parameters consistent with the above rates. It is observed that both programs again produced very similar alignments and, moreover, they were all able to identify most indels correctly.

Table 2 below shows the average speeds of CAT and Context-free CAT on SPARC Ultra II Model 1300. The speed-up of Context-free CAT over CAT is illustrated in Figure 7. The speed-up deceases with the length of sequences because the "atomic" codon alignments (*i.e.* the ones that cannot be further reduced, such as the codon alignments shown in Figure 1 and Figure 5(c)) for CAT are more complicated and require more time to compute than the ones for Context-free CAT, and the percentage of time spent by each program on setting up the atomic codon alignment table decreases with the length. We expect the speed-up to approach $\frac{16644}{292} = 57$ (but never goes below 57) when the sequences get really long.

Table 2. The average speeds (in seconds) of CAT and Context-free CAT.

length	102	201	300	402	501	600	702	801	900	1002	1200	1500
CAT	898.5	1872	2496	3032.5	3486	3463	4166.5	4490	4820	5414.5	6177	8138
C.f. CAT	1	2.5	5.5	9	13.5	17.5	26.5	33.5	40	50.5	68	104

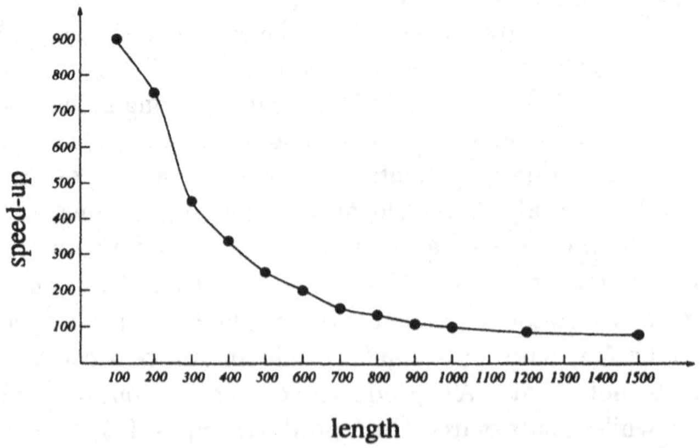

Fig. 7. The speed-up of Context-free CAT over CAT.

6 Concluding Remarks

Both programs CAT and Context-free CAT spend a significant amount of time in computing the costs of atomic codon alignments, even though a table is used to avoid duplicate computations for the same atomic codon alignment. When the same set of cost parameters $c_p, c_d, g_{open}, g_{ext}$ are used again and again, it is possible to speed up the programs by recycling the atomic codon alignment table.

We learned very recently that C. Pedersen has developed a quadratic time algorithm for Hein's model independently [9]. His algorithm requires tables of at least 24968 entries.

Acknowledgement. We are grateful to Jotun Hein for many valuable discussions on the subject. We would also like to thank Xian Zhang for providing real and simulated test data.

References

1. M. Dayhoff, et al., *Atlas of Protein Sequence and Structure*, 5 suppl. 3, pp. 345-352, 1978.
2. O. Gotoh, An improved algorithm for matching biological sequences, *J. Mol. Biol.* 162, pp. 705-708, 1981.
3. D. Gusfield, *Algorithms on Strings, Trees, and Sequences*, Cambridge University Press, 1997.
4. J. Hein, An algorithm combining DNA and protein alignment, *Journal of Theoretical Biology* 167, pp. 169-174, 1994.
5. J. Hein and J. Støvlbæk, Genomic alignment, *J. Mol. Evol.* 38, pp. 310-316, 1994.
6. J. Hein and J. Støvlbæk, Combined DNA and protein alignment, *Methods in Enzymology* 266, pp. 402-418, 1996.
7. Y. Hua, An improved algorithm for combined DNA and protein alignment, *M. Eng. Thesis*, Department of Computer and Electrical Engineering, McMaster University, 1997.
8. S. Needlemann and C. Wunsch, A general method applicable to the search for similarities in the amino acid sequences of two proteins, *J. Mol. Biol.* 48, pp. 443-453, 1970.
9. C. Pedersen, Computational analysis of biological sequences, *Manuscript*, 1997.
10. D. Sankoff, Matching sequences under deletion/insertion constraints, *Proc. Nat. Acad. Sci.* 69(1), pp. 4-6, 1972.
11. D. Sankoff, R. Cedergren and G. Lapalme, Frequency of insertion-deletion, transversion, and transition in the evolution of 5S ribosomal RNA, *J. Mol. Evol.* 7, pp.133-149, 1976.
12. P. Sellers, On the theory and computation of evolutionary distances, *SIAM J. Appl. Math.* 26, pp. 787-793, 1974.
13. D. States, et al., *Methods: A companion to methods in Enzymology* 3, pp. 66-70, 1991.

Genome Halving

Nadia El-Mabrouk[1] and Joseph H. Nadeau[2] and David Sankoff[3]

[1] Centre de recherches mathématiques, Université de Montréal, CP 6128 succursale Centre-Ville, Montréal, Québec, H3C 3J7. elmabrou@crm.umontreal.ca

[2] Department of Genetics, Case Western Reserve University, 10900 Euclid Avenue, Cleveland, Ohio 44106-4955. jhn4@po.cwru.edu

[3] Centre de recherches mathématiques, Université de Montréal, CP 6128 succursale Centre-Ville, Montréal, Québec, H3C 3J7. sankoff@ere.umontreal.ca

Abstract. Genome duplication is an important source of new gene functions and novel physiological pathways. In the course of evolution, the nucleotide sequences of duplicated genes tend to diverge through mutation, so that one copy loses function (becomes a pseudogene) or develops a new function, encoding a distinct but similar product. Originally a duplicated genome contains two identical copies of each chromosome, but through inversion or other intrachromosomal movement, the gene orders in each pair of chromosomes change independently, and through reciprocal translocation, parallel linkage patterns between the two copies are disrupted. Eventually, all that can be detected are several chromosome segments of greater or lesser length (*blocks*), each of which appears twice in the genome, containing many paralogous genes in parallel orders. The study of genome duplication based on block data includes the inference of the synteny or linkage structure of the pre-duplication genome, the nature of the post-duplication rearrangement events, and the statistics of gene loss versus functional divergence. We propose a suite of *Genome halving* problems for algorithmic solution, some of which address the evolution of gene order, and others which deal with relations of synteny only. We present an efficient and accurate heuristic for the latter type of problem, and apply it to the genome duplication which has been described for *Saccharomyces cerevisiae*.

1 Genome duplication

In the course of evolution, there are a number of mechanisms which result in the appearance of two or more identical or nearly identical genes in the same genome – a gene family. Mechanisms such as unequal crossing over give rise to adjacent or closely linked copies on the same chromosome. Others, such as reverse transcription, can produce copies on different chromosomes. Whatever the mechanism, the existence of duplicate genes can lead to important changes in the genetic makeup of an organism. Over a relatively short period of time, evolutionary speaking, all but one of the genes in a family may accumulate random mutations and lose their function, i.e. become pseudogenes. Sometimes, however, one or more of the genes in a family incur mutations that lead them to encode for similar, but not identical, products. This is an evolutionary "opportunity" for the organism; it allows for specialization of gene products to function optimally under differing conditions in the life cycle or in different tissues. It can generally speed up evolution in that while one copy of a gene is free to "explore" mutational space to find some new function, the original function, necessary for survival, is preserved by the other copy.

Perhaps the rarest, but surely the most spectacular cause of gene duplication, is tetraploidization of the genome. Normally a lethal accident of meiosis or other reproductive step, if this doubling of the genome can be resolved in the organism and eventually fixed as a normalized diploid state in a population, it represents a simultaneous duplication of the entire genetic complement. It transcends other mecanisms for gene duplication in that not only is one copy of each gene free to evolve its own function, but it can evolve in concert with any subset of the thousands of other extra gene copies (cf [6] for accounts of gene family coevolution). Whole new physiological pathways may emerge, involving novel functions for many of these genes. Genome duplication is thus a likely source of rapid and far-reaching evolutionary progress. Its rarity does not detract from its importance.

Evidence for its effects has shown up across the eukaryote spectrum. More than two hundred million years ago the vertebrate genome underwent two duplications [2, 8, 15]. Although numerous chromosome rearrangements such as inversions and reciprocal translocations have subsequently occurred, the number of rearrangements has been sufficiently modest that hundreds of conserved paralogous segments can be detected in the human genome since the ancient duplications; similar observations hold for the murine genome [13, 14] and for less intensively mapped vertebrate genomes. More recent genome duplications are known to have occurred in some vertebrate lines, such as the frogs [22], the salmoniform fish [15] and zebrafish [17].

Comparison of chromatin-eliminating *Ascaridae* with other nematodes suggest that somatic cells of these worms have discarded a good proportion of the genes present in germ cells, possible because these are redundant duplicates arising through genomic doubling some 200 million years ago [10].

Genome duplication is particularly prevalent in plants. Comparison of the well-studied rice [1], oats (wild and domestic), corn [1, 7] and wheat [12] genomes indicate several occurrences in the cereal lineage. Soybeans [20], rapeseed [19], and other cultivars have genome duplications in their ancestry. Paterson *et al.* have presented convincing evidence that one or more genome duplications also occurred much earlier in plant evolution [16].

Recently, following the complete sequencing of all *Saccharomyces cerevisiae* chromosomes, the prevalence of gene duplication [9, 3] has led to the conclusion that this yeast genome is also the product of an ancient doubling [21].

Subsequent to genome duplication, duplicated genes tend to diverge through mutation, so that one copy loses function (becomes a pseudogene) or develops a new function, encoding a distinct but similar product. Originally a duplicated genome contains two identical copies of each chromosome, but through inversion or other intrachromosomal movement, the gene orders in each pair of chromosomes change independently, and through reciprocal translocation, parallel linkage patterns between the two copies are disrupted. Eventually, all that can be detected are several chromosome segments of greater or lesser length (*blocks*), each of which appears twice in the genome, containing many paralogous genes in parallel orders. The study of genome duplication based on block data includes the inference of the synteny or linkage structure of the pre-duplication genome, the nature of the post-duplication rearrangement events, and the statistics of gene loss versus functional divergence [14]. In this paper we propose a suite of *Genome halving* problems for algorithmic solution, one of which addresses the evolution of gene order, and another which deals with relations of synteny only. We present an efficient and accurate heuristic for the latter problem, and apply it to the genome duplication which has been postulated for *Saccharomyces cerevisiae.*

2 Genome halving with ordered chromosomes

As part of their detailed study of the post-duplication evolution of the *S.cerevisiae*, Seoighe and Wolfe [18] ask how many translocations are necessary to account for the present configuration of paralogous segments in this 16-chromosome genome, starting from a tetraploidization of an ancestral 8-chromosome genome.

The *Genome halving on ordered chromosomes* problem can be formalized most simply as follows. We are given a genome containing n chromosomes $\{c_1, \cdots, c_n\}$, where each chromosome c can be represented as a string of blocks (or segments) $s_{i_1(c)}s_{i_2(c)} \cdots s_{i_{k_c}(c)}$ and where each block occurs exactly twice in the genome, either on two different chromosomes or twice on the same chromosome. Since there is no biologically meaningful reason for assigning a particular left-to-right orientation to a chromosome, the reverse string $\rho(s_{i_1(c)} \cdots s_{i_{k_c}(c)}) = s_{i_{k_c}(c)} \cdots s_{i_1(c)}$ is considered identical to the original string as a representation of the chromosome.

A translocation consists of dividing the string of blocks on a chromosome a into a prefix a_1 and a suffix a_2, and similarly for $b = b_1b_2$, and from them creating two strings a_1b_2 and b_1a_2 (or $a_1\rho(b_1)$ and $\rho(a_2)b_2$) to replace the former two chromosomes. (Only one of a_1, a_2, b_1, b_2 may be null.) How many translocations does it take to replace the n given chromosomes with n reconstructed ones, such that there are $n/2$ pairs of identical chromosomes?

There are a number of variants of this problem. First, in the *Genome halving on ordered chromosomes with centromeres* problem each string is punctuated with a special symbol s_{ce} (the centromere). The translocations are constrained so that each new string created must also contain exactly one s_{ce}.

Second, in the *Genome halving with variable chromosome number* problem, the number n' of reconstructed chromosomes is not fixed at n, but is free to vary among the even integers, in the search for a minimizing number of translocations. All that is required is that the reconstructed genome contain $n'/2$ pairs of identical chromosomes.

Third, where the centromere is taken into account, we can define *Genome halving with oriented blocks* problem. Here the data for each block indicates whether it has positive polarity (oriented away from the centromere) or negative polarity (oriented towards the centromere). Since translocation does not change the polarity of blocks, to solve this problem it is necessary to allow reversals of substrings of a chromosome. For strings not containing the centromere (paracentric reversals), the reverse string of $s_1 \cdots s_m$ is $R(s_1 \cdots s_m) = -s_m \cdots - s_1$, where the minus sign indicates a change of polarity. For strings containing the centromere (pericentric reversals), the reverse string is $R(s_1 \cdots s_{ce} \cdots s_m) = s_m \cdots s_{ce} \cdots s_1$. The objective function to be minimized is the number of translocation plus the number of reversals, possibly with different cost coefficients for the two types of operation. Note that in this problem, we still have the property that the reversal of the entire chromosome $R(s_{i_1(c)} \cdots s_{i_{k_c}(c)}) = \rho(s_{i_1(c)} \cdots s_{i_{k_c}(c)})$ $= s_{i_{k_c}(c)} \cdots s_{i_1(c)}$ since this is necessarily a pericentric reversal.

It is possible to define oriented genome halving problems without taking into account the centromere, but these are not directly interpretable biologically. The reason for this is that the left-to-right labeling of the blocks in a chromosome is no longer innocuous. In contrast to all the previous versions of the problem, reversing a chromosome here changes its nature; every block changes polarity, there being no pericentric reversals. The unfortunate effects of this can be seen in the results of a translocation – newly adjoining blocks from the two contributing chromosomes will have the same or different polarities depending on the arbitrary choice of which ends of these chromosomes were left and which were right. This is not realistic, since whether or not two newly adjoining blocks have the same polarity (i.e. are on the same DNA strand) is predictable from knowledge of their polarity in the original chromosomes.

Thus in genome halving problems incorporating increasing degrees of biological information, centromere location should be included before orientation.

Seoighe and Wolfe devised a heuristic algorithm to solve the ordered genome halving problem, though it is not clear in [18] which version of the problem is addressed. When applied to the yeast data involving 55 segments each in two copies, distributed over 16 chromosomes, the best solution they found was 41 translocations. It is important to note that they did not feel their solution was directly interpretable in terms of evolutionary history, since it severely underestimated the amount of evolution suggested by statistical analysis of ancillary data. Nevertheless, the optimization type of analysis, aside from its intrinsic algorithmic interest, provides an important baseline for evaluating more statistically-oriented approaches.

3 Genome halving with unordered chromosomes

The significance of an analysis based on the order of the blocks on the chromosomes depends on the extent that this order is affected solely by translocation. If in the course of evolution, these blocks were repeatedly shuffled by processes of inversion (reversals) and transposition, a possibility that Seoighe and Wolfe discount (based on meaningful but not overwhelming evidence), then block order might be only indirectly related to translocational history. Then it might be of interest to analyze the data based only on synteny data: which blocks are on which chromosome; a set-theoretical formulation rather than the string theory formulation in the previous section.

The problem we will discuss is *Unordered genome halving with variable chromosome number*, though the algorithm we propose produces a solution with fixed n. It would be both meaningful and feasible to include centromere considerations to the problem with unordered chromosomes, but not questions of block orientation, since reversal operations require taking into account block adjacency and other order information.

Thus we consider a genome G to be a collection of n subsets $S_1, \cdots, S_n$ of a set $B = \{b_1, \cdots, b_k\}$ containing k elements (blocks of genes). We suppose that each block in B appears exactly twice among these n subsets (including the possibility that a block has "multiplicity" 2 within a single subset and appears in no other subset). We refer here to these subsets both as chromosomes and as synteny sets.

The problem is to find the minimum number of translocations necessary to transform G into a genome G' made up of two identical copies of $n'/2$ chromosomes, where n' is even.

Notation :

- Let S_1, S_2, T_1, T_2 be four sets such that $S_1 \bigcup S_2 = T_1 \bigcup T_2$ and at most one of these sets is empty. An operation of form $(S_1, S_2) \longrightarrow (T_1, T_2)$ is called a *translocation* of S_1 and S_2, in the sense of [5].
  - If none of the sets is empty, it is a *strict translocation*.
  - If S_1 or S_2 is empty, it is a *fission*.
  - If T_1 or T_2 is empty, it is a *fusion*.
- We define a *duplicated genome* to be made up of two identical copies of $n'/2$ chromosomes.
- For a genome G, we define $D(G)$ to be the *halving cost* of G, the minimal number of translocations necessary to transform G into some duplicated genome.
- Given a genome G, we call a sequence $(\sigma_1, \cdots, \sigma_t)$ of $t = D(G)$ translocations, which transform G into a duplicated genome, an *optimal sequence of translocations* for G.

We define the *intersection graph induced by* G to be $S_G = (U, E)$ where the vertex set is $U = \{1, \cdots, n\}$ and the edge set E satisfies $(i, j) \in E$ if and only if $S_i \bigcap S_j \neq \emptyset$ (for $i \neq j$), or S_i contains a block of multiplicity 2 (for the case $i = j$).

For the purposes of the genome halving problem, the intersection graph $\mathcal{S}_G$ is equivalent to G. In fact, if two synteny sets contain more than one block in common, then it suffices to designate any one of these blocks as a "representative" for the purposes of analyzing potential translocations.

For all i, $1 \leq i \leq n$, let B_i be the subset of B defined by $b \in B_i$ if and only if for some $j \neq i$, the designated element of $S_i \bigcap S_j$ is b, or else b has multiplicity 2 within S_i. We label each vertex i of $\mathcal{S}_G$ with B_i.

Then, let $\sigma = (S_i, S_j) \longrightarrow (S'_i, S'_j)$ be any translocation that operates on two chromosomes S_i and S_j of G, and transforms them to chromosomes S'_i and S'_j. This translocation corresponds to moving a number of blocks from B_i to B_j, and from B_j to B_i, and subtracting and adding appropriate edges to the graph. Henceforth, we need consider only the sets B_i, rather than the synteny sets S_i.

3.1 Properties of translocations

For each vertex (chromosome) i of $\mathcal{S}_G$, we define the set $C_i = \{j, (i,j) \in E\}$ to contain those chromosomes j which have at least one block in common with i.

Let e be the number of edges in $\mathcal{S}_G$. Suppose $\sigma = (\{B_{i_1}, B_{i_2}\}, \{B_{j_1}, B_{j_2}\}) \longrightarrow (\{B_{i_1}, B_{j_1}\}, \{B_{i_2}, B_{j_2}\})$ is a translocation of chromosomes i and j. Let $\mathcal{S}'_G = \sigma(\mathcal{S}_G)$ be the graph obtained from $\mathcal{S}_G$ by applying σ, and e' the new number of edges. Then $e' \leq e$; a translocation can only reduce or maintain the number of edges. This is a consequence of our statement of the problem where we considered the blocks of genes in B as the smallest units to be manipulated. In general, in a description of genome evolution through translocation, we cannot define in advance which blocks of genes will be uninterrupted by translocation, so that a translocation may well increase the number of edges in an intersection graph representation, by subdividing a block. In the present context, however, every time a block is transferred from one chromosome to another, the whole block is transferred, and any edge induced by this block, say between a third chromosome l and either chromosome i or j, becomes an edge between l and one of the new chromosomes formed by the translocation.

If the third chromosome l was connected to both i and j, it could be that the number of edges is actually reduced by the translocation. In effect, $e' < e$ if and only if there are two blocks $b_1 \in \{B_{i_1}, B_{i_2}\}$, $b_2 \in \{B_{j_1}, B_{j_2}\}$ satisfying $\{b_1, b_2\} \in \{B_{i_1}, B_{j_1}\}$ or $\{b_1, b_2\} \in \{B_{i_2}, B_{j_2}\}$ and there is some chromosome $l \in C_i \cap C_j$ such that $\{b_1, b_2\} \in B_l$. In other words, a translocation which reduces the number of edges is one which increases the size of intersections between chromosomes.

The translocation σ maximally decreases the number of edges only if $C_i \cap C_j$ is maximal.

3.2 Some bounds

Let G be a genome with n chromosomes. The following is a trivial consequence of the fact that a fusion reduces the number of chromosomes by 1.

Lemma 1. *Any genome G' obtainable from G through $D(G)$ translocations contains $n' \geq n - D(G)$ chromosomes.*

Thus for small values of $D(G)$, n' cannot be much smaller than n. For biological realism, we expect n' to be close to n. This is guaranteed on the upper side by:

Lemma 2. *There exists a duplicated genome G' obtainable from G through a sequence of $D(G)$ translocations, such that $n' \leq n$.*

Proof. We proceed by induction on $D(G)$. If $D(G) = 1$, let σ be the translocation that transforms G into a duplicated genome G'. If σ is a fusion or a strict translocation, then $n' \leq n$ since these do not increase the number of chromosomes. If σ is a fission, then to arrive at a duplicated genome we must have three chromosomes of form $S_1 = (T_1, T_2), S_2 = T_1, S_3 = T_2$, and σ must be the fission $(T_1, T_2) \longrightarrow T_1, T_2$ operating on S_1 and dividing it into two chromosomes containing T_1 and T_2, respectively. We can replace this fission by fusion $T_1, T_2 \longrightarrow (T_1, T_2)$ operating on the two chromosomes S_2 and S_3. This produces a duplicated genome without increasing the number of chromosomes.

Suppose now that the induction hypothesis is true up to $t - 1$, and suppose that $D(G) = t$.

DasGupta *et al.* [4] proved that if there exists an optimal sequence of translocations for transforming one genome into another, then there exists another sequence, containing the same number of fusions, fissions and strict translocations, such that the fissions are ordered after all the translocations and fusions.

Then there exists an optimal sequence of translocations $\sigma = (\sigma_1, \cdots, \sigma_t)$ which transforms G into a duplicated genome G', such that σ_1 is a fusion or a translocation. Let $G_1 = \sigma_1(G)$ and let n_1 be the number of chromosomes de G_1. We have $n_1 \leq n$, and by the induction hypothesis $n' \leq n_1$. Thus $n' \leq n$.

An upper bound for the minimum number of translocation $D(G)$ follows directly from the observation that a trivial duplicated genome can be obtained through $n - 1$ fusions followed by a single fission. Thus

Lemma 3.

$$D(G) \leq n$$

Define the C_i as above. Then a lower bound is given by

Lemma 4. *Let e be the number of edges of S_G and p the size of the largest intersection $C_i \cap C_j$. Then*

$$D(G) \geq \left\lceil \log_2 \left(\frac{e - \frac{n}{2}}{p} + 1 \right) \right\rceil$$

Proof. The maximum number of edges in the graph of a duplicated genome is $n/2$, so that at least $e - n/2$ edges must be removed.

Suppose there exists optimal sequence of translocations $(\sigma_1, \cdots, \sigma_t)$, where $t = D(G)$, such that at each step i, σ_i is a translocation whch removes a maximum number of edges. For each i, let r_i ne the number of edges removed by σ_i. By definition, there are no translocations removing more than r_i edges. Thus σ_i may add edges so that the next translocation removes at most $2r_i$ edges. Thus $r_{i+1} \leq 2r_i$. Note that σ_1 must remove p edges.

Let $r \geq 0$ the smallest integer $2^r - 1 \geq \frac{e-\frac{n}{2}}{p}$. Then it follows from the above that $D(G) \geq r$. Now, $r = \left\lceil \log_2 \left(\frac{e-\frac{n}{2}}{p} + 1 \right) \right\rceil$, and our result follows immediately.

For the yeast genome with $n = 16$ chromosomes, based on the 55 blocks found by Seoighe et Wolfe [18], the corresponding graph $\mathcal{S}_G$ contains $e = 40$ edges and $p = 5$. Lemmas 3 and 4 assure us that $3 \leq D(G) \leq 16$.

4 An algorithm

In this section, we will deal with the case where n is even, and we wish the duplicated genome to contain n chromosomes also. The data typically contain many more than $n/2$ duplicated blocks, and may be represented by a graph whose vertices represent chromosomes and whose edges connect vertices where the two chromosomes share at least one block. The goal is the efficient transformation of the graph to a matching bipartite graph (with $n/2$ edges), by eliminating (and occasionally adding) appropriate edges through translocation.

4.1 The data

Let $CH = \{c_1, \cdots, c_n\}$ be the set of chromosomes and $BL = \{b_1, \cdots, b_k\}$ be the set of gene blocks, each belonging to exactly two chromosomes (or to a single chromosome, but with multiplicity 2). Let $S_1, \cdots, S_n$ be the synteny sets of chromosomes $c_1, \cdots, c_n$, respectively.

4.2 Initialization

Construct the set BR of subsets of BL, such that each element of BR is either the intersection of two synteny sets, or a subset of form $\{b\}$ where b is of multiplicity 2 within a single chromosome. A block (any one) of an element of BR is chosen as its representative. In the course of the algorithm, some elements of BR will be amalgamated, and representatives designated anew.

Construct the set E of edges of the intersection graph. Each edge e of E is defined by two chromosomes c_i, c_j and an element of BR represented by some block b. We write $e = (c_i, c_j, b)$

For all $c \in CH$, let $C_c = \{c_j \in CH \,|\, e = (c, c_j, b) \in E \text{ for some } b \in BR\}$.

The intersection graph is completely determined by E. During the execution of the algorithm, we will denote by $\mathcal{G}(E)$ the current graph.

4.3 Results

When the algorithm stops, the results are contained in the variables F, BR and T, where F is the edge set of the output graph containing $n/2$ independent edges, BR is the final set of intersections and T is the sequence of translocations which have been used. Each translocation is represented by a quadruplet $t = (c, c', \mathcal{B}, \mathcal{B}')$, where $c, c' \in CH$, $\mathcal{B}, \mathcal{B}' \subset BR$, indicating the movement of $\mathcal{B}$ from c to c', and the movement of $\mathcal{B}'$ from c' to c. Only one of the two sets of blocks $\mathcal{B}$ or $\mathcal{B}'$ can be missing. In this case, we write $\mathcal{B} = \emptyset$ or $\mathcal{B}' = \emptyset$.

4.4 Description of the algorithm

After initializing $F = \emptyset$, the first part of the algorithm tries to maximally reduce the number of edges in $\mathcal{G}(E)$. At each step we search for the translocation which removes the most edges. By a **fan** of size r, we denote a variable of form $f = (\{c, c'\}, \{c_1, c_2, \cdots, c_r\})$, where $r > 1$, and $c, c', c_1, \cdots, c_r$ are vertices of $\mathcal{G}(E)$ such that $\{c_1, \cdots, c_r\} \subset C_c \cap C_{c'}$. For all i, $1 \leq i \leq r$, suppose $e_i = (c, c_i, b_i)$ and $e'_i = (c', c_i, b'_i)$ the two edges of E linking c to c_i and c' to c_i, respectively.

A translocation removes a maximum number of edges if and only if it operates on a fan $f = (\{c, c'\}, \{c_1, c_2, \cdots, c_r\})$ satisfying $\{c_1, \cdots c_r\} = C_c \cap C_{c'}$ and $|C_c \cap C_{c'}|$ is maximal, and if it is of form $t = (c, c', \mathcal{B}, \mathcal{B}')$, where $\{b_1, \cdots, b_r\} \subset \mathcal{B} \cup \mathcal{B}'$. Such a translocation removes exactly r edges. We will only consider translocations such that $\mathcal{B} \cup \mathcal{B}' = \{b_1, \cdots, b_r\}$.

In particular, for all i, $1 \leq i \leq r$, one of the two edges e_i, e'_i is removed. In order to keep as many fans as possible, we keep the edge which is involved in the structure of the largest number of fans. Suppose that the edge e'_i is removed (i.e. the edge e_i is maintained). This means that the translocation t moves block b'_i from chromosome c' to chromosome c. In this case, we amalgamate the subsets corresponding to blocks b_i and b'_i in BR and we designate b_i to be the representative of this new subset (see Figure 1).

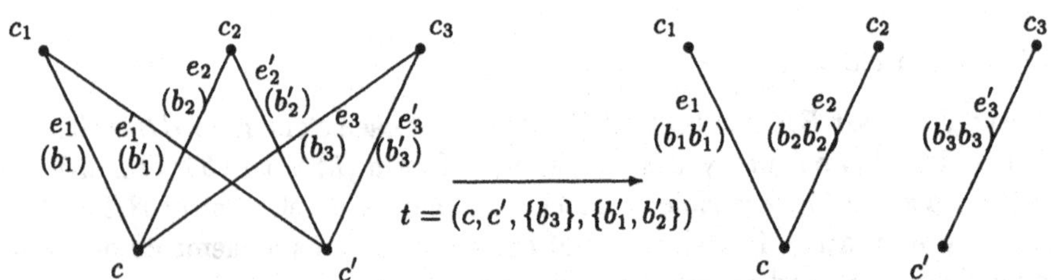

Fig. 1. Translocation of type 1 removing three edges.

As long as $\mathcal{S}(E)$ still contains fans, we choose one of maximal size. If there are two such, two other choice criteria come into play:

Since we are trying to construct a matching bipartite graph of size $n/2$, we choose a translocation which maintains perfect matching (assuming perfect matching already holds). A corollary of Tutte's Theorem is that a necessary condition for a graph to constitute a perfect matching is that it have no connected component with an odd number of vertices. We thus try to find a fan such that the corresponding translocation results in a graph $S(E)$ containing no connected component of odd size.

If several fans of maximal size satisfy the previous condition (or none do), we choose one such that the corresponding translocation maintains as many fans as possible. To implement this, for each fan $f = (\{c, c'\}, \{c_1, c_2, \cdots, c_r\})$ and each $i \leq r$, we calculate the scores s_i and s'_i counting the number of fans other than p, containing the edges e_i and e'_i, respectively. A score s_p for the whole fan p is derived by summing the scores of all edges it would maintain and subtracting the scores of all it would remove. We choose one with the highest score.

A translocation on a fan of size r will be termed a type 1 translocation removing r edges.

Up to now in the algorithm, we have been primarily concerned with reducing the number of edges as quickly as possible. In addition, our reduction of fans is designed so that the remaining edges are potentially (though not necessarily) elements in the ultimate solution to the problem. When there are no more fans, we are left only with translocations which decrease the number of edges of the graph by at most 1. Indeed, it is always possible to find a translocation which will remove an edge and not create any more. However, since our final goal is a graph with exactly $n/2$ independent edges, translocations which create new edges between chromosomes may be necessary to set up pairings for the final duplicated genome.

The second part of the algorithm consists of trying to identify as many edges as possible in $S(E)$ destined to be contained in the final graph, in order that a minimum of further translocations will be required. This is essentially the classical maximum matching problem of graph theory. Edmonds [11] gave the first polynomial algorithm for finding a maximum matching in a non-bipartite graph, based on the technique of "shrinking" certain odd cycles.

We use this method to find the largest possible number of edges, which are then added to F, the set of edges in the final graph. Each edge $e = (c, c', b)$ so added to F automatically establishes the pairing of chromosomes c and c'.

For the remainder of the algorithm, we will need a set CC consisting of all currently paired chromosomes in CH. At the outset of the second part of the algorithm, $CC = \emptyset$. At the end of the algorithm, we require $CC = CH$. For all $c \in CC$, we define $\bar{c}$ to be the chromosome paired with c. We have $\bar{\bar{c}} = c$

After this step, the edges $e = (c, c', b)$ still in E/F are of three types:

1. $c = c'$ and $c \notin CC$;
2. $c \neq c'$ and only one of the two chromosomes c and c' is already paired, i.e. belongs to CC;
3. The two chromosomes c and c' are already paired but not with each other, i.e. $c, c' \in CC$ and $\bar{c} \neq c'$.

The first two types of edge allow us to pair the as yet unpaired chromosomes.

We group (arbitrarily) all the edges of type (1) and (2), e.g. $e_1 = (c_{11}, c_{21}, b_1)$ and $e_2 = (c_{12}, c_{22}, b_2)$, where $c_{11}, c_{12} \notin CC$. Because n is even, there are either zero, or an even number, of such edges. We carry out the translocation $t = (c_{12}, c_{21}, b_2, b_1)$. This translocation removes edges e_1 and e_2 and creates edges $e = (c_{11}, c_{12}, b_1)$ and $e_0 = (c_{21}, c_{22}, b_2)$. At the same time, we pair chromosomes c_{11} and c_{12} (Figure 2). We call this a translocation of type 2.

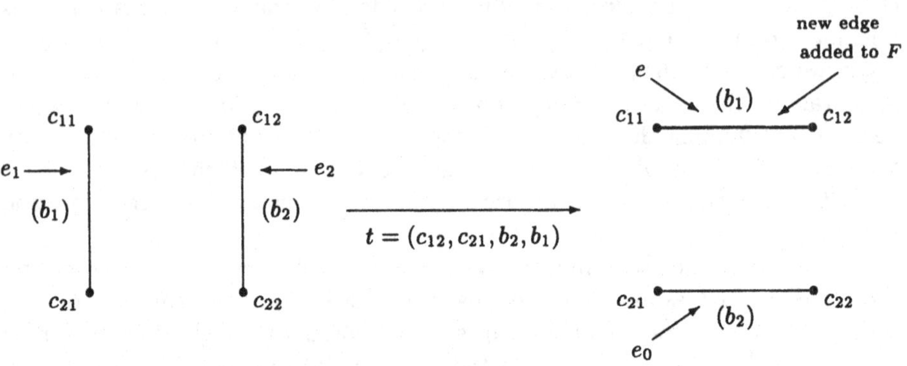

Fig. 2. Translocation of type 2. Two edges are removed, two edges are created, and one of these new edges is added to the final edge set F.

Finally, only edges of type (3) remain to be removed. To do this, we apply translocations which remove one edge of $\mathcal{G}(E)$ at a time (Figure 3). Let $e_1 = (c_{11}, c_{21}, b_1) \in E$ be an edge of type (3) and let $e = (c_{11}, c_{12}, b)$ be the edge such that $\overline{c_{11}} = c_{12}$. Then, the translocation $t = (c_{12}, c_{21}, \emptyset, b_1)$ removes the edge e_1. Here, subsets corresponding to blocks b and b_1 have to be amalgamated. We call this a translocation of type 3.

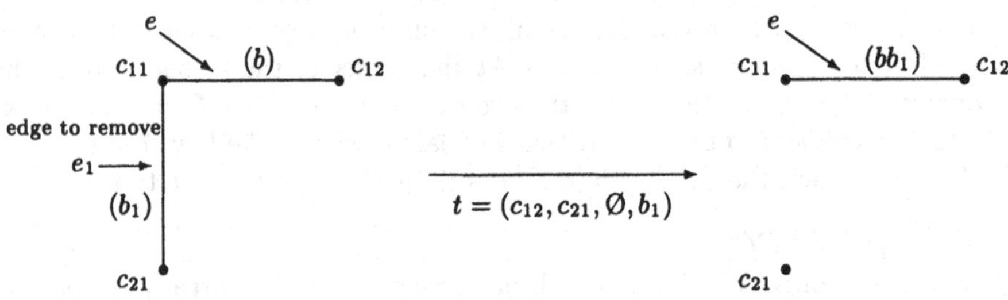

Fig. 3. Translocation of type 3. One edge is removed.

4.5 Procedures

In our implementation, the following routines carry out the steps described above.

Decomposition into fans : The procedure **fan_max** decomposes the graph $\mathcal{G}(E)$ into fans and returns, if possible, an appropriate fan of maximal length.

Processing a fan : The fan $p = (\{c, c'\}, \{c_1, \cdots, c_r\})$ output by the procedure **fan_max** is processed by the procedure **analyze_fan** as in Figure 1.

Choosing a subgraph : The procedure **edmonds** chooses a maximal matching of the graph $\mathcal{G}(E)$.

Pair remaining chromosomes : The procedure **pair_rem_chro** completes the pairing of the as yet unpaired chromosomes as in Figure 2.

Remove remaining edges : The procedure **remove_rem_edges** completes the removal of edges belonging to E/F as in Figure 3.

4.6 The main program

The procedures described in the previous section are used in the main program below. The procedure **fan_max** returns the value T if $\mathcal{G}(E)$ contains a fan.

```
fan_exist = T;
while fan_exist do
        fan_exist = fan_max(fan);
        if fan_exist = T do
                analyze_fan(fan);
end while
edmonds;
pair_rem_chro;
remove_rem_edges;
```

5 Analyzing the yeast data

The data we analyzed, drawn from [21], are listed in Table 1. The Roman numerals are standard notation for the 16 *S. cerevisiae* chromosomes, and the lists of blocks present in each chromosome are numbered according to the Wolfe and Shields notation. It should be noted that the fact that 55 blocks were detected is a function of the criteria and procedures used for assessing similarities between genes and defining blocks (i.e. BLASTP scores $\geq$ 200 for each paralogous gene pair, at least three genes per block, no two more than 50 kilobases apart, and conservation of gene order and orientation, aside from some short reversals). Conceivably a good number of additional blocks could be added by relaxing some of these.

Our heuristic found a number of solutions to unordered genome halving using only 13 translocations. One of these solutions is described in Tables 2 and 3.

I : 2 1	IX : 38 39 27
II : 4 3 7 8 5 6	X : 10 40 41 28 42
III : 9 10 11	XI : 42 40 43 35 41 52 38
IV : 20 12 12 54 15 21 3 13 16 17 24 22 14 23 19 18 9	XII : 53 53 31 55 16 18 17 45 30 15 44
V : 28 25 27 4 26 13	XIII : 46 44 19 43 54 48 47 46
VI : 55 36	XIV : 49 20 37 50 39 11
VII : 36 25 26 32 6 33 5 30 34 31 29	XV : 49 21 22 52 50 23 45 51 47 2
VIII : 35 14 37 29 1	XVI : 48 32 33 51 8 24 7 34

Table 1. Lists of blocks corresponding to each of the 16 chromosomes of the yeast genome.

Table 2 contains a possible ancestral genome (note that the numbering of the chromosomes is only very indirectly related, through the algorithm, to the labels used for modern *S.cerevisiae*), and Table 3 traces the algorithm in reconstructing the actual yeast genome from the duplicated genome. The different solutions arose through different choices in the algorithm when there were several possible, such as when several fans had equivalent scores.

Identical Chromosome Pair	*Blocks Contained*
1, 8	1 2 45 44 15 16 17 18 30 31
2, 16	7 8 24 32 33 34 51 48
3, 4	9 11 10
5, 10	28 40 41 42 13 25 26 4 27 38
6, 7	36 55
9, 14	39 20 49 50 37 12 21 22 23 19 54 14 29 3 5 6
11, 12	35 52 43 53
13, 15	47 46

Table 2. One possible duplicated genome. The actual yeast genome is obtained from this one using 13 translocations.

To be as certain as possible that our solution is optimal, we carried out the following tests: at each step, from among all the translocations which remove the maximum possible number of edges, choose one *randomly* instead of using our choice criteria. Table 4 gives the results in terms of the number of translocations required in 1000 trials on the yeast data set.

Chro. A	*Chro. B*	*Blocks A*	*Blocks B*	*Type of translocation*
7	12	55		3
11	12	53		3
1	15	45 44 15 16 17 18 30 31		3
8	12	45 44 15 16 17 18 30 31	35 52 43	2
10	4	13 25 26 4 27 38	10	1, removes 2 edges
9	4	20 49 50 37 12 21 22 23 19 54 14 29 3 5 6	27 38	1, removes 2 edges
5	11	38 40 41 42		1, removes 2 edges
14	4	12 21 22 23 19 54 14 29 3 5 6	11	1, removes 2 edges
4	2	3 5 6 4	24 32 33 34 51 48	1, removes 3 edges
15	4	15 16 17 18 30 31	21 22 23 19 54 14 29 49 50 37 51 48	1, removes 4 edges
15	8	14 29 37	2 52 43	1, removes 4 edges
15	13	19 54 43 44 46 48		1, removes 5 edges
4	7	5 6 25 26 29 30 31 32 33 34		1, removes 5 edges

Table 3. Translocations applied in reconstructing the current yeast genome from the duplicated genome of table 2. The first line should be read as Block 55 transferred from current chromosome number 7 (initialized as *S.cerevisiae* chromosome VII), to chromosome number 12, with nothing transferred in the other direction, according to a translocation of type 3.

Translocations required	*Frequency*
13	35
14	194
15	268
16	418
17	85

Table 4. Number of translocations required in 1000 trials, using randomized choice of maximal translocation at each step.

6 Discussion and Conclusions

How are we to interpret our solution of 13 translocations, compared to that of [18] with 41? There is little danger in assuming that both results are optimal for their respective problems, which though not mathematically guaranteed, is likely given that they are each based on many runs of a locally optimal procedure. The major reason for the different scores is of course the difference between the unordered and ordered versions of the problem.

This again brings up the question of which problem is more appropriate. Seoighe and Wolfe argue that there is little evidence that reversals have played a major role in scrambling the yeast chromosome. However, if there were as many as 70 or 80 translocations, plus a few reversals, we could expect the blocks on any given chromosome to be paired with blocks from a random sequence of the 16 chromosomes. If this were the case, it would be vanishingly unlikely to find a pattern such as the four blocks out of eleven on Chromosome XII paired with blocks on a single other chromosome (IV), and three occurrences of three blocks in common between pairs of chromosomes. Another way of looking at this is that under the random model, we could expect about six cases where two blocks occur on the same pair of chromosomes. In fact, there are 21 such pairs.

These results suggest the need for unbiased methods to evaluate the relative rate of translocations versus reversals. Perhaps the *Ordered genome halving with oriented blocks* problem would be useful here, with some way of optimizing the relative costs of the two rearrangement operations. In the same way as unordered genome halving is related to synteny distance [5, 4], the ordered genome halving problems have obvious relationships with minimal rearrangement distance problems, and techniques for solving the latter may have some implications for the problems enunciated here.

7 Acknowledgements

Research supported by grants to DS from the Natural Sciences and Engineering Research Council of Canada (NSERC) and the Canadian Genome Analysis and Technology program, and a CRM postoctoral fellowship to N E-M. DS is a Fellow of the Canadian Institute for Advanced Research.

References

1. Ahn, S., Tanksley, S.D.: Comparative linkage maps of rice and maize genomes. Proc. Natl. Acad. Sci. USA **90** (1993) 7980-7984.
2. Atkin, N. B., Ohno, S.: DNA values of four primitive chordates. Chromosoma **23** (1967) 10–13
3. Coissac, E., Maillier, E., Netter, P.:A comparative study of duplications in bacteria and eukaryotes: the importance of telomeres. Molecular Biology and Evolution **14** (1997) 1062–1074
4. DasGupta,B., Jiang, T., Kannan, S., Li, M., Sweedyk, Z.: On the complexity and approximation of syntenic distance. RECOMB 97. Proceedings of the First Annual International Conference on Computational Molecular Biology (1997) ACM Press, 99-108
5. Ferretti, V., Nadeau, J.H., Sankoff, D.: Original synteny. In Combinatorial Pattern Matching. Seventh Annual Symposium (D. Hirschberg and G .Myers, ed.) Lecture Notes in Computer Science **1075** (1996) Springer Verlag, 159–167
6. Fryxell, K.J.: The coevolution of gene family trees. Trends in Genetics **12** (1996) 364–369.

7. Gaut, B.S., Doebley,J.F.: DNA sequence evidence for the segmental allotetraploid origin of maize. Proc. Natl. Acad. Sci., U.S.A. **94** (1997) 6809– 6814.
8. Hinegardner, R.: Evolution of cellular DNS content in teleost fishes. American Naturalist **102** (1968) 517–523
9. Mewes, H.W., Albermann, K., Bähr, M., Frishman, D., Gleissner, A., Hani, J., Heumann, K., Kleine, K., Maierl, A., Oliver, S.G., Pfeiffer, F., Zollner, A.: Overview of the yeast genome. Nature **387**(suppl.) (1997) 7–65
10. Muller, F., Bernard, V., Tobler, H.: Chromatin diminution in nematodes. Bioessays **18** (1996) 133–138
11. Lovász, L., Plummer, M.D.: Matching Theory. Annals of discrete mathematics **121** (1986) 357–369
12. Moore, G., Devos, K. M., Wang, Z., Gale, M. D.: 1995. Grasses, line up and form a circle. Current Biology **5** (1995) 737–739.
13. Nadeau, J. H.: Genome duplication and comparative mapping. In Advanced Techniques in Chromosome Research (ed. Adolph, K.T.) (1991) (Marcel Dekker, New York) 269–296
14. Nadeau, J.H., Sankoff, D.: Comparable rates of gene loss and functional divergence after genome duplications early in vertebrate evolution. Genetics **147** (1997) 1259–1266
15. Ohno, S., Wolf, U., Atkin, N. B.: Evolution from fish to mammals by gene duplication. Hereditas **59** (1968) 169–187
16. Paterson, A.H., Lan, T.-H., Reischmann, K.P., Chang, C., Lin, Y.-R., Liu, S.-C., Burow, M.D., Kowalski, S.P., Katsar, C.S., DelMonte, T.A., Feldmann, K.A., Schertz, K.F., Wendel, J.F.: Toward a unified genetic map of higher plants, transcending the monocot-dicot divergence. Nature Genetics **14** (1996) 380–382
17. Postlethwait, J.H, Yan, Y.-L., Gates, M.A., Horne, S., Amores,A., Brownlie, A., Donovan, A., Egan, E.S., Force, A., Gong, Z., Goutel, C., Fritz, A., Kelsh, R., Knapik, E., Liao, E., Paw, B., Ransom, D., Singer, A., Thomson,T., Abduljabbar, T.S., Yelick, P., Beier, D., Joly, J.-S., Larhammar, D., Rosa, F., Westerfield, M., Zon, L.I., and Talbot, W.S.: Vertebrate genome evolution and the zebrafish gene map. Nature Genetics 18 (1998) 345–349.
18. Seoighe, C., Wolfe, K.H.: Extent of genomic rearrangement after genome duplication in yeast. Proceedings of the National Academy of Sciences USA 95 (1998) 4447–4452.
19. Scheffler, J. A., Sharpe, A.G., Schmidt, H., Sperling, P., Parkin, I.A.P., Lühs, W., Lydiate, D.J., Heinz, E.: Desaturase multigene families of Brassica napus arose through genome duplication. Theoretical and Applied Genetics **94** (1997) 583–591
20. Shoemaker, R.C.,Polzin, K., Labate, J., Specht, J., Brummer, E.C., Olson, T., Young, N., Concibido, V., Wilcox, J., Tamulonis, J.P., Kochert, G. Boerma, H.R.: Genome duplication in soybean (Glycine subgenus soja). Genetics **144** (1996) 329–228
21. Wolfe, K.H., Shields, D.C.: Molecular evidence for an ancient duplication of the entire yeast genome. Nature **387** (1997) 708–713
22. Xu, R-H., Kim, J.,Taira, M., Lin, J.J., Zhang, C.-H., Sredni, D., Evans, T., Kung, H.-F.: Differential regulation of neurogenesis by the two Xenopus GATA-1 genes. Molecular and Cellular Biology **17** (1997) 436–443

Author Index

Springer and the environment

At Springer we firmly believe that an international science publisher has a special obligation to the environment, and our corporate policies consistently reflect this conviction.

We also expect our business partners – paper mills, printers, packaging manufacturers, etc. – to commit themselves to using materials and production processes that do not harm the environment. The paper in this book is made from low- or no-chlorine pulp and is acid free, in conformance with international standards for paper permanency.

Lecture Notes in Computer Science

For information about Vols. 1–1361

please contact your bookseller or Springer-Verlag

Vol. 1362: D.K. Panda, C.B. Stunkel (Eds.), Network-Based Parallel Computing. Proceedings, 1998. X, 247 pages. 1998.

Vol. 1363: J.-K. Hao, E. Lutton, E. Ronald, M. Schoenauer, D. Snyers (Eds.), Artificial Evolution. XI, 349 pages. 1998.

Vol. 1364: W. Conen, G. Neumann (Eds.), Coordination Technology for Collaborative Applications. VIII, 282 pages. 1998.

Vol. 1365: M.P. Singh, A. Rao, M.J. Wooldridge (Eds.), Intelligent Agents IV. Proceedings, 1997. XII, 351 pages. 1998. (Subseries LNAI).

Vol. 1366: Z. Li, P.-C. Yew, S. Chatterjee, C.-H. Huang, P. Sadayappan, D. Sehr (Eds.), Languages and Compilers for Parallel Computing. Proceedings, 1997. XII, 428 pages. 1998.

Vol. 1367: E.W. Mayr, H.J. Prömel, A. Steger (Eds.), Lectures on Proof Verification and Approximation Algorithms. XII, 344 pages. 1998.

Vol. 1368: Y. Masunaga, T. Katayama, M. Tsukamoto (Eds.), Worldwide Computing and Its Applications — WWCA'98. Proceedings, 1998. XIV, 473 pages. 1998.

Vol. 1370: N.A. Streitz, S. Konomi, H.-J. Burkhardt (Eds.), Cooperative Buildings. Proceedings, 1998. XI, 267 pages. 1998.

Vol. 1371: I. Wachsmuth, M. Fröhlich (Eds.), Gesture and Sign Language in Human-Computer Interaction. Proceedings, 1997. XI, 309 pages. 1998. (Subseries LNAI).

Vol. 1372: S. Vaudenay (Ed.), Fast Software Encryption. Proceedings, 1998. VIII, 297 pages. 1998.

Vol. 1373: M. Morvan, C. Meinel, D. Krob (Eds.), STACS 98. Proceedings, 1998. XV, 630 pages. 1998.

Vol. 1374: H. Bunt, R.-J. Beun, T. Borghuis (Eds.), Multimodal Human-Computer Communication. VIII, 345 pages. 1998. (Subseries LNAI).

Vol. 1375: R. D. Hersch, J. André, H. Brown (Eds.), Electronic Publishing, Artistic Imaging, and Digital Typography. Proceedings, 1998. XIII, 575 pages. 1998.

Vol. 1376: F. Parisi Presicce (Ed.), Recent Trends in Algebraic Development Techniques. Proceedings, 1997. VIII, 435 pages. 1998.

Vol. 1377: H.-J. Schek, F. Saltor, I. Ramos, G. Alonso (Eds.), Advances in Database Technology - EDBT'98. Proceedings, 1998. XII, 515 pages. 1998.

Vol. 1378: M. Nivat (Ed.), Foundations of Software Science and Computation Structures. Proceedings, 1998. X, 289 pages. 1998.

Vol. 1379: T. Nipkow (Ed.), Rewriting Techniques and Applications. Proceedings, 1998. X, 343 pages. 1998.

Vol. 1380: C.L. Lucchesi, A.V. Moura (Eds.), LATIN'98: Theoretical Informatics. Proceedings, 1998. XI, 391 pages. 1998.

Vol. 1381: C. Hankin (Ed.), Programming Languages and Systems. Proceedings, 1998. X, 283 pages. 1998.

Vol. 1382: E. Astesiano (Ed.), Fundamental Approaches to Software Engineering. Proceedings, 1998. XII, 331 pages. 1998.

Vol. 1383: K. Koskimies (Ed.), Compiler Construction. Proceedings, 1998. X, 309 pages. 1998.

Vol. 1384: B. Steffen (Ed.), Tools and Algorithms for the Construction and Analysis of Systems. Proceedings, 1998. XIII, 457 pages. 1998.

Vol. 1385: T. Margaria, B. Steffen, R. Rückert, J. Posegga (Eds.), Services and Visualization. Proceedings, 1997/1998. XII, 323 pages. 1998.

Vol. 1386: T.A. Henzinger, S. Sastry (Eds.), Hybrid Systems: Computation and Control. Proceedings, 1998. VIII, 417 pages. 1998.

Vol. 1387: C. Lee Giles, M. Gori (Eds.), Adaptive Processing of Sequences and Data Structures. Proceedings, 1997. XII, 434 pages. 1998. (Subseries LNAI).

Vol. 1388: J. Rolim (Ed.), Parallel and Distributed Processing. Proceedings, 1998. XVII, 1168 pages. 1998.

Vol. 1389: K. Tombre, A.K. Chhabra (Eds.), Graphics Recognition. Proceedings, 1997. XII, 421 pages. 1998.

Vol. 1390: C. Scheideler, Universal Routing Strategies for Interconnection Networks. XVII, 234 pages. 1998.

Vol. 1391: W. Banzhaf, R. Poli, M. Schoenauer, T.C. Fogarty (Eds.), Genetic Programming. Proceedings, 1998. X, 232 pages. 1998.

Vol. 1392: A. Barth, M. Breu, A. Endres, A. de Kemp (Eds.), Digital Libraries in Computer Science: The MeDoc Approach. VIII, 239 pages. 1998.

Vol. 1393: D. Bert (Ed.), B'98: Recent Advances in the Development and Use of the B Method. Proceedings, 1998. VIII, 313 pages. 1998.

Vol. 1394: X. Wu. R. Kotagiri, K.B. Korb (Eds.), Research and Development in Knowledge Discovery and Data Mining. Proceedings, 1998. XVI, 424 pages. 1998. (Subseries LNAI).

Vol. 1395: H. Kitano (Ed.), RoboCup-97: Robot Soccer World Cup I. XIV, 520 pages. 1998. (Subseries LNAI).

Vol. 1396: E. Okamoto, G. Davida, M. Mambo (Eds.), Information Security. Proceedings, 1997. XII, 357 pages. 1998.

Vol. 1397: H. de Swart (Ed.), Automated Reasoning with Analytic Tableaux and Related Methods. Proceedings, 1998. X, 325 pages. 1998. (Subseries LNAI).

Vol. 1398: C. Nédellec, C. Rouveirol (Eds.), Machine Learning: ECML-98. Proceedings, 1998. XII, 420 pages. 1998. (Subseries LNAI).

Vol. 1399: O. Etzion, S. Jajodia, S. Sripada (Eds.), Temporal Databases: Research and Practice. X, 429 pages. 1998.

Vol. 1400: M. Lenz, B. Bartsch-Spörl, H.-D. Burkhard, S. Wess (Eds.), Case-Based Reasoning Technology. XVIII, 405 pages. 1998. (Subseries LNAI).

Vol. 1401: P. Sloot, M. Bubak, B. Hertzberger (Eds.), High-Performance Computing and Networking. Proceedings, 1998. XX, 1309 pages. 1998.

Vol. 1402: W. Lamersdorf, M. Merz (Eds.), Trends in Distributed Systems for Electronic Commerce. Proceedings, 1998. XII, 255 pages. 1998.

Vol. 1403: K. Nyberg (Ed.), Advances in Cryptology – EUROCRYPT '98. Proceedings, 1998. X, 607 pages. 1998.

Vol. 1404: C. Freksa, C. Habel. K.F. Wender (Eds.), Spatial Cognition. VIII, 491 pages. 1998. (Subseries LNAI).

Vol. 1405: S.M. Embury, N.J. Fiddian, W.A. Gray, A.C. Jones (Eds.), Advances in Databases. Proceedings, 1998. XII, 183 pages. 1998.

Vol. 1406: H. Burkhardt, B. Neumann (Eds.), Computer Vision – ECCV'98. Vol. I. Proceedings, 1998. XVI, 927 pages. 1998.

Vol. 1407: H. Burkhardt, B. Neumann (Eds.), Computer Vision – ECCV'98. Vol. II. Proceedings, 1998. XVI, 881 pages. 1998.

Vol. 1409: T. Schaub, The Automation of Reasoning with Incomplete Information. XI, 159 pages. 1998. (Subseries LNAI).

Vol. 1411: L. Asplund (Ed.), Reliable Software Technologies – Ada-Europe. Proceedings, 1998. XI, 297 pages. 1998.

Vol. 1412: R.E. Bixby, E.A. Boyd, R.Z. Ri ' 'os-Mercado (Eds.), Integer Programming and Combinatorial Optimization. Proceedings, 1998. IX, 437 pages. 1998.

Vol. 1413: B. Pernici, C. Thanos (Eds.), Advanced Information Systems Engineering. Proceedings, 1998. X, 423 pages. 1998.

Vol. 1414: M. Nielsen, W. Thomas (Eds.), Computer Science Logic. Selected Papers, 1997. VIII, 511 pages. 1998.

Vol. 1415: J. Mira, A.P. del Pobil, M.Ali (Eds.), Methodology and Tools in Knowledge-Based Systems. Vol. I. Proceedings, 1998. XXIV, 887 pages. 1998. (Subseries LNAI).

Vol. 1416: A.P. del Pobil, J. Mira, M.Ali (Eds.), Tasks and Methods in Applied Artificial Intelligence. Vol.II. Proceedings, 1998. XXIII, 943 pages. 1998. (Subseries LNAI).

Vol. 1417: S. Yalamanchili, J. Duato (Eds.), Parallel Computer Routing and Communication. Proceedings, 1997. XII, 309 pages. 1998.

Vol. 1418: R. Mercer, E. Neufeld (Eds.), Advances in Artificial Intelligence. Proceedings, 1998. XII, 467 pages. 1998. (Subseries LNAI).

Vol. 1419: G. Vigna (Ed.), Mobile Agents and Security. XII, 257 pages. 1998.

Vol. 1420: J. Desel, M. Silva (Eds.), Application and Theory of Petri Nets 1998. Proceedings, 1998. VIII, 385 pages. 1998.

Vol. 1421: C. Kirchner, H. Kirchner (Eds.), Automated Deduction – CADE-15. Proceedings, 1998. XIV, 443 pages. 1998. (Subseries LNAI).

Vol. 1422: J. Jeuring (Ed.), Mathematics of Program Construction. Proceedings, 1998. X, 383 pages. 1998.

Vol. 1423: J.P. Buhler (Ed.), Algorithmic Number Theory. Proceedings, 1998. X, 640 pages. 1998.

Vol. 1424: L. Polkowski, A. Skowron (Eds.), Rough Sets and Current Trends in Computing. Proceedings, 1998. XIII, 626 pages. 1998. (Subseries LNAI).

Vol. 1425: D. Hutchison, R. Schäfer (Eds.), Multimedia Applications, Services and Techniques – ECMAST'98. Proceedings, 1998. XVI, 532 pages. 1998.

Vol. 1427: A.J. Hu, M.Y. Vardi (Eds.), Computer Aided Verification. Proceedings, 1998. IX, 552 pages. 1998.

Vol. 1430: S. Trigila, A. Mullery, M. Campolargo, H. Vanderstraeten, M. Mampaey (Eds.), Intelligence in Services and Networks: Technology for Ubiquitous Telecom Services. Proceedings, 1998. XII, 550 pages. 1998.

Vol. 1431: H. Imai, Y. Zheng (Eds.), Public Key Cryptography. Proceedings, 1998. XI, 263 pages. 1998.

Vol. 1432: S. Arnborg, L. Ivansson (Eds.), Algorithm Theory – SWAT '98. Proceedings, 1998. IX, 347 pages. 1998.

Vol. 1433: V. Honavar, G. Slutzki (Eds.), Grammatical Inference. Proceedings, 1998. X, 271 pages. 1998. (Subseries LNAI).

Vol. 1434: J.-C. Heudin (Ed.), Virtual Worlds. Proceedings, 1998. XII, 412 pages. 1998. (Subseries LNAI).

Vol. 1435: M. Klusch, G. Weiß (Eds.), Cooperative Information Agents II. Proceedings, 1998. IX, 307 pages. 1998. (Subseries LNAI).

Vol. 1436: D. Wood, S. Yu (Eds.), Automata Implementation. Proceedings, 1997. VIII, 253 pages. 1998.

Vol. 1437: S. Albayrak, F.J. Garijo (Eds.), Intelligent Agents for Telecommunication Applications. Proceedings, 1998. XII, 251 pages. 1998. (Subseries LNAI).

Vol. 1438: C. Boyd, E. Dawson (Eds.), Information Security and Privacy. Proceedings, 1998. XI, 423 pages. 1998.

Vol. 1439: B. Magnusson (Ed.), System Configuration Management. Proceedings, 1998. X, 207 pages. 1998.

Vol. 1441: W. Wobcke, M. Pagnucco, C. Zhang (Eds.), Agents and Multi-Agent Systems. Proceedings, 1997. XII, 241 pages. 1998. (Subseries LNAI).

Vol. 1443: K.G. Larsen, S. Skyum, G. Winskel (Eds.), Automata, Languages and Programming. Proceedings, 1998. XVI, 932 pages. 1998.

Vol. 1444: K. Jansen, J. Rolim (Eds.), Approximation Algorithms for Combinatorial Optimization. Proceedings, 1998. VIII, 201 pages. 1998.

Vol. 1445: E. Jul (Ed.), ECOOP'98 – Object-Oriented Programming. Proceedings, 1998. XII, 635 pages. 1998.

Vol. 1446: D. Page (Ed.), Inductive Logic Programming. Proceedings, 1998. VIII, 301 pages. 1998. (Subseries LNAI).

Vol. 1448: M. Farach-Colton (Ed.), Combinatorial Pattern Matching. Proceedings, 1998. VIII, 251 pages. 1998.

Vol. 1456: A. Drogoul, M. Tambe, T. Fukuda (Eds.), Collective Robotics. Proceedings, 1998. VII, 161 pages. 1998. (Subseries LNAI).